DEATH AND ANTI-DEATH
(VOLUME 8)

DEATH AND ANTI-DEATH, VOLUME 8:
Fifty Years
After Albert Camus
(1913-1960)

Charles Tandy, Editor

Ria University Press

www.ria.edu/rup

2010

Printed in the United States of America

Ria University Press   **Palo Alto, California**

Death And Anti-Death, Volume 8:
Fifty Years After Albert Camus (1913-1960)

Charles Tandy, Editor

FIRST PUBLISHED IN HARDBACK AND PAPERBACK 2010

PUBLISHED BY
Ria University Press
PO Box 20170 at Stanford
Palo Alto, California 94309 USA

www.ria.edu/rup

Distributed by Ingram
Available from most bookstores and all Espresso Book Machines

DEDICATED TO

Glen Harold Stassen

Dr. Glen Stassen
first introduced the thought of Albert Camus to me (Charles Tandy)
more than four decades ago

CONTENTS

PREFACE AND ACKNOWLEDGEMENTS

Death And Anti-Death, Volume 8:
Fifty Years After Albert Camus (1913-1960)

Volume 8, as indicated by the anthology's subtitle, is in honor of Albert Camus (1913-1960). The chapters do not necessarily mention him (but some chapters do). The chapters (by professional philosophers and other professional scholars) are directed to issues related to death, life extension, and anti-death, broadly construed. Most of the contributions consist of scholarship unique to this volume. As was the case with all previous volumes in the Death And Anti-Death Series By Ria University Press, the anthology includes an Index as well as an Abstracts section that serves as an extended table of contents. (Volume 8 also includes a BRIEF COMMUNICATIONS section.)

The editor gratefully acknowledges support and assistance from the following:

- Center for General Education, Fooyin University (Taiwan)

- Research Center for Medical Humanities, Fooyin University (Taiwan)

- R. Michael Perry, Ph.D., Society for Universal Immortalism (USA)

BRIEF COMMUNICATIONS

Death And Anti-Death, Volume 8:
Fifty Years After Albert Camus (1913-1960)

◆◆◆

Invitation To Communicate

The Editor invites email messages to him at <tandy@ria.edu> for possible inclusion in the BRIEF COMMUNICATIONS section of future volumes.

◆◆◆

On Changing One's Mind: The Bayesian Way

Nick Bostrom*

Belief is not an all-or-nothing thing—believe or disbelieve, accept or reject. Instead, I have degrees of belief, a subjective probability distribution over different possible ways the world could be. This means I am constantly changing my mind about all sorts of things, as I reflect or gain more evidence. While I don't always think explicitly in terms of probabilities, I often do so when I give careful consideration to some matter. And when I reflect on my own cognitive processes, I must acknowledge the graduated nature of my beliefs.

The commonest way in which I change my mind is by concentrating my credence function on a narrower set of possibilities than before. This occurs every time I learn a new

---

* An earlier version of this paper appeared in response to the 2008 EDGE Question: "What Have You Changed Your Mind About?" The Edge Foundation, Inc., <http://ww.edge.org/q2008/q08_2. html#bostrom>.

piece of information. Since I started my life knowing virtually nothing, I have changed my mind about virtually everything. For example, not knowing a friend's birthday, I assign a 1/365 chance (approximately) of it being August 11. After she tells me that August 11 is her birthday, I assign that date a probability of close to 100 percent. (Never exactly 100 percent, for there is always a non-zero probability of miscommunication, deception, or other error.)

It can also happen that I change my mind by smearing out my credence function over a wider set of possibilities. I might forget the exact date of my friend's birthday but remember that it is sometime in the summer. The forgetting changes my credence function from being almost entirely concentrated on August 11 to being spread out more or less evenly over the summer months. After this change of mind, I might assign a 1-percent probability to my friend's birthday being August 11.

My credence function can become more smeared out not only by forgetting but also by learning—learning that what I previously took to be strong evidence for some hypothesis is in fact weak or misleading evidence. (This type of belief change can often be mathematically modeled as a narrowing rather than a broadening of credence function, but the technicalities of this are not relevant here.) For example, over the years I have become moderately more uncertain about the benefits of medicine, nutritional supplements, and much conventional health wisdom. This belief change has come about as a result of several factors. One of the factors is that I have read some papers that cast doubt on the reliability of the standard methodological protocols used in medical studies and their reporting. Another factor is my own experience of following up on MEDLINE some of the exciting medical findings reported in the media—almost always, the search of the source literature reveals a much more complicated picture, with many studies showing a positive effect, many showing a negative effect, and many showing no effect. A third factor is the arguments of a health economist friend of mine, who takes a dim view of the marginal benefits of medical care.

Typically, my beliefs about big issues change in small steps. Ideally, these steps should approximate a random walk, like the stock market. It should be impossible for me to predict how my beliefs on some topic will change in the future. If I believed that a year hence I would assign a higher probability to some hypothesis than I do today—why, in that case, I could raise the probability right away. Given knowledge of what I will believe in the future, I would defer to the beliefs of my future self, provided I think my future self will be better informed than I am now and at least as rational.

I have no crystal ball to show me what my future self will believe. But I do have access to many other selves, who are better informed than I am on many topics. I can defer to experts. Provided they are unbiased and are giving me their honest opinion, I should perhaps always defer to people who have more information than I do—or to some weighted average of expert opinion if there is no consensus. Of course, the proviso is a big one: Often I have reason to disbelieve that other people are unbiased or that they are giving me their honest opinion. However, it is also possible that I am biased and self-deceiving. An important unresolved question is how much epistemic weight a wannabe Bayesian thinker should give to the opinions of others. I'm looking forward to changing my mind on that issue, hopefully by my credence function becoming concentrated on the correct answer.

◆◆◆

CONTRIBUTORS TO THIS VOLUME

Death And Anti-Death, Volume 8:
Fifty Years After Albert Camus (1913-1960)

Giorgio Baruchello, Ph.D.

Born in Genoa, Italy, Giorgio Baruchello is Professor of Philosophy at the Faculty of Humanities and Social Sciences of the University of Akureyri, Iceland. He read philosophy in Genoa and Reykjavík, Iceland, and holds a Ph.D. in philosophy from the University of Guelph, Canada. His publications encompass several different areas, especially social philosophy, value theory, and history of philosophy. Since 2005 he edits *Nordicum-Mediterraneum* <http://nome.unak.is>, the first Icelandic scholarly journal in Nordic and Mediterranean studies.

Benjamin P. Best

Ben Best, President and CEO of the Cryonics Institute (USA), is a well-known activist in cryonics and life extension advocacy. He holds undergraduate degrees in pharmacy; in physics and computing science; and, in finance. His monograph *Mechanisms of Aging* was reprinted in the *Anti-Aging Clinical Protocols 2004-2005* of the American Academy of Anti-Aging Medicine. Best's review "Nuclear DNA damage as a direct cause of aging" challenges the OncoSENS claim that nuclear DNA damage only matters for aging because of cancer. Best has published articles on his website <http://www.benbest.com> on more than 150 diverse topics ranging from science and medicine to history and philosophical musings.

Thomas O. Buford, Ph.D.

Tom Buford (1932-) holds the Louis G. Forgione Chair of Philosophy at Furman University (USA) and has been an adherent of the Boston Personalism branch of philosophy. He received his Doctorate of Philosophy at Boston University in 1963, where he studied under Peter Anthony Bertocci (1910-1989), who had studied under Edgar Sheffield Brightman (1884-1954), who had studied under Borden Parker Bowne (1847-1910). Bowne, a friend of the philosopher William James, founded the Boston Personalist tradition in philosophy. In addition to editing three collections and coediting two – most recently *Personalism Revisited* – Buford has authored four books, perhaps the most notable of which is *In Search of a Calling* (1995).

Gregory M. Fahy, Ph.D.

Dr. Gregory Fahy is the Chief Scientific Officer and Vice President of 21st Century Medicine, Inc. (USA), a company specializing in advanced methods of preserving complex living systems at either cryogenic or hypothermic temperatures. A California native, he received his B.S. from the University of California at Irvine and his Ph.D. in pharmacology from the Medical College of Georgia (USA). He developed organ banking technology at the American Red Cross research facilities and at the U.S. Navy in Maryland and continues this work presently in California. Known as the founder of vitrification research, Dr. Fahy is widely published and holds numerous patents in low temperature biotechnology and aging intervention.

Terry Grossman, M.D.

Dr. Grossman is founder and medical director of Frontier Medical Institute in Denver, Colorado (USA), a leading longevity clinic. Certified in antiaging medicine, he lectures internationally on longevity and antiaging strategies. In the words of Arline Brecher, coauthor of *Forty Something Forever*, "I've met good writers and good doctors, but seldom are they one and the same. Dr. Terry Grossman breaks the mold and sets a new standard for physicians." His numerous publications include these books: *The Baby Boomers' Guide to Living Forever* (2000), and, with Ray Kurzweil, *Fantastic Voyage: How to Live Long Enough to Live Forever* (2004) and *TRANSCEND: Nine Steps to Living Well Forever* (2009).

William James, M.D.

Born January 11, 1842 in New York City. 1869-Received M.D. from Harvard. 1875-Began teaching psychology at Harvard. 1882-Death of William's father, Henry James Sr. 1890-Published The Principles of Psychology. 1892-Turned lab over to Hugo Munsterberg. 1897-Published Will to Believe and Other Essays. 1907-Published Pragmatism and officially resigned from Harvard. Died August 26, 1910 at the age of 68.

John Randolph LeBlanc, Ph.D.

John Randolph LeBlanc (Ph.D., Louisiana State University, USA, 1997) is Associate Professor of Political Science at the University of Texas at Tyler (USA) where he teaches political philosophy and public law. He is author of *Ethics and Creativity in the Political Thought of Simone Weil and Albert Camus* (2004) as well as several articles on the ethical and political implications and ethical possibilities of artistic creativity in Camus and his readers, most notably the Palestinian-American literary and cultural critic Edward Said. He is currently working on a book-length study of Said's political theory. He may be reached at <randy_leblanc@uttyler.edu>.

Jack Lee, Ph.D.

Dr. Jack Lee received his Ph.D. in Philosophy from the University of Queensland (Australia) in 2000. Presently he is an Assistant Professor at National Central University (Taiwan). Dr. Lee is author of numerous publications, including *Can Death Be a Harm to the Person Who Dies* (Kluwer Academic Publishers, 2002); and, as Editor and contributor, *Sustainability and Quality of Life* (Ria University Press, 2010).

J. R. Lucas

Fellow of the British Academy. Fellow of Merton College, University of Oxford, UK. Author: *The Principles of Politics*, 1966, 1985; *The Concept of Probability*, 1970; *The Freedom of the Will*, 1970; (jointly) *The Nature of Mind*, 1972; (jointly) *The*

Development of Mind, 1973; *A Treatise on Time and Space*, 1973; *Essays on Freedom and Grace*, 1976; *Democracy and Participation*, 1976 (Portuguese, 1985); *On Justice*, 1980; *Space, Time and Causality*, 1985; *The Future*, 1989; (jointly) *Spacetime and Electromagnetism*, 1990; *Responsibility*, 1993; (jointly) *Ethical Economics*, 1997; *The Conceptual Roots of Mathematics*, 2000; (jointly) *An Engagement with Plato's Republic*, 2003; *Reason and Reality*, 2009. Also see: <http://users.ox.ac.uk/~jrlucas>.

Ludwig A. Minelli, LL.M.

Ludwig A. Minelli (born 1932) holds an LL.M. (Master of Laws) from Zurich University (Switzerland). He is a Swiss lawyer and human rights activist. He is the founder of Dignitas <www.dignitas.ch>, an organization that helps permanently ill people to end life in a manner which relieves pain and suffering; Dignitas has helped hundreds of people with an accompanied, self-determined end of life. Minelli is also the founder and general secretary of the Swiss Society for the European Convention on Human Rights. Minelli supports physician-assisted suicide, believing that it should be available to all suffering individuals.

R. Michael Perry, Ph.D.

Dr. Perry (Ph.D., computer science) has worked at Alcor Foundation, a cryonics organization now in Scottsdale, Arizona, since 1987. In spare time he completed a book, *Forever for All*, which deals with scientific and moral issues connected with physical immortality. He is currently working on a revised edition of his book, and on a deeper investigation of the philosophical issues connected with personal identity and survival. Meanwhile, an interest in computers, mathematical programming, and artificial intelligence continues. Dr. Perry is a cofounder and member of the Society for Universal Immortalism which is devoted to one day solving the problem of death in its entirety through a scientific approach.

John Searle, D. Phil.

John Rogers Searle (born 1932) is an American philosopher – currently the Slusser Professor of Philosophy at the University of California, Berkeley. Searle began his college education at the University of Wisconsin-Madison, and subsequently became a Rhodes Scholar at Oxford University (UK) where he earned an undergraduate degree and a doctorate in philosophy and ethics. Widely noted for his contributions to the philosophy of language, philosophy of mind and social philosophy, he began teaching at Berkeley in 1959. He received the Jean Nicod Prize in 2000, and the National Humanities Medal in 2004. He has written over 20 books, and over 200 articles. His work is translated into 21 languages.

Asher Seidel, Ph.D.

Asher Seidel is an associate professor of philosophy at Miami University (Ohio, USA). His specialties include metaphysics, epistemology, and philosophy of mind. While the foregoing have been his primary areas, he has also published in the areas of philosophy of language, philosophy of mathematics, and social and political philosophy. He has lately focused on philosophical aspects of posthumanism, and has published two monographs in this area.

David Simpson, Ph.D.

David Simpson serves on the visiting faculty of The School for New Learning, DePaul University (USA), where he teaches courses in moral philosophy and cosmopolitan ethics and global community. His main area of interest is intellectual history, and he has a particular interest in writers, like Albert Camus, who reside at the crossroads of literature and philosophy. He received his Ph.D. in English and Comparative Literature from Columbia University (USA), and has published articles and reviews on a variety of topics, from ancient philosophy and poetry to modern jazz, slang, and popular culture. He is the author of the article on Camus for the Internet Encyclopedia of Philosophy.

Charles Taliaferro, Ph.D.

Charles Taliaferro, Professor of Philosophy, St. Olaf College, is the author or editor of thirteen books, including *Evidence and Faith: Philosophy and religion since the seventeenth century* (Cambridge University Press). He is on the editorial board of the American Philosophical Association, Philosophy Compass, and other journals.

Charles Tandy, Ph.D.

Dr. Charles Tandy received his Ph.D. in Philosophy of Education from the University of Missouri at Columbia (USA) before becoming a Visiting Scholar in the Philosophy Department at Stanford University (USA). Tandy is author or editor of numerous publications, including the *Death And Anti-Death* set of anthologies from Ria University Press. Tandy, along with Nobel Laureates and others, is a member of the Board of Advisors of the Lifeboat Foundation. Tandy is Associate Professor of Humanities, and Senior Faculty Research Fellow in Bioethics, at Fooyin University (Taiwan). In 2010, Dr. Tandy authored *21st Century Clues: Essays in Ethics, Ontology, and Time Travel* (Ria University Press, 2010). Also see: <www.segits.com>.

James Yount

James R. ("Jim") Yount is the Chief Operating Officer of the American Cryonics Society, Inc. (a nonprofit charity) where he also serves as a Governor. He is a past President, and has been a Board member since 1974. Mr. Yount writes for "Long Life Magazine," where he serves as a Contributing Editor. He is the author of a supplemental article in the Ria University edition of R.C.W. Ettinger's *Prospect of Immortality* (Ria University Press, 2005). He is a contributing writer to the anthology *Being Human - The Technological Extensions of the Body* (Agincourt/Marsilio, 1999), edited by Houis, Mieli, and Stafford.

ABSTRACTS FOR THIS VOLUME

Death And Anti-Death, Volume 8:
Fifty Years After Albert Camus (1913-1960)

Abstract Of Pages 33-52
CHAPTER ONE
Homer, Heroes And Humanity:
Vico's *New Science* On Death And Mortality
Giorgio Baruchello
The past decades have witnessed a renewed and growing interest in the thought of Neapolitan philosopher Giambattista Vico (1668—1744), particularly with regard to the third edition of his most famous work, *The New Science* (1744). However, as this book is concerned, Vico's understanding of death and mortality has not been the subject of any scholarly study. It is therefore the aim of the present chapter to provide one, thus revealing how the topics of death and mortality recur in, and allow for the explanation of, three key-themes in his philosophy: poetic wisdom, the providential unfolding of ever-changing civilizations, and the universal unchanging hallmarks of actual humanity.
KEYWORDS: barbarism; civilization; hero(es/ic/ism); history; law; philosophy; poetry; rhetoric; Roman law; trope; Vico

Abstract Of Pages 53-78
CHAPTER TWO
Cryonics: A Scientific Challenge To Death
Benjamin P. Best
Very low temperatures create conditions that can preserve tissue for centuries, possibly including the neurological basis of the human mind. Through a process called vitrification, brain tissue can be cooled to cryogenic temperatures without ice formation. Damage associated with this process is theoretically reversible in the same sense that rejuvenation is theoretically possible by specific foreseeable technology. Injury to the brain due to stopped blood flow is now known to result from a complex series of processes that take much longer to run to completion than the six minute limit of ordinary resuscitation technology.
KEYWORDS: brain; cryobiology; mind; stasis; vitrification

Abstract Of Pages 79-90
CHAPTER THREE
Primary Institutions
Thomas O. Buford

Are any institutions primary? If yes, are they made or found? The ontological status of general social patterns of societies (e.g., education or family), within which individuals formulate their own specific ones, was raised in John Searle's *The Construction of Social Reality*. Here we use radical empiricism (Borden Parker Bowne), not methodological skepticism (Descartes), to search for structures that are implicit in experience and manifest themselves within it: Education as an enduring and stable social structure is creatively-found. Thus our answers to the two questions raised by Searle are that there are primary institutions and that they are neither found nor made; they are both found and made.

KEYWORDS: Bowne [Borden Parker Bowne]; Descartes; education; family; methodological skepticism; radical empiricism; social ontology; social reality

Abstract Of Pages 91-120
CHAPTER FOUR
Physical And Biological Aspects Of Renal Vitrification
Gregory M. Fahy et al.

This paper reviews the general problems of preserving kidneys at cryogenic temperatures without ice crystal damage and describes the results of an experiment in which a rabbit kidney was vitrified, rewarmed, and transplanted with long term survival of the recipient, the vitrified kidney being the sole renal support. Analysis of cryoprotective agent distribution within the kidney identified noninvasive means of determining when a kidney is stable enough to vitrify and recover without freezing upon rewarming (devitrification). Armed with this information, it may be possible to design methods that will consistently allow kidneys to be vitrified and rewarmed without ice crystal damage.

KEYWORDS: cryopreservation; freezing, alternative to; survival, after vitrification; cryoprotectant toxicity; kidney, rabbit; transplantation, rabbit kidney, after vitrification and rewarming; organ banking; bioartificial organs; organ freezing; ice, avoidance of

Abstract Of Pages 121-146
CHAPTER FIVE
Latest Advances In Antiaging Medicine
Terry Grossman

Rapid progress is being made in our ability to modify the aging process. The leading causes of death (cardiovascular disease, cancer, lung disease, diabetes) are the end result of the aging process. Emerging genomics technology will allow individuals to establish personalized antiaging programs, while early detection of heart disease and cancer will contribute to longevity. Biotechnological therapies involving stem cells, recombinant DNA, proteomics, therapeutic cloning and gene-based therapies are expected to play major roles in promoting successful aging. These future therapies have the potential to greatly extend human longevity.

KEYWORDS: aging; genomics; caloric restriction; longevity; hormone replacement; stem cells; molecular nanotechnology; nanobots; artificial intelligence

Abstract Of Pages 147-170
CHAPTER SIX
The Will To Believe
William James

This is a reprint of the classic essay by William James. It is sometimes concluded that its primary purpose is meant to be a defense of religion. It seems, however, that James is really emphasizing that there are particular conditions for particular kinds of beliefs properly held, but that the details will vary from individual to individual. Moreover, "a rule of thinking which would absolutely prevent me from acknowledging certain kinds of truth if those kinds of truth were really there, would be an irrational rule." Indeed, contrary to religious or philosophical absolutism, James argues that we should view our beliefs as tentative or uncertain, always open to revision.

KEYWORDS: absolutism; avoidable option; dead option; forced option; genuine option; living option; momentous option; religion; trivial option; truth

Abstract Of Pages 171-198
CHAPTER SEVEN
Politics, Death, And Camus's Late Anarchic Style
John Randolph LeBlanc

This essay argues that a consciousness of death shapes Albert Camus's response to his political environment. He demonstrates a nuanced understanding of the technology of death at work in his world, but is not always able to resist it. After embracing the deadly political methodology of modernity during the postwar purges in France, Camus refused to continue that affiliation when his native Algeria was engulfed in a bitter war of decolonization. Turning away from politics as usual, he sought a new politics characterized by a more humane form of interaction, a creative, life-affirming form analogous to that of nonviolent anarchism.

KEYWORDS: creativity; Comfort [Alex Comfort]; aphoristic politics; Algerian Civil War; postwar purges; politics and modern science; modernism; modernity; ethics; justice

Abstract Of Pages 199-210
CHAPTER EIGHT
Can One Be Harmed Posthumously?
Jack Lee

I argue (contrary to the Epicureans) that one can be harmed by posthumous events: 1. Some of a person's interests can survive the death of the person. 2. The interests which survive the death of the person may be impaired by posthumous events. 3. Harm is the impairment of interest. 4. Therefore, a person can be harmed by posthumous events – that is, one can be harmed posthumously. I also show that if a person is harmed posthumously, it is the *ante-mortem* person who is the subject of the posthumous harm. And he is harmed *before* his death although the posthumous harm-event occurs after he dies.

KEYWORDS: posthumous harm; interest; Epicurus; ante-mortem person

Abstract Of Pages 211-224
CHAPTER NINE
The Gödelian Argument: Turn Over The Page
J. R. Lucas

The Gödelian Argument uses Gödel's theorem to show that minds cannot be explained in purely mechanist terms. It has been put forward, in different forms, by Gödel himself, by Roger Penrose, and by me. Most of the critics have not read either of my, or either of Penrose's, expositions carefully, and seek to refute arguments we never put forward, or else propose as a fatal objection one that had already been considered and countered in our expositions of the argument. Hence my title. To be reasonable is not necessarily to be rule-governed, and actions not governed by rules are not necessarily random. If belatedly, let's turn over the page.

KEYWORDS: Gödel [Kurt Gödel]; machines; minds; Penrose [Roger Penrose]; randomness; reason; rules; Turing test

Abstract Of Pages 225-234
CHAPTER TEN
The Function Of Assisted Suicide
In The System Of Human Rights
Ludwig A. Minelli
The courts are beginning to recognize one's right to **suicide**. Nevertheless, in such decisions up until the end of 2010, no court has considered the problem of attempted yet failed suicides, which by far exceeds the number of finalised suicides. It is not at all easy to end one's own life. As long as residents of other countries have to travel to Switzerland for **assisted** suicide because the law of their own country does not allow them to ask for **assisted** suicide at home, neither their freedom of choice nor their right to suicide can be said to correspond with the guarantees of the European Convention on Human Rights.

KEYWORDS: assisted suicide; assisted suicide law; end of life; euthanasia; failed suicides; human rights; right to assisted suicide; right to suicide; suicide; suicide law; Switzerland; voluntary end of life

Abstract Of Pages 235-292
CHAPTER ELEVEN
Death, Resurrection, And Immortality:
Some Mathematical Preliminaries
R. Michael Perry

A mathematical formalism is developed that, it is argued, has relevance to the nature of personhood, and addresses certain fundamental problems including death, resurrection and immortality. An argument from principle is given that resurrection, even after total obliteration or "information death," is logically possible and not in violation of generally accepted physics. The argument is not based on the necessity of recovering a "hidden past" but, as a last resort, on the eventual likelihood of a duplicate of a lost person recurring fortuitously, supposing time and space are overall unlimited.

KEYWORDS: death; resurrection; immortality; topology; personhood; multiverse; observer; mathematics of resurrection; mathematics of immortality

Abstract Of Pages 293-302
CHAPTER TWELVE
The Chinese Room Argument
John Searle

Strong Artificial Intelligence is answered by a simple thought experiment. If computation were sufficient for cognition, then any agent lacking a cognitive capacity could acquire that capacity simply by implementing the appropriate computer program for manifesting that capacity. The thought experiment is underlain by a deductive proof, herein articulated. The Chinese Room Argument is incidentally also a refutation of the Turing Test and other forms of logical behaviorism. The brain is a causal mechanism, not a mere simulation. The Argument is **NOT** meant to show that computers or machines can't think, that it is impossible to build a thinking machine, or that minds or thinking machines are substrate dependent.

KEYWORDS: artificial intelligence; brain; cause; computation; consciousness; logical behaviorism; minds; simulation; substrate; Turing test

Abstract Of Pages 303-332
CHAPTER THIRTEEN
What's Best For Us
Asher Seidel

I urge that the best future course for humanity is to migrate to a posthuman existence. I consider three reasons in support of this recommendation, settling on the reason that life-extending discoveries will likely be announced in the near future. I speculate that the haphazard implementation of these discoveries occasions suffering, but that if humanity survives the likely tumult there awaits posthuman possibilities of greatly enhanced cognitive and creative powers. Other possible consequences of this transition to posthumanity include childlessness and sexlessness. Concomitant with, and as consequences of, this discussion are philosophical implications regarding ethics, metaphysics, and epistemology. Various objections to my overall suggestion are considered.

KEYWORDS: posthumanism; transhumanism; ethics; extended life; cognitive enhancement

Abstract Of Pages 333-362
CHAPTER FOURTEEN
Camus, Plague Literature, And The Apocalyptic Tradition
David Simpson

The Plague and *The State of Siege* represent Camus's innovative reworking of traditional apocalyptic literature. Inspired by Melville's *Moby Dick* and building on previous examples of plague literature, including the accounts of Thucydides, Lucretius, and Defoe, *The Plague* enlarges upon its sources by adding a sustained pattern of allegory and symbolism. In *The State of Siege*, Camus radically adapts the conventions of the medieval morality play and blends them with elements of modern theatre and dystopian satire and with themes from Kafka and Orwell. Both works illustrate the author's philosophy of rebellion and also represent important contributions to Holocaust literature.

KEYWORDS: Melville; Kafka; Dostoyevsky; Orwell; Black Death (bubonic plague); allegory; Holocaust; totalitarianism; Nazi occupation of France; morality play; Daniel Defoe

Abstract Of Pages 363-378
CHAPTER FIFTEEN
The Absurd Walls Of Albert Camus
Charles Taliaferro

An analysis and defense of the reasons behind Camus' conclusion that, given secular naturalism, human life is absurd because of the impersonal (non-teleological) nature of the cosmos, the one-directional nature of time, and the probable, final annihilation of human life. Employing recent work by Thomas Nagel, the paper also articulates conditions when the feeling that life is absurd can provide reasons for pursuing a worldview in which human life has meaning.

KEYWORDS: Thomas Nagel; Bertrand Russell; absurdity; meaning; naturalism; theism; one-directional nature of time; teleology; death; reason

CHAPTER SIXTEEN
Camusian Thoughts About The Ultimate Question Of Life
Charles Tandy

Albert Camus said he lived in a time of absurdity. I argue we still live there. Then I look at his two major philosophical works: *The Myth of Sisyphus* asks "Is my life worth living?"; *The Rebel* asks "How do I live a meaningful life?" Like Camus, we can embrace absurdity for the purpose of rebelling against it and giving each person their due. "All may indeed live again," says Camus, if in solidarity we will continue to continue to refuse to give in to literal death and ideological death. Our Camusian common task – the struggle against death and the birthing of Meridiankind – is an occasion of strange joy.

KEYWORDS: absurdity; Camus [Albert Camus]; death; ideology; meaning of life; power; reason; rebellion; religion; science; solidarity

CHAPTER SEVENTEEN
The UP-TO Project:
How To Achieve World Peace, Freedom, And Prosperity
Charles Tandy

Neither nature nor human nature prevents our achieving a free and prosperous world at stable peace in the 21st century. That does not mean it will be easy. I try to show that two new politically feasible organizations, working in concert, can do much to help us in the task. I outline a structure for an UP (**U**nion of Well-ordered

Peoples) and for a TO (Treaty Organization Acting for a Better Cosmos). At this doubly-unique terrestrial-extraterrestrial point in history, UP-TO is politically doable. A window of opportunity has opened. It is an open invitation for the world to achieve peace, freedom, and prosperity for the first time in history.

KEYWORDS: European Union; extinction; extraterrestrial; intentional communities; O'Neill [Gerard K. O'Neill]; political philosophy; prisoner's dilemma; Rawls [John Rawls]; Taiwan; transparency; well-ordered peoples

Abstract Of Pages 419-448
CHAPTER EIGHTEEN
Life And Death, And The Identity Problem
James Yount

Experiments, mostly "thought experiments" proposed by Robert Ettinger in *The Prospect of Immortality* (first published in 1964) to determine the bounds of Personal Identity, are examined. Additional thought experiments are posed. The determination that the individual "revived" from a frozen or vitrified state is the same individual who was frozen or vitrified is very important to one undergoing cryonic suspension. A distinction is made between *inside observer* (person frozen or vitrified) and *outside observer* (everyone else). Use of information that made up the memories may be acceptable to "reconstruct" an individual from the viewpoint of the outside observer but is not acceptable to the inside observer.

KEYWORDS: cryonics; cyborg; Ettinger [Robert Ettinger]; immortality; technology; information theory; memory; restoration; singularity; cryonic suspension

CHAPTER ONE

Homer, Heroes And Humanity:
Vico's *New Science* On Death And Mortality

Giorgio Baruchello

The past decades have witnessed a renewed and growing scholarly interest in the thought of Neapolitan philosopher Giambattista Vico (1668—1744), particularly with regard to the third edition of his most famous work, *The New Science* (1744). Perhaps entangled in the loops of recurring history or in the book's obscure prose, fame took a long time to catch up with it. In Vico's lifetime, *The New Science* was received, reviewed and circulated poorly, not to mention that Vico himself paid for the publication of its first edition in 1725 (cf. Capongiri, 1953).

Even in Italy, any major acclaim, if not the widespread knowledge of Vico's main work, remained limited for longer than a century and a half, i.e. until Benedetto Croce (1911) initiated a substantial revival of Vichian studies in the 20th century.[1] In this perspective, just as much is owed to Croce's close associate and friend, Fausto Nicolini (e.g. 1932, 1955, 1991), whose lifelong laborious recovery of archival materials, sources and historical information has allowed for an articulate and deeper understanding of Vico's personal vicissitudes, intellectual formation and socio-cultural milieu.

As far as the Anglophone academe is concerned, Collingwood's 1913 translation of Croce's study on Vico, Adams' *Life and Writings of Giambattista Vico* (1935) and the 1948 English translation of the third edition of Vico's *New Science* by Thomas Goddard Bergin and Max Harold Fisch constitute the three crucial steps in the 20th-century renaissance of Vichian studies in the British Isles and North America.

The recognition of Vico's originality by liberal icon Isaiah Berlin (1960, 1976, 2000) was also important, and so was the use that James Joyce made of Vico's ideas in designing his most experimental literary works (cf. Verene, 1987; Robischaud, 2003). Moreover, as of the 1980s, new translations and essays were written and promoted by Giorgio Tagliacozzo and Donald Phillip Verene, editors of *New Vico Studies*, which by now have succeeded in presenting to the English-speaking academic community the majority of Vico's works (Bayer & Verene, 2009: 199-204).

This is not to imply that Vico had been ignored during the 19th century. Italian scholarship was neither oblivious nor indifferent to him (cf. Giannantonio, 2009). Occasional references can be retrieved in the works of leading French and German intellectuals. For instance, Auguste Comte praised Vico for his insights in the stages of human progress (Pickering, 2009: 297) and dedicated to him a day in his positivist calendar, i.e. the 17th of the 11th month or "Descartes" (Comte, 1849). Karl Marx cited Vico while discussing the historical nature of man and the human nature of history (1970: 372n3). Jules Michelet saluted Vico as a distant prophet of "the terrible creative power of ordinary people" (Grafton, 1993: 51). Nevertheless, it was only in the 20th century that Vico became a full member of the Western philosophical canon.[2]

Despite the recent wealth of Vichian studies, as far as I can discover, Vico's comprehension of death and mortality in *The New Science* has not been yet the subject of any scholarly study.[3] Allusions to the metaphorical "death of a nation" appear in works dealing with Vico's theory of historical cycles (e.g. Pompa, 1990: 203; this "death" too will be addressed here), but not to death and mortality as such. It is the aim of the present chapter to provide one, revealing how the topics of death and mortality recur in, and allow for the explanation of, three essential themes of Vico's philosophy, i.e. poetic wisdom, the providential unfolding of ever-changing civilizations, and the universal unchanging hallmarks of actual humanity.[4]

Homer

Death pervades the works of Homer. References to, and discussions of, Homer pervade Vico's *New Science*. As a result, death pervades Vico's *New Science*. But why was Vico interested in Homer?

As he set out to review as much ancient material evidence as possible—the activity of "philology" (§ 7)—and extract from it the providential logic of the course of human history—the activity of "philosophy" (§ 7)—Vico could not avoid referring to the Homeric poems, which are the founding stone of Greek culture and, arguably, Western culture at large.[5] Besides, Vico and his Baroque colleagues were bound to literary documents to a higher extent than today's historians and social scientists, who have benefitted enormously from the widespread technological application of scientific knowledge to the investigation of the remote past of humankind (cf. Capongiri, 1953).

Concerning the received view of Homer in the times of Vico, there was a widespread belief that both the *Iliad* and the *Odyssey* displayed a profound philosophical character. As Plato himself had argued, these poems were supposed to be the work of a true philosophical genius, who had cast profound metaphysical thoughts in marvellous poetic formulations (§ 780).

Vico took issue with this belief, and death played a role in leading him to do so.

According to Vico, no philosopher would have ever been so complacent *vis-à-vis*, not to say keen on, the "cruel and fearful... battles and deaths" (§ 894) with which Homer's poems are rife, and the *Iliad* in particular. Similarly, the leading character of this poem, the fearless demigod Achilles, is far too merciless, vindictive and self-indulgently fickle to be regarded as a model of virtue under whatever philosophical perspective one may opt for.

For instance, Vico observes that Achilles "is pleased—he who carries with him the fate of Troy—to see all the Greeks fall to ruin and suffer miserable defeat at Hector's hands." (§ 786) As morally disturbing as it may sound, Achilles is so obdurate and self-centred that "not even in death is he placated for the loss of his Briseis until the unhappy beautiful royal maiden Polyxena... has been sacrificed before his tomb" (§ 786). Vico's measure is not full though, given that "what is really past understanding" for him is "that a philosopher's gravity and propriety of thought could have been possessed by a man who amused himself by inventing so many fables worthy of old women entertaining children, as those with which Homer stuffed his other poem, the Odyssey" (§ 786).

Homer's delight in portraying repeatedly and unashamedly "[s]uch crude, coarse, wild, savage, volatile, unreasonable or unreasonably obstinate, frivolous and foolish customs" does not suit mature intellects that are capable of philosophising, but "men who are like children in the weakness of their minds, like women in the vigor of their imaginations and like violent youths in the turbulence of their passions" (§ 787). Consequently, Vico's conclusion is unequivocal: "we must deny to Homer any kind of esoteric wisdom" and rather step up to "seeking out the true Homer." (§ 787)

This is not to say that there is no genius in Homer's epic. Vico did believe the *Iliad* and the *Odyssey* to be magnificent poems. He praises openly Homer's inventiveness and, as further explained in the next subsection ("Heroes"), there is more to Vico's appreciation of poetry than sheer style (§§ 404-8, 806-8). For him, Homer's poems were genial works, but of the literary sort; hence they had to be grasped in their own terms, i.e. *qua* tokens of "poetic wisdom" (§ 364). In them lurks no hidden form of "sublime" theoretical knowledge that can be excavated by some subtle application of philosophical reasoning (§ 780).

Uniquely for his time, Vico thought the *Iliad* and the *Odyssey* to be highly imaginative, synthetic depictions of reality that were produced—indeed the only ones producible—by a fascinating primitive civilization, somehow closer to the "American Indians"

36

of his day than to him (§ 375). For Vico, Homer's Greece was an archipelago of order-conscious, slave-owning societies, which were headed by verse-reciting murderous warriors (§ 81). They were not the home of long-bearded, toga-robed sages surrounded by respectful disciples and sun-basked white colonnades (cf. Grafton, 1993).

What is more, the two Homeric poems seemed so dissimilar to Vico, as well as the tales and stylistic devices deployed in each of them, that *The New Science* posits the audacious and still-debated claim that they were composed by two distinct authors (§ 880), if not even by a multitude of "Homers." Vico's "true Homer" is finally revealed to be an "idea or a heroic character of Grecian men insofar as they told their histories in song." (§ 873) In other words, the legendary blind poet is actually the *a posteriori* personification of the "history of the natural law of the gentes of Greece" (§ 904).[6]

Homer's works are not the only ancient literary sources of Vico's *New Science* that are redolent with death. A plethora of dramatic, gruesome, and sometimes morbid myths are cited throughout his book. For example, we read of: "modest maiden Daphnes", whom Apollo "pursues to the point of death" (§ 80); Cadmus' slaying of the dragon and the fight of his ghost-soldiers (§ 81); Aeneas' murder of "his *socius* Misenus when it is needful for a sacrifice" (§ 558); "Hercules [who], stained by the blood of the ugly centaur Nessus… go[es] forth in madness and die[s]" (§ 802); and the cruel flaying of Marsyas by Apollo (§ 1021).

Death lies all around the gods and the mythical heroes of old.

Heroes

Death informs and surrounds the other major body of "philological" references harvested in Vico's *New Science* for the "philosophical" understanding of history: Roman law.

The *New Science* focuses especially upon the early stages of the Roman republic. Contrary to its common albeit misleading representation as a precocious experiment in democratic self-rule,

Vico describes Rome's senatorial republic as a harsh oligarchic regime: "forty years after the expulsion of Tarquinius Superbus, in the comfortable assurance of his death, the nobility had again begun to be insolent toward the unhappy plebs." (§ 619) Seen from the side of Rome's *populus*, the republic was no better than the previous tyrannical regime: "the liberty instituted by Brutus was not popular (the freedom of the people from their lords) but aristocratic (the freedom of the lords from the Tarquin tyrants). (§ 664; see also § 26)

The celebrated leaders of the young Roman republic, its "heroes", are depicted as intense, ruthless, and unyieldingly elitist. "Brutus... Scaevola... Manlius... Curtius... Fabricius and Curius... Atilius Regulus" either faced a brutal death or brought it onto others, sometimes even onto their own children (§ 668). Still, "what did any of them do for the poor and unhappy Roman plebs? Assuredly they did but increase their burdens by war, plunge them deeper in the sea of usury, in order to bury them to a greater depth in the private prisons of the nobles, where they were beaten with rods on their bare backs like abject slaves." (§ 668) Furthermore, "if anyone in this period of Roman virtue attempted to relieve the lot of the plebs with some sort of agrarian or grain law, he was accused of treason and sent to his death." (§ 668; see also § 115)

As a matter of "philological" fact, Vico notes that in republican Rome the death penalty was a standard feature of the judicial system (§§ 115, 997, 1021).

The deadly severity of the republic's aristocracy is, in Vico's view, just another instance of a recurrent historical phenomenon, to which Greek myths and Homeric poems had given ample though unnoticed testimony. The recurrent historical phenomenon, which applies equally to "Assyrians, Chaldeans, Phoenicians and Egyptians" (§ 317[7]), is the creation of oligarchies by "heroic fathers" or "heroes" (§ 521). These were men who had overcome the earliest condition of quasi-animal life in the wildernesses of the world (e.g. Vico's "Patagonian giants" or Homer's "cyclopes"; § 338) and the successive family-wide farming gatherings (§ 296), so as to establish the first real civil communities or "first cities" (§ 16). They were heroic examples of *pater familias*, who were thought of initially as "mortal gods" (§ 449), then as priests (e.g. §§ 25, 110, 549, 627) and eventually as monarchs (e.g. §§ 513, 582-3, 964).

Death, once more, is a crucial element in the picture.

The *pater familias* was the one who had the supreme right of life and death (*ius vitae necisque*) over his children, wives, slaves and animals, i.e. all of his animate possessions (e.g. §§ 517, 582, 670). Ancient oligarchies extended the deadly power of the "heroic fathers" onto larger and larger populations. Even so, the fundamental social criterion remained analogous to that between father and child, i.e. master and slave or, in the early days of Rome, patrician and plebeians or "plebs" (§§ 25, 1069, 1079).

These two groups' mutual relations were forged in acts of supreme domination and abject submission, which *The New Science* collects and examines extensively. Again, these acts of domination and submission were reflected not solely in Roman religious and legal formulations, but also and most originally in much older myths, e.g. the "Herculean knot by which clients were said to be *nexi* or tied to the lands they had to cultivate for the nobles... [or] how Ulysses is on the point of cutting off the head of Antinous [i.e. Eurylochus], the chief of his *socii*, just for a word which, though well meant, does not please him" (§ 558).

Clarifying further the notion of "poetic wisdom" addressed in the previous subsection ("Homer"), Vico's "philosophical" interpretation of the ancient myths reveals itself as being socio-political. As he writes: although "philosophers later found all these fables convenient for the meditation and exposition of their moral and metaphysical doctrines... poets had... political ideas" in mind (§§ 720-1).

Equipped with this novel hermeneutical perspective, Vico could explain the tragic death of Orpheus as follows: "the founder of Greece, with his lyre or cord or force, which signify the same thing as the knot of Hercules (the knot with which the Petelian law was concerned), met his death at the hands of the Bacchantes (the infuriated plebs), who broke his lyre to pieces (the lyre being the law...)... so that already in Homer's time the heroes were taking foreign women to wife, and bastards were coming into royal successions" (§ 659).

Today, Vico's novel hermeneutical perspective is part of anthropology's methodological armoury. In his day, it allowed *The New Science* to make sense of a vast number of myths, thus engendering an inventive historical consciousness and, as I shall explain, to regard humanity as unfolding in stages according to an unfaltering inner criterion (cf. Bayer, 2004).

The very deadly time of the "heroic fathers" was not the end of history, as the mythical end of Orpheus implies. It was a step in a divinely inspired chain of events that, *mutatis mutandis*, are to take place in all civilization (e.g. §§ 5, 29). In the case at hand, the oligarchs' subjects, whether *qua* citizens of Pericles' Athens (§ 592) or plebs of Quintus Publilius Philo's Rome, did eventually break free from the chains of slavery. They established a "commonwealth" of "popular liberty." (§ 112) As tokens of the changed humanity of these novel, free communities, habits and institutions became milder, and the death penalty far less common than before (§ 1022).

Vico's *New Science* embraces an enormous array of sources and historical events. Human civilizations have been many and they have been changing all the time. Nonetheless, the logic according to which human civilizations have been unfolding can be described as a consistent triadic process. This process manifests itself in all civilizations, in a contextually particular and never identical fashion. Henceforth, according to Vico, no historiography had been able to grasp it fully before his "new science", which he defined as "a rational civil theology of divine providence" (§ 2).

Specifically, after God's providence allowed humankind to overcome an otherwise insuperable animal-like condition, societies are to advance through three stages: an "age of the gods" (i.e. rural theocratic gatherings), an "age of the heroes" (i.e. civil oligarchic regimes), and an "age of men" (i.e. "popular commonwealths" and related anarchy-correcting humane "monarchies"; § 31; see also § 630).

From the peaks of refinement of the third stage, societies can relapse subsequently into a novel form of "barbarian times" and refuel another cycle of stages, thus instantiating "the recourse the nations take in natures and customs." (§ 699) This "recourse" preserves however some of the wisdom accumulated up to that point, so that the new barbaric times are shorter and somewhat less barbarous (§ 159).[8]

Vico scholar Leon Pompa (1990: 203) calls the recurrence of barbarism the "death of a nation". Yet, for Vico, this "death" was literally a godsend, for it guarantees that history may move forward in spite of human errors. Vico's "recourse" of natures and customs is not the eternal recurrence of the same. Falling back into "primitive simplicity" is a "last remedy of providence" (§ 1106), which rescues ungodly populations from the socially devastating evils of selfish "premeditated malice" and proud scepticism, i.e. the two causes of falsely civilised "liv[ing] like... beasts in... solitude of spirit and will, scarcely any two being able to agree since each follows his own pleasure or caprice." (§ 1106)

Such are the bitter fruits of the irreligious and over-refined "barbarism of reflection", which lets the human being lose touch with its natural, corporeal and imaginative faculties and with each society's particular forms of traditional wisdom, including the poetic one (§§ 1106-11). Therefore, for Vico, returning to a novel form of "primitive simplicity" means that human beings are able to regain "piety, faith and truth" (§ 1106). Simon (1981: 317) rightly dubs this point of Vico's thought a "phoenix-like rejuvenation."

All aspects of human civilizations are guided by this triadic providential logic, e.g. the notions of quantity ("weight... measure... number"; § 713), the main psychological dispositions or "customs" of the age (§§ 919-21), the graphic signs or "characters" standardly utilised (§§ 932-6), the valid sources of legal "authority" (§§ 942-6), the penal chastisements ordinarily employed (§§ 1021-2), the spirit of the times or "sects of time" ("fashion"; §§ 975-9), and even the rhetorical "tropes" that make conceptualisation and comprehension possible (§§ 402-8).

Vico taught rhetoric for many years and his *New Science* does not reduce it to style, as it is still widely done today (cf. Mazzotta, 1999). Instead, Vico takes rhetorical tropes to be the mental-linguistic "poetic logic" enabling us to conceive of reality and try to grasp it (§ 404).[9]

In this context, Vico refers to "mortality" (§ 407), which was conceptualised initially by "metonymy of cause for effect", such as the "little fable" of a female "pale Death" that seized human beings (§ 406). It was only later, "as particulars were elevated into universals or parts united... [to] make up... wholes" by increasingly evolved human intellects, that the poetic founders of civilization stepped to the more abstract trope of "metaphor", such as "mortals" for "men" (§ 407). Needless to say, "irony" was amongst the last tropes to be developed, for it requires a high degree of "reflection" to be understood, since "it is fashioned of falsehood" (§ 408)—try saying that rhetoric is, at least for Vico, nothing but empty talk.

Humanity

Vico's *New Science* accounts for both diversity and similarity in human history. On the one hand, his work amasses an incredible amount of incredibly detailed—and sometimes blatantly incredible—information dealing with specific events and specific moments of specific human civilizations. On the other hand, *The New Science* detects consistent trends and patterns across this variety of phenomena, but without positing them as though they were devoid of unique circumstances of manifestation. An admirer and, at times, a devoted misinterpreter of Plato, Vico sought for the unifying harmony that could be discerned behind a vast tapestry of intimately particular historical events (cf. Du Bois Marcus, 2001).

Once again, death is relevant, also in connection with this Platonic aim.

Amongst the consistent trends and patterns across historical phenomena, Vico selects funerary rites or "burial" as a universal indication of civilization as such (e.g. §§ 12-3, 40, 333). This is another discovery that Vico claims to have derived from the proper interpretation of ancient poetic wisdom, as in the following example: "Aeneas... proceeds through the underworld to the Elysian fields (for the heroes, having settled in the cultivated fields, enjoyed eternal peace in death if they had proper burial). Here he beholds his ancestors and those who are to come after him (for on the religion of the graves, called the underworld by the poets, were founded the first genealogies, from which history took its beginning)." (§ 721)

Vico claims that "among all peoples the civil world began with religion" (§ 8) which sanctions the institutions of "marriage" (§ 11) and "burial" (§ 12). In this manner, death lies at the heart of humanity's establishment *qua* humanity: "Indeed *humanitas* in Latin comes first and properly from *humando*, burying." (§ 12; see also § 537)

Vico employs vivid images to make his point: "[to realize] what a great principle of humanity burial is, imagine a feral state in which human bodies remain unburied on the surface of the earth as food for crows and dogs. Certainly this bestial custom will be accompanied by uncultivated fields and uninhabited cities. Men will go about like swine eating the acorns found amidst the putrefaction of their dead." (§ 337)[10]

Religion, marriage and burial are for Vico the principles upon which humanity was established and, even in the case of recurrence of barbarism, preserved, for: "all nations, barbarous as well as civilized, though separately founded because remote from each other in time and space, keep these three human customs: all have some religion, all contract solemn marriages, all bury their dead." (§ 333) Significantly, "burial grounds were called by the Latins religious places par excellence." (§ 529)

As the dramatic event that it undoubtedly is in human affairs, remote as much as present, death is at the heart of civilization in another way, since: "by the graves of their buried dead the giants

43

showed their dominion over their lands, and Roman law called for burial of the dead in a proper place to make it religious (§ 531). Notions of territory, borders, property, ethnic identity, even statehood, emerge from sacred graveyards (§§ 529, 531, 637, 1051).

Additionally, poetic wisdom suggests that death informs religion and *a fortiori* civilization itself in a third way. The primitive, animal-like giants, from whom we all descend, were afraid of death. They lived in treacherous wilderness; they were exposed to the elements and to their most bewildering of expressions, that is to say, the lightening. As Vico explains, the first religious beliefs sprang out of a genuine "terror of Jove [or any other deification of the sky (e.g. §§ 62, 377, 383, 689, 712)], whom they feared as the wielder of the thunderbolt." (§ 502)[11] Modern languages contain the memory of such archaic beginnings in exclamations like *"moure bleu*! [or *morbleu*!]... *parbleu*!... [in which] *bleu* [is] the sky; and, as the gentile nations used 'sky' for Jove, the French must have used *bleu* for God" (§ 482).

Out of the same fear of death developed morality, in the form of conjugal love, which Vico believes to be inextricably linked to religion: "the virtue of the spirit began likewise to show itself among them, restraining their bestial lust from finding its satisfaction in the sight of heaven, of which they had a mortal terror." (§ 504)

Upon these first conjugal unions grew slowly civil society as well, in accordance with God's plan: "divine providence initiated the process by which the fierce and violent were brought from their outlaw state to humanity and by which nations were instituted... It did so by awaking in them a confused idea of divinity, which they in their ignorance attributed to that to which it did not belong." (§ 178; see also § 518)

Interestingly, Vico adds that morality still relies upon this fear, at least as concerns the giant-equivalents of fully human societies, i.e. children: "Hence came the eternal property among all nations, that piety is instilled in children by the fear of some divinity" (§ 503).[12]

Burials, in their connection with the notion of the underworld, reveal another token of the triadic logic that pervades the development of human history. Initially, as Vico writes, "the first lower world must not have been any deeper than the source of the springs." (§ 714) Then, "[w]ith the practice of burial the idea of the underworld was extended, and the poets called the grave the underworld... Thus the lower world was no deeper than a ditch, like that in which Ulysses, according to Homer, sees the underworld and the souls of the dead Heroes" (§ 715) albeit "[l]ater... the underworld had the depth of a furrow. It is to this underworld that Ceres (the same as Proserpine, the seed of the grain) is carried off by the god Pluto" (§ 716). "Finally", so as to emphasise the social divide of the heroic stage of civilization, "the underworld was taken to be the plains and the valleys, as opposed to the lofty heaven set on the mountain tops." (§ 717)[13]

The heroes' fatal severity discussed in the previous subsection ("Heroes") extends to this domain of human life too: the death of the hero's other, the enemy, the inferior. Vico notes that it was a widespread heroic custom "that of denying burial to enemies slain in battle, leaving their unburied bodies instead as a prey to dogs and vultures (on which account the unhappy Priam found so costly the ransom of his son's body, though the naked corpse of Hector had already been dragged by Achilles's chariot three times around the walls of Troy)." (§ 781)

Once more, Homer's heroic poetry discloses for Vico the true origins of humanity, which so much owes to the fearful awareness of mortality, the political usages of death, and death's sacralisation in religious funerary rites.[14]

Bibliography

Adams, H.P. (1935). *The Life and Writings of Giambattista Vico*. London: Allen and Unwin.

Baruchello, G. (2003). Pietist Prejudice in Gadamer's Misreading of Vico. *Filosofia Oggi 26*(3): 291-306.

Baruchello, G. (2004). Giambattista Vico. *Great Thinkers A-Z.* Edited by J. Baggini and J. Stangroom. London: Continuum: 242-244.

Bayer, T.I. (2004). History as Symbolic Form: Cassirer and Vico. *Idealistic Studies 34*(1): 49-65.

Bayer, T.I. and Verene, D.P. (2009). *Giambattista Vico. Keys to the* New Science. *Translations, Commentaries, and Essays.* Ithaca and London: Cornell University Press.

Berlin, I. (1960). The Philosophical Ideas of Giambattista Vico. *Art and Ideas in Eighteenth-Century Italy.* Rome: Edizioni di Storia e Letteratura.

Berlin, I. (1976). *Vico and Herder: Two Studies in the History of Ideas.* New York: Viking.

Berlin, I. (2000). *Three Critics of the Enlightenment: Vico, Hamann, Herder.* Edited by Henry Hardy. London: Pimlico.

Capongiri, A.R. (1953). *Time and Idea: The Theory of History in Giambattista Vico.* Chicago: Henry Regnery.

Collingwood, R.G. (1913). *The Philosophy of Giambattista Vico.* London: Howard Latimer.

Comte, A. (1849). The Positivist calendar. Retrieved from: http://personal.ecu.edu/mccartyr/pos-cal.html

Croce, B. (1911/1965). *La filosofia di G.B. Vico.* 6th edition. Bari: Laterza.

Di Miele, A. (2007). La cifra nel tappeto. Note su Paci interprete di Vico. *Bollettino del Centro di Studi Vichiani 37*: 87-103.

Du Bois Marcus, N. (2001). *Vico and Plato.* New York: Peter Lang.

Feldman, B. and Richardson, R.D. (1972). *The Rise of Modern Mythology, 1680-1860*. Bloomington: Indiana University Press.

Giannantonio, V. (2009). *Oltre Vico. L'identità del passato a Napoli e Milano tra '700 e '800*. Lanciano: Carabba.

Grafton, A. (1993). Fear and Loathing in Naples, *The New Republic 209*(12): 51-7 [Book review of *G.B. Vico: The Making of an Anti-Modern* by Mark Lilla]

Heidegger, M. (1927/1996). *Being and Time*. Translated by Joan Stambaugh. New York: State of New York University Press.

Lilla, M. (1993). *G.B. Vico: The Making of an Anti-Modern*. Cambridge, Mass.: Harvard University Press.

Marini, C. (1852). *Giambattista Vico al cospetto del XIX secolo*. Napoli: Stamperia Strada Salvatore.

Marx, K. (1867/1970). *Capital. Vol. 1*. Edited by Friedrich Engels. Translated by S. Moore and E. Aveling. London: Lawrence and Wishart.

Mazzotta, G. (1999). *The New Map of the World: The Poetic Philosophy of Giambattista Vico*. Princeton: Princeton University Press.

McMurtry, J. (2009-10). What is good? What is bad? The value of all values across time, place and theories. In *Encyclopedia of Life Support Systems*. Retrieved from http://www.eolss.net

Nicolini, F. (1932). *La giovinezza di Vico*. Bari: Laterza.

Nicolini, F. (1955). *Saggi vichiani*. Napoli: Giannini.

Nicolini, F. (1991). *Giambattista Vico nella vita domestica: la moglie, i figli, la casa*. Venosa: Ossana.

Pickering, M. (2009). *Auguste Comte. An Intellectual Biography. Volume II*. Cambridge: Cambridge University Press.

Pompa, L. (1990). *Vico. A Study of the 'New Science'*. 2nd edition. Cambridge: Cambridge University Press.

Robischaud, P. (2003). Joyce, Vico, and National Narrative, *James Joyce Quarterly 41*(1-2): 185-96.

Simon, L.H. (1981). Vico and Marx: Perspectives on Historical Development. *Journal of the History of Ideas 42*(2): 317-31.

Verene, D.P. (Ed., 1987). *Vico and Joyce*. Albany: State University of New York Press.

Verene, D.P. and Black, M. (1994). *Vico: A Bibliography of Works in English from 1884 to 1994*. Bowling Green: Philosophy Documentation Center.

Vico, G.B. (1708/1965). *On the Study Methods of Our Time*. Translated by E. Gianturco. New York: Bobbs-Merrill.

Vico, G.B. (1709/1988). *On the Most Ancient Wisdom of the Italians Unearthed from the Origins of the Latin Language*. Translated by L.M. Palmer. Ithaca: Cornell University Press.

Vico, G.B. (1744/1987). *La scienza nuova*. La Spezia: Fratelli Melita.
Vico, G.B. (1744/1988). *The New Science of Giambattista Vico*. Translated by T.G. Bergin and M.H. Fisch. 3rd printing. Ithaca and London: Cornell University Press.

Villemaire, D. (1994). What Kuhn Really Said. *New Vico Studies 12*: 75-80.

Endnotes

[1] After testifying to the oblivion of Vico's work in 18th-century Italy, Marini (1852) tries ardently to show how nearly all major philosophical systems of the first half of the 19th century are indebted to Vico's *New Science*. The results of his efforts are perplexing, to say the least, for they are marred by the jingoistic presumption that nearly all valuable modern Western philosophy

and historiography is somehow a variation on tunes played by "the great solitary of Vatolla, the immortal genius of the Italian peninsula" (p. 127). Marini does identify a few tokens of actual knowledge of Vico's theories in non-Italian authors such as Condorcet, Michelet, De Maistre and Chateaubriand, which however fall short of proving his much more ambitious point. Rather, it can be evinced from Marini's book that Vico's focus on the theoretical discernment of human history had been a precursor of a philosophical trend that emerged forcefully only in the 19th century, well after Vico's death.

As for any large-scale engagement with Vico, that was left mainly to the scholars of the 20th century, when he became "the buzzword of a thousand academics" (Grafton, 1993: 52).

[2] Most 19th-century admirers of Vico commended specific notions of his *New Science* (e.g. the empathetic understanding of other nations, the popular nature of Homeric poetry). However, none of them seemed to adhere to the *New Science* as a project. Even less did they seem to grasp or care about the Catholic, conservative spirit of Vico's enterprise. His *New Science* included significant innovation as to the means of investigation that it employed, but it had also deeply anti-modern ends. Rejecting abstract ideals of social reform and Cartesian rationalism, Vico's work wished to show how Divine Providence had been guiding human history since its remotest origins and how reason, left to its own devices, hence deprived of the aid of tradition, is bound to fall into scepticism and relativism (cf. Lilla, 1993).

On the one hand, Vico claims: "To be useful to the human race, philosophy must raise and direct weak and fallen man, not rend his nature or abandon him in his corruption." (§ 129) On the other hand, philosophers have often displayed too proud an over-reliance upon reason and consequently forgotten about the religious truths that lie at the foundation of any cohesive human community. Thus, Vico "dismisses from the school of our Science the Stoics who seek to mortify the senses and the Epicureans who make them the criterion. For both deny providence, the former chaining themselves to fate, the latter abandoning themselves to chance... Both should be called monastic or solitary philosophers." (§ 130)

[3] I should emphasise that this book chapter is not dealing with the related but distinct topic of immortality in Vico's *New Science*.

[4] This book chapter assumes no previous knowledge of Vico's work on the reader's part. Therefore, I endeavour to explain succinctly the main issues pertaining to *The New Science*, while addressing the three key-themes highlighted *via* death and mortality.

Hopefully, the reader who does not have any previous knowledge of Vico will find this book chapter a useful introduction to *The New Science*; whilst the reader who does have some previous knowledge will deepen it with regard to Vico's understanding of death and mortality.

[5] Vico's *New Science* offers a chronology of world's history and discusses it (§§ 43-118). This way, Vico does address prehistory and several ancient civilizations, but eventually *The New Science* focuses on the Greek, Roman and, to a lesser extent, Israelite civilizations. This is due first of all to the limitations of his personal knowledge and competences, but also to the material available to him and to 18th-century Western scholars in general. Besides, he could have never managed to provide as exhaustive an account of the history of all cultures, nor even of just all the Mediterranean cultures. Focussing upon two well-known and reasonably evidence-endowed cases was a much more realistic strategy.

[6] Vico's original interpretation of Homer found positive acceptance amongst Europe's Romantics (cf. Feldman & Richardson, 1972).

[7] Vico's references to these pagan nations and to the "Patagonian giants" (§ 338) in the same paragraph of the present chapter appear in the Italian text and in the first printing of the English translation of the *New Science* cited in my bibliography, but not in the third printing of the latter.

[8] Towards the end of the returned barbaric times of Medieval Europe, a new heroic poet appears, i.e. the Florentine Dante Alighieri, who collects the wisdom of the age like Homer did before him (§ 564). Still, hence indicating the reduced savagery of the recourse of history, Medieval Europe was capable of producing also tokens of masterly philosophical reflection such as "Peter Lombard" in Paris and his "subtlest scholastic theology" (§ 159).

This imperfect cyclicality is probably Vico's most original take on a notion of development in three stages that he claims to derive from archaic Egyptian wisdom and Roman historian Marcus Terentius Varro (§ 40). Although he does address to some extent metaphysical issues, Vico's focus is social and historical. Perhaps he was afraid of getting in trouble with Napoli's very active Inquisition, or more simply he wished to show that history does not make anything; we, with God's help, do.

[9] The originally mute savages were led by providence into gesturing (§ 402) and then into communicating by few, simple, sung words prompted by "violent passions" (§ 230). Afterwards, slowly, they progressed into "univocal", true poetry (§ 403). Prose, or "articulate speech", with its armoury of synonyms, analogies and even means of deceptions, came last (§ 929).

[10] From religion and, in particular, from burials originates also the belief in the immortality of the soul (e.g. § 360).

[11] Since fear of death expresses a desire to live, it is implied that life, rather than death *per se*, is the ultimate onto-axiological springboard of religion (cf. McMurtry, 2009-10).

[12] Vico's interest in pedagogical matters is the result of his long professional experience *qua* university lecturer and mentor of youngsters from well-off families, as well as of profound epistemological concerns (cf. Mazzotta, 1999).

First of all, Vico (1709) was a humanist who tried to defend literary and historical studies from the rampant scientism of his age. He reminded his colleagues that the knowledge of the natural sciences was bound to be more superficial than the knowledge of the formal and human sciences, insofar as we can know better that which we make (e.g. geometry, law, poetry) than that which we have not made (e.g. planets, animals, ourselves). As he famously wrote: *"verum factum convertuntur"* [truth and made are mutually convertible].

Secondly, Vico (1708) did never forget that learning begins in childhood and that children must be approached in their own terms, just like the poets of ancient times. Their most developed faculties are not rational, but sensorial and imaginative. More to the point, these faculties are still essential in reflective adulthood, e.g. as imagination transcends given categories of thought in the natural sciences (cf. Villemaire, 1994). For Vico, the human mind is always and anyhow a "mind in a body" (Di Miele, 2007: 98).

[13] There is not always an exact correspondence between the three ages of historical development and each and every area of human life that Vico subdivides in three parts. Thus, the three forms of underworld that he discusses in §§ 715-21 would seem to apply to the first two ages only.

[14] Vico achieves historically the awareness of the importance of mortality in humanity's self-understanding, which Martin Heidegger (1927) also achieves, yet existentially, by focussing on the anxiety-ridden experience of the individual.

CHAPTER TWO

Cryonics: A Scientific Challenge To Death

Benjamin P. Best[*]

§1. Cryonics Overview
§2. Cooling
§3. Vitrification and Cryogenic Storage
§4. "Reversible Death"?
§5. Cryonics Procedures
§6. Science and Indirect Evidence
§7. Conclusions

§1. Cryonics Overview

Cryonics is the practice of preserving humans and animals at cryogenic temperatures in the hope that future science can restore them to a healthy living condition as well as rejuvenate them. At present cryonics can only be performed after pronouncement of legal death of the cryonics subject.

The scientific justification for the practice of cryonics is based on several key concepts: (1) Low temperature can slow metabolism. Sufficiently low temperature can virtually stop chemical changes for centuries. (2) Ice formation can be reduced or even eliminated by the use of vitrification mixtures. (3) Legally dead does not mean "irreversibly dead". Death is a process, not an event -- and the process takes longer than is commonly believed. (4) Damage associated with low temperature preservation and clinical death that is not reversible today is theoretically reversible in the future.

* This is a newly modified version of the 2008 version available at *Rejuvenation Research* 11(2): 493-503.

Pronouncement of legal death is necessary before cryonics procedures can begin because cryonics is not yet a proven, recognized medical procedure. Following legal death, a cryonics team can then begin preservation procedures immediately. Cryonics preservation procedures are intended to protect the tissues of cryonics subjects while cooling them to temperatures below −120°C with minimal alteration of tissue structure after cardiac arrest.

In the first stages the circulation and respiration of the cryonics subject is mechanically restored, the subject is administered protective medicines and is rapidly cooled to a temperature between 10°C and 0°C. The subject's blood is washed out and a significant amount of body water is replaced with a cryoprotectant mixture to prevent ice formation. The subject is cooled to a temperature below −120°C and held in cryostasis. When and if future medicine has the capability, the subject will be re-warmed, cryoprotectant will be removed, tissues will be repaired, diseases will be cured, and the subject will be rejuvenated (if required).

§2. Cooling

Preservation of food in refrigerators and freezers is based on the principle of lowering temperature to reduce the rate of biochemical degradation. Cooling to reduce metabolic rate (and ultimately to bring chemical processes to a virtual halt) is at the heart of cryonics practice. Initial cooling after pronouncement of death involves placing the cryonics subject in a bath of ice water. Cardiopulmonary support with mechanical active compression/decompression also speeds cooling because of heat transfer from flowing blood.

Cryonics subjects are cooled with convection, a combination of conduction and fluid motion. In convection, a solid object (such as a cryonics subject) is cooled by a fluid (liquid or gas) that is rapidly circulated, such that the fluid can carry heat away from the conduction layer around the solid object. In cooling a cryonics subject from human body temperature (37°C) to 10°C, cooling by rapid circulation of ice-water is far more effective because of the convection effect than cooling by ice-packs or by standing water.

The formula governing convection is **Newton's Law of Cooling** which equates the rate of heat transfer to $hA(T_s-T_f)$ where T_s is the starting temperature (head or body of a cryonics subject), T_f is the final temperature (temperature of the cooling medium, e.g., ice water or cold nitrogen gas), A is the surface area of the solid, and h is a variable which is dependent upon the rate of fluid motion as well as the thermal conductivity and heat capacity of the cooling medium. Faster fluid motion and higher thermal conductivity will increase the value of h. Newton's Law of Cooling predicts that cooling rate is greatest at the start of cooling when T_s is much greater than T_f. The cooling rate declines exponentially thereafter.

Reduction in temperature can considerably extend the time without blood flow before irreversible damage occurs. Many people, especially children, have been reported to survive 20 minutes to an hour or more of cardiac arrest with complete neurological recovery after hypothermic accidents, such as drowning in cold water[1,2]. Metabolic rate can be dramatically reduced by cooling.

Duration of ischemic time necessary to cause 50% neuronal damage in gerbils has been shown to increase exponentially with lowering of brain temperature from 37°C to 31°C[3]. Six out of six experimental hypothermic dogs having tympanic temperature of 10°C were shown to endure 90 minutes of cardiac arrest without subsequent neurological damage, and two out of seven endured 120 minutes without evident neurological damage[4]. A review of animal models of ischemic stroke found that infarct volume was reduced by about one-third by cooling to 35°C within 90 to 180 minutes of initiation of permanent ischemia[5].

Reducing core body temperature 3 to 5°C has been shown to significantly improve survival and neurological outcome of human patients suffering cardiac arrest due to ventricular fibrillation[6]. Humans have been subjected to deep hypothermic cardiac arrest for aortic surgery for over an hour without gross neurological deficits. The subjects reached complete electrocerebral silence (zero electroencephalographic bispectral index) between temperatures of 16°C and 24°C[7] and were re-warmed without neurological deficit, confirming that dynamic brain activity can be lost and regained without loss of personal identity.

The extension of hypothermic protection from ischemic injury to subzero temperatures is seen in the northern wood frog (*Rana sylvatica*) which can survive in a semi-frozen state without heartbeat for months at temperatures as low as −3°C to −6°C with full recovery upon re-warming[8]. In 1966 a Japanese researcher replaced blood with glycerol to reduce ice formation in cat brains cooled to −20°C. After 45 days with no blood circulation at −20°C the revived cat brains demonstrated normal-looking EEG activity[9].

The relationship between reaction rate (**k**) of chemical reactions (including metabolism and the processes of ischemic injury) and temperature (**T**) can be described by the Arrhenius equation[10]:

$$k = Ae^{-E_a/RT}$$

where **T** is in Kelvins, E_a is the **activation energy**, **R** is the **universal gas constant** (8.314 Joules/mole-Kelvin) and **A** is the **frequency factor** (related to frequency of molecular collisions and the probability that collisions are favorably oriented for reaction). Taking the natural logarithm of both sides of this equation gives:

$$\ln k = (-E_a/RT) + \ln A$$

For each of two different temperatures, T_1 and T_2, there will be a different temperature-dependent reaction rate, k_1 and k_2:

$$\ln k_1 = (-E_a/RT_1) + \ln A$$
$$\ln k_2 = (-E_a/RT_2) + \ln A$$

Subtracting **ln k_2** from **ln k_1** gives a single equation for the four variables:

$$\ln k_1 - \ln k_2 = ((-E_a/RT_1) + \ln A) - ((-E_a/RT_2) + \ln A)$$

which can be simplified to:

$$\ln (k_1/k_2) = (E_a/R)*(1/T_2 - 1/T_1)$$

or $k_1/k_2 = e^{(Ea/R)*(1/T2 - 1/T1)}$

The reaction rates of enzymes at various temperatures give a close approximation to the relationship between temperature and metabolic rate. Lactate dehydrogenase from rabbit muscle -- which has an activation energy (E_a) of 13,100 calories/mole[11] -- can be taken as a representative enzyme. Using one thermochemical calorie equal to 4.184 Joules gives 54,810 Joules/mole.

Comparing the reaction rate (k_1) for lactate dehydrogenase at 40°C (313 Kelvins) (T_1) to the reaction rate (k_2) at 30°C (303 Kelvins) (T_2) gives:

$$k_1/k_2 = e^{((54,810 \text{ J/mol})/(8.314 \text{ J/mol-K}))*(1/303 \text{ K} - 1/313 \text{ K})} = 2.004$$

The reaction rate at 40°C is almost exactly twice the reaction rate at 30°C or, conversely, dropping the temperature 10°C has the effect of cutting the reaction rate nearly in half. This is in agreement with the Q_{10} rule, a rule of thumb that between 0°C and 40°C reaction rates are reduced by one-half to one-third for every 10°C drop in temperature[12].

This exponential drop in reaction rates with declining temperature means that reaction rates would become infinitesimally small at cryogenic temperatures (temperatures below −100°C) if chemical reactions were possible at those temperatures.

If lactate dehydrogenase reaction rate was representative of metabolism in general, the metabolism at 37°C would be 18 times faster than at 0°C. Experimentally it has been observed that the rate of oxidative phosphorylation at 4°C is about one-twentieth the rate at 37°C[13], a figure roughly in agreement with the value just calculated.

A reaction rate that is 9 octillion times faster at human body temperature than at −196°C would indicate essentially no reaction for millennia at the lower temperature. At this rate it would take 100 sextillion years for the ischemic biochemical reactions that occur at 37°C in six minutes to occur at liquid nitrogen temperature. But even these figures understate chemical inertness at lower cryogenic temperatures because the Arrhenius equation is based on the assumption of a fluid or gas medium in which normal chemistry is possible. Below −130°C even vitrified mammalian tissues are in a solid state, with a viscosity in excess of 10^{13} poise[14,15], a viscosity about 10^{15} (one quadrillion) times greater than the viscosity of water at 20°C[16]. The resulting diffusion rates are insignificant over geological time spans. At liquid nitrogen temperature mammalian tissues would even be stable against background radiation over periods of many centuries poise[14,17].

It is a misconception that freezing mammalian tissue typically results in ice formation within cells, causing the cells to burst. As mammalian tissues are cooled water leaves cells osmotically to form extracellular pure water-ice crystals. The unfrozen solution will contain increasing concentrations of toxic electrolytes. Ultimately enough extracellular ice will form to crush cells in the remaining unfrozen channels[14]. Whether mechanical crushing or toxic electrolytes is the cause of damage following ice formation during slow cooling remains a subject of debate among cryobiologists[18]. Cryonics practice is based on efforts to reduce or eliminate freezing, however.

§3. Vitrification and Cryogenic Storage

Cryonics practice has long sought to minimize ice formation by perfusing cryonics subjects with anti-freeze compounds known as cryoprotectants, traditionally glycerol. As of 2007 both of the major cryonics organizations doing cryoprotectant perfusions (Alcor Life Extension Foundation and the Cryonics Institute) claim to have eliminated ice formation in the brain by the use of vitrification solution, but make no such claim for other organs or tissues[19,20].

Vitrification is solidification to an amorphous (glassy) state which is distinct from the crystalline state characteristic of ice. Amber is a familiar example of a vitreous (amorphous, non-crystalline) solid. Pure water can be made to vitrify if cooled not more slowly than three million Kelvins per second[21], a cooling rate impractical for animal tissues. Sucrose can be cooled rapidly enough to be vitrified into "cotton candy", but with slower cooling will form a "rock candy" crystal. Adding corn syrup to sucrose allows it to be cooled slowly to the non-crystalline solid used in lollipops. Silicon dioxide can be rapidly cooled to vitreous silica or can be slowly cooled to the crystalline form (quartz). Common glassware and window panes are made by adding sodium and calcium oxides to silicon dioxide to produce a molten liquid that can cool slowly as an increasingly viscous syrup to an amorphous (non-crystalline) solid. In the absence of a phase transition from liquid to solid crystal at melting/fusion temperature, there is a great increase in viscosity (characterized as solidification) which occurs at a glass transition temperature (T_g) that is determined by cooling rate.

Cryoprotectants most frequently used in cryobiology include dimethylsulfoxide (DMSO) as well as the polyols ethylene glycol (an automobile anti-freeze), propylene glycol (once used to reduce ice crystals in ice cream) and glycerol (used since the 1950s to cryopreserve sperm and blood cells). All of these compounds are capable of hydrogen bonding with water to prevent water molecules from organizing themselves into ice. These cryoprotectants also act by colligative interference that hinders water molecules from forming the ice lattice. Mixtures of cryoprotectants can be less toxic than the pure cryoprotectants, and can completely eliminate ice formation. The use of ice blockers (non-cryoprotectant substances such as anti-freeze proteins that chemically block ice crystal growth) in vitrification mixtures can further reduce toxicity and concentration needed to vitrify[22].

Difficulty in achieving sufficiently high cryoprotectant concentration to eliminate ice formation, while at the same time minimizing cryoprotectant toxicity, has been the limiting factor preventing better recovery of biological systems from cryopreservation. Rapid cooling can permit the use of lower cryoprotectant concentrations to prevent ice formation, but rapid cooling becomes increasingly difficult for increasingly larger tissues. Cryoprotectant toxicity varies inversely with temperature, so the use of less viscous cryoprotectant mixtures can speed tissue penetration and thereby reduce the tissue cryoprotectant exposure time at higher temperatures before cooling.

A number of possible explanations for cryoprotectant toxicity have been proposed, but the exact molecular mechanisms remain elusive[23]. Insofar as cryoprotectants do not destroy molecules, the damage they cause may not be irreparable. Moreover, considerable success has been made in reducing the toxicity of vitrification mixtures[24,25], and there is no reason to believe that further toxicity reductions cannot be made.

The mammalian organ which has been studied by the most researchers attempting organ vitrification is the ovary. Variable success has been achieved with the ovaries of a number of species, but the greatest success has been with the mouse ovary. Vitrified mouse ovaries cryopreserved at −196°C have been re-warmed to produce live-pup birth rates comparable to that seen with fresh ovaries[26].

A study on rat hippocampal slices showed that it is possible for vitrified slices cooled to a solid state at −130°C to have viability upon re-warming comparable to that of control slices that had not been vitrified or cryopreserved. Ultrastructure of the CA1 region (the region of the brain most vulnerable to ischemic damage) of the re-warmed slices is seen to be quite well preserved compared to the ultrastructure of control CA1 tissue[27]. Cryonics organizations perfuse brains with vitrification solution until saturation is achieved.

Tissues which have been vitrified and cryopreserved are assessed for viability as well as for ultrastructure. Intracellular K^+/Na^+ ratio is a commonly used method of assessing viability, although other methods (such as measurement of intracellular ATP content) could be useful in the future. The sodium pump which maintains membrane potential will not function without binding to ATP and Na^+ inside the membrane and K^+ outside the membrane.

Although a cell can maintain a membrane potential for several hours without a functional sodium pump, the slow leak of Na^+ into the cell and consequent leak of K^+ out of the cell will result in a complete loss of membrane potential after several hours. Similarly, if the cell dies in the sense of no longer being capable of producing energy (ATP) in the mitochondria, the sodium pump will cease to operate. Thus, normal intracellular K^+/Na^+ ratios indicate functioning sodium pumps and intact cell membranes.

To assay the intracellular K^+/Na^+ ratio tissues are placed in mannitol to wash away extracellular ions. Then Trichloroacetic acid is used to rupture cell membranes and release intracellular ions. A flame photometer or atomic absorption spectrometer can be used to determine the relative concentrations of sodium and potassium ions. Viability studies of vitrified hippocampal slices using intracellular K^+/Na^+ ratios indicated viability in excess of 90% normal[27].

A rabbit kidney has been vitrified, cooled to −135°C, re-warmed and transplanted into a rabbit. The formerly vitrified transplant functioned well enough as the sole kidney to keep the rabbit alive indefinitely[28]. Some people imagine a need to understand brain function as being essential for brain cryopreservation. But in simple terms, a kidney produces urine and a brain produces consciousness. A re-warmed brain that is physiologically restored should be able to produce consciousness no less than a re-warmed vitrified kidney can produce urine. Preservation of structure and restoration of physiology should result in restoration of function, irrespective of the organ or tissue.

The vitrification mixture used in preserving the rabbit kidney is known as M22. M22 is used by the cryonics organization Alcor for vitrifying cryonics subjects. Perfusion of rabbits with M22 has been shown to preserve brain ultrastructure without ice formation[29].

Cooling from 0°C to −130°C should be rapid to minimize the possibility of ice formation. When cooling from −130°C to −196°C thermal stress on large solid vitrified samples can cause cracking and fracturing[30]. Although it should theoretically be possible to cool to −196°C slowly enough to avoid cracking, the requisite cooling rates are unknown and may be too slow to be practical. Annealing a vitrified sample near glass transition temperature can reduce thermal stress[31], but this may not be adequate. Due to its more well-defined nature, cracking damage may be much easier to repair than freezing damage.

§4. "Reversible Death"?

As recently as the 1950s it was believed that death is irreversible when the heart stops. Today it is established that CardioPulmonary Resuscitation (CPR) in combination with Automated External Defibrillators (AEDs) can restore many people to life who were clinically dead because of cardiac arrest[32]. But it is still widely believed that after about six minutes of cardiac arrest without circulation irreparable brain damage has already occurred.

In 1976 Peter Safar (the "father of CPR") showed that dogs could be subjected to twelve minutes of cardiac arrest without neurological damage by the use of elevated arterial pressure, norepinephrine, heparin, and hemodilution with dextran 40[33]. Over a decade later an experiment showed that spontaneous EEG activity returned in 50% of cats subjected to one hour of global cerebral ischemia followed by reperfusion and treatment with norepinephrine (or dopamine), heparin, insulin, and acidosis buffers. Six out of fifteen of the cats submitted to intensive care regained spontaneous respiration, and one of those cats survived a full year with normal neurological function (except slight ataxia)[34]. The six minute limit is not mainly a neurological phenomenon, it is a problem of increased vascular resistance that can be overcome (in part) by increasing perfusion pressure[35].

Reperfusion injury refers to the tissue damage inflicted when blood flow is restored after an ischemic period of more than about twenty minutes. The re-supply of blood after an excessive period of ischemia initiates inflammatory processes and causes oxygen to form toxic free radicals (reactive oxygen species) such as superoxide[36]. Xanthine oxidase-produced superoxide damages the endothelium far more than the parenchyma[37]. In inflammatory conditions, such as occur in reperfusion, inducible nitric oxide synthetase can increase nitric oxide concentration to thousands of times normal levels[38]. During reperfusion, abnormally high amounts of superoxide convert almost all available nitric oxide to perxoynitrite -- regarded as the agent causing most of the damage to brain capillary endothelial cells[39].

Despite the damaging effects of excitotoxicity[40] brain structure is normally retained post-mortem much longer than is commonly appreciated. In the cerebral cortex of rats subjected to occlusion of blood flow to the cortex (cerebral ischemia) only 15% of neurons were necrotic after 6 hours. Most neurons (65%) did not become necrotic until 12 hours after the cessation of blood flow[41]. Neurons isolated from the brains of autopsied elderly humans an average of 2.6 hours post-mortem showed 70-90% viability after two weeks *in vitro*[42].

Part of the reason why more than six minutes of cardiac arrest currently leads to neurological damage is because the ischemia starts a process of neuron self-destruction (apoptosis) which takes many hours to complete. But therapies are on the horizon which may interfere with apoptosis. Neurons in the CA1 sector of the hippocampus are much more vulnerable to becoming necrotic following ischemia than neurons elsewhere in the brain[43]. But cell death in the hippocampus following ischemia can be significantly reduced by the use of caspase inhibitors that arrest the apoptotic process[44]. Caspase inhibitors have also been used to block apoptosis in cryopreserved hematopoietic cells re-warmed from cryogenic temperatures[45]. Bag−1 protein which binds pro-apoptotic members of the Bcl−2 protein family has demonstrated powerful anti-apoptotic effects on rat livers subjected to ischemia/reperfusion injury[46].

Most neuroscientists agree that the anatomical basis of mind is encoded in physical structures of the brain, particularly neuropil connectivity and synaptic strengths[47] and possibly neuronal epigenetic structure[48]. The fact that complete absence of electrical activity in the brain does not prevent full neurological recovery[7,49] supports the proposition that the ultimate basis of consciousness is structural rather than dynamic and can therefore be preserved at cryogenic temperatures.

Possibly associated with redundancy of information storage in the brain, significant recovery of cerebral cortex function is possible following stroke[50,51,52]. Neural stem cell transplantation therapies have the potential to further augment brain recovery from damage due to ischemia, toxins and cryopreservation injury[53]. These considerations increase the amount of damage that may be tolerable for restoration of a human cryonically preserved under suboptimal conditions.

Preservation of brain structure and restoration of brain function are essential to cryonics. Other organs and tissues are not as important because artificial organs and tissue regeneration by stem cells should be easily accomplished by future medicine. Appendage regeneration in salamanders is already being used as a guideline for mammalian regenerative medicine[54]. Biodegradable 3-dimensional braided fibrous scaffolds[55] can potentially be used for construction of organs, if not whole bodies.

Conservative cryonics strives to minimize damage and minimize reliance on future molecular repair technologies. In many cases cryonics subjects have experienced less than a minute of cardiac arrest before circulation has been restored. Evidence that the neurological basis of mind is preserved long beyond the six minute limit gives hope that molecular medicine to reverse apoptosis and repair damaged blood vessels may allow for recovery of cryonics subjects who did not benefit from prompt treatment. It is unlikely that cryonics is worthless after six minutes without blood circulation. Many tissues are alive when the heart stops, and take hours to die.

Under the best circumstances, cryonics subjects experience virtually no ice formation in the brain. Repairs to vitrified brain tissue that had experienced little ischemic damage could be performed above cryogenic temperatures along with curing diseases and rejuvenating.

Although there are sound legal requirements for drawing a distinct line between life and death, biological and psychological reality point to a continuum rather than discrete binary states. Consciousness emerges gradually from embryo to fetus to child to adult, and consciousness can diminish gradually in neurodegenerative diseases. After the heart stops the anatomical basis of mind (the brain) decomposes over a period of hours and days at a temperature-dependent rate. The non-binary character of consciousness would also become evident in revived cryonics subjects whose brains had been partially destroyed and then repaired, resulting in partial amnesia and partial restoration of original identity.

§5. Cryonics Procedures

Pre-treatment of terminal cryonics subjects to reduce ischemia/reperfusion injury is advisable, but is too rarely done in practice. For example, intravenous injection of the alpha-tocopherol form of Vitamin E (20 mg/kg) 30 minutes prior to ischemia has been shown to significantly reduce lipid peroxidation and neurological damage[56]. It is better to include both alpha-tocopherol and gamma-tocopherol because gamma-tocopherol removes peroxynitrite whereas alpha-tocopherol does not[57]. Vitamin E pre-treatment for cryonics patients has the additional advantage of reducing blood clotting -- and does not have the risk of gastric bleeding associated with aspirin. Many fish oils (especially salmon oil) afford the same benefit, in addition to reducing the risk of cardiac arrest[58]. Reduced clotting in a cryonics patient is usually a great benefit. But for patients undergoing surgery, Vitamin E and fish oils may be prohibited because of the danger of excessive bleeding.

Cryonics procedures are generally only practiced on subjects who have made contractual and funding arrangements in advance with a cryonics organization (such as Alcor Life Extension Foundation, the American Cryonics Society or the Cryonics Institute). In optimal circumstances a cryonics subject will be pronounced legally dead very quickly after their heart stops. Only after legal death has been pronounced can the cryonics procedures begin.

Once cardiac arrest has occurred and death has been pronounced, a cryonics subject can be given medications to maintain sedation, reduce cerebral metabolism, prevent/reverse blood clotting, increase blood pressure, stabilize pH against acidosis, and protect against ischemia/reperfusion injury.

Cryonics procedures involve restoring blood circulation and respiration as soon as possible to keep tissues alive. In cryonics, this is called CardioPulmonary Support (CPS) rather than CardioPulmonary Resuscitation (CPR) because resuscitation after death has been pronounced is not desired (a DNR, Do Not Resuscitate, condition). Propofol (2,6-diisopropylphenol) is given partly because its sedative action can prevent resuscitation, with the added benefit that it can be neuroprotective[59]. Propofol has been shown to inhibit the neural cell apoptosis that can occur as a consequence of ischemia/reperfusion injury[60].

Heparin is used to prevent blood clotting. Streptokinase is the usual thrombolytic used to break up blood clots. THAM (Tris-Hydroxymethyl AminoMethane) is a buffer that maintains arterial pH without producing carbon dioxide and also maintains intracellular pH because it readily crosses cell membranes[61].

When the equipment is available, cryonics teams restore circulation and respiration with mechanical devices capable of restoring circulation on the down-stroke (compression) as well as the up-stroke (decompression). Active Compression-DeCompression (ACDC) and interposed abdominal compression can improve CPS perfusion considerably[62]. Epinephrine has commonly been used to supplement CPS by maintaining blood pressure, although vasopressin may also be used[63].

While the cryonics subject is receiving ACDC CPS, he or she is in a bath of circulating ice water. Cooling is much more rapid in water than in air[66] and is much more rapid in flowing water than in still water due to Newton's Law of Cooling. The cooling of the cryonics subject is also considerably hastened by the blood circulation resulting from ACDC CPS.

Once the cryonics subject is cooled to below 10°C perfusion with vitrification solution can begin. Vitrification of the brain is achieved by cryoprotectant exposure times which are considerably longer than was seen for vitrification of hippocampal slices. Cryoprotectants are toxic, so increased cryoprotectant exposure time means increased toxicity to the exposed tissues. But large organs and body tissues cannot be cooled as rapidly as tissue slices, therefore higher cryoprotectant concentrations must be used to prevent ice formation.

Suggestions have been made for systems to store cryonics subjects at temperatures closer to −130°C to eliminate cracking due to thermal stress in cooling to −196°C. Subjects stored just below −130°C would still be in the solid state, insofar as vitrification solutions have a glass transition temperature just below −120°C[23]. Storage just below −130°C has yet to be implemented on all but a few cryonics subjects.

Cryonics subjects are stored in thermos-bottle-like containers of liquid nitrogen, a storage method which is both inexpensive and not dependent upon electricity (not so vulnerable to power failure).

§6. Science and Indirect Evidence

Many critics assert that cryonics is not science and has no capability of becoming science until a mammal has been re-warmed and reanimated after having been cryopreserved at cryogenic temperatures. But model-building based on extrapolations from indirect evidence is central to science.

No one has ever seen the core of the Earth. Models have been constructed of the state of the universe in the first millionth of a second following the Big Bang. Scientists describe the state of the Earth after years of global warming. Models have been constructed of the state of contained nuclear waste hundreds of thousands of years in the future -- models upon which considerable reliance is placed in the disposal of nuclear waste. Past landings by humans on the moon provide indirect evidence that future landings of humans on Mars may be possible.

Many people cryopreserve umbilical cord stem cells from their newborn baby on the basis of potential future scientific developments rather than on the basis of current science. Germ cells and DNA from endangered species are similarly cryopreserved in anticipation of future technology. Influenza vaccine only has a 50% chance of protecting a person over 65[65,66]. It is not unscientific to risk modest or experimental medical treatments that are justified by indirect evidence for some probability of success, rather than absolute guarantee of success.

If there are plausible models for the repair and reanimation of cryopreserved cryonics subjects[67,68] it seems reasonable to rely upon them when deciding upon human cryopreservation as a long-term treatment which may or may not succeed. To wait until a cryopreserved mammal has been reanimated before cryopreserving humans could mean that many lives will be lost. It would be like waiting tens or hundreds of thousands of years to ensure that nuclear waste can be contained before containing those wastes. Indirect evidence exists to support the claim that revival of a cryopreserved mammal is not essential for cryonics practice to be scientifically justified.

§7. Conclusions

Low temperatures slow biological time, effectively stopping time at liquid nitrogen temperature. Cryoprotectants greatly reduce damage caused by tissue cryopreservation, and effective vitrifications can prevent ice formation completely. Cryoprotectant toxicity is likely a reparable injury, and less toxic means of cryopreservation continue to be discovered. Applied to humans and animals, cryopreservation is a means of achieving a stable biological state that is reversible in principle.

Dying is a process that begins, not ends, when heartbeat and blood circulation stop. When legal death is declared based on cardiopulmonary arrest, cryonics procedures aim to minimize further injury by artificially restoring blood circulation and rapidly reducing temperature. The duration of clinical death at warm temperatures beyond which the brain information that defines a human being is lost may be many hours.

The proposition that aging is a disease that can be treated and perhaps eventually be reversed (rejuvenation), is based on the general understanding that aging consists of a multitude of specific pathologies on cellular and molecular levels that can be studied, understood, and reversed with foreseeable tools. Pathologies caused by global cerebral ischemia (clinical death), by cryopreservation, and those of other presently-incurable diseases are similarly amenable to analysis and possible future repair. If aging damage can be repaired at some future time, it is not unreasonable to think that damage due to cryonics procedures can also be repaired. And if aging damage can be repaired at some future time, cryonics may be the only way for many people living today to reach future medicine that can cure presently-incurable diseases and rejuvenate.

Added Note From The Editor (Charles Tandy)

Another fact seemingly worth mentioning is that there are many human adults alive and well today due to their cryogenic suspended-animation when they were mere embryos.

REFERENCES

1. Eich C, Brauer A, Kettler D. Recovery of a hypothermic drowned child after resuscitation with cardiopulmonary bypass followed by prolonged extracorporeal membrane oxygenation. Resuscitation. 2005 Oct;67(1):145-8.

2. Bolte RG, Black PG, Bowers RS, Thorne JK, Corneli HM. The use of extracorporeal rewarming in a child submerged for 66 minutes. JAMA. 1988 Jul 15;260(3):377-9.

3. Takeda Y, Namba K, Higuchi T, Hagioka S, Takata K, Hirakawa M, Morita K.
Quantitative evaluation of the neuroprotective effects of hypothermia ranging
from 34 degrees C to 31 degrees C on brain ischemia in gerbils and determination
of the mechanism of neuroprotection. Crit Care Med. 2003 Jan;31(1):255-60.

4. Behringer W, Safar P, Wu X, Kentner R, Radovsky A, Kochanek PM, Dixon CE,
Tisherman SA. Survival without brain damage after clinical death of 60-120 mins in dogs using suspended animation by profound hypothermia. Crit Care Med. 2003 May;31(5):1523-31.

5. van der Worp HB, Sena ES, Donnan GA, Howells DW, Macleod MR. Hypothermia in
animal models of acute ischaemic stroke: a systematic review and meta-analysis.
Brain. 2007 Dec;130(Pt 12):3063-74.

6. Bernard S. Hypothermia after cardiac arrest: expanding the therapeutic scope.
Crit Care Med. 2009 Jul;37(7 Suppl):S227-33.

7. Hayashida M, Sekiyama H, Orii R, Chinzei M, Ogawa M, Arita H, Hanaoka K, Takamoto S. Effects of deep hypothermic circulatory arrest with retrograde cerebral perfusion on electroencephalographic bispectral index and suppression ratio. J Cardiothorac Vasc Anesth. 2007 Feb;21(1):61-7.

8. Costanzo JP, Lee RE Jr, DeVries AL, Wang T, Layne JR Jr. Survival mechanisms of vertebrate ectotherms at subfreezing temperatures: applications in cryomedicine. FASEB J. 1995 Mar;9(5):351-8.

9. Suda I, Kito K, Adachi C. Viability of long term frozen cat brain in vitro. Nature. 1966 Oct 15;212(5059):268-70.

10. CHEMISTRY:The Central Science (Seventh Edition); pp.511-512; Brown,TL, et. al., Editors; Prentice Hall (1997).

11. Low PS, Bada JL, Somero GN. Temperature adaptation of enzymes: roles of the free energy, the enthalpy, and the entropy of activation. Proc Natl Acad Sci U S A. 1973 Feb;70(2):430-2.

12. Davidson EA, Janssens IA. Temperature sensitivity of soil carbon decomposition and feedbacks to climate change. Nature. 2006 Mar 9;440(7081):165-73.

13. Dufour S, Rousse N, Canioni P, Diolez P. Top-down control analysis of temperature effect on oxidative phosphorylation. Biochem J. 1996 Mar 15;314 (Pt 3):743-51.

14. Mazur P. Freezing of living cells: mechanisms and implications. Am J Physiol. 1984 Sep;247(3 Pt 1):C125-42.

15. Angell CA. Liquid fragility and the glass transition in water and aqueous solutions. Chem Rev. 2002 Aug;102(8):2627-50.

16. CRC Handbook of Chemistry and Physics (76th Edition), p.6-10, David R. Lido, Editor-in-Chief CRC Press (1995-1996).

17. Glenister PH, Whittingham DG, Lyon MF. Further studies on the effect of
radiation during the storage of frozen 8-cell mouse embryos at -196 degrees C.
J Reprod Fertil. 1984 Jan;70(1):229-34.

18. Mazur,P, "Principles of Cryobiology" in Life In the Frozen State, pp.37-51, Barry J. Fuller, et. al., Editors; CRC Press (2004).

19. Chamberlain,F, Cryonics 21(4):4-9 (2000).

20. Best,B, Long Life: Longevity through Technology, 39(3-4):5-6 (2007).

21. Bald WB. On crystal size and cooling rate. J Microsc. 1986 Jul;143(Pt 1):89-102.

22. Wowk B, Leitl E, Rasch CM, Mesbah-Karimi N, Harris SB, Fahy GM. Vitrification enhancement by synthetic ice blocking agents. Cryobiology. 2000 May;40(3):228-36.

23. Fahy GM, Lilley TH, Linsdell H, Douglas MS, Meryman HT. Cryoprotectant toxicity and cryoprotectant toxicity reduction: in search of molecular mechanisms.
Cryobiology. 1990 Jun;27(3):247-68.

24. Fahy GM, Wowk B, Wu J, Paynter S. Improved vitrification solutions based on the predictability of vitrification solution toxicity. Cryobiology. 2004 Feb;48(1):22-35.

25. Fahy GM, Wowk B, Wu J, Phan J, Rasch C, Chang A, Zendejas E. Cryopreservation of organs by vitrification: perspectives and recent advances.
Cryobiology. 2004 Apr;48(2):157-78.

26. Hasegawa A, Mochida N, Ogasawara T, Koyama K. Pup birth from mouse oocytes in preantral follicles derived from vitrified and warmed ovaries followed by in vitro growth, in vitro maturation, and in vitro fertilization. Fertil Steril. 2006 Oct;86 Suppl 4:1182-92.

27. Pichugin Y, Fahy GM, Morin R. Cryopreservation of rat hippocampal slices by vitrification. Cryobiology. 2006 Apr;52(2):228-40.

28. Fahy GM, Wowk B, Pagotan R, Chang A, Phan J, Thomson B, Phan L. Physical and biological aspects of renal vitrification. Organogenesis. 2009 Jul;5(3):167-75.
[This is reprinted as Chapter Four in the present volume.]

29. Lemler J, Harris SB, Platt C, Huffman TM. The arrest of biological time as a bridge to engineered negligible senescence. Ann N Y Acad Sci. 2004 Jun;1019:559-63.

30. Fahy GM, Saur J, Williams RJ. Physical problems with the vitrification of large biological systems. Cryobiology. 1990 Oct;27(5):492-510.

31. Baudot,A., Odagescu,V, "An automated cryostat designed to study the vitrification of cryoprotective solutions without any function", CRYOBIOLOGY 53:442 (2006) [Abstract].

32. Cobb LA, Fahrenbruch CE, Walsh TR, Copass MK, Olsufka M, Breskin M, Hallstrom AP. Influence of cardiopulmonary resuscitation prior to defibrillation in patients with out-of-hospital ventricular fibrillation. JAMA. 1999 Apr 7;281(13):1182-8.

33. Safar P, Stezoski W, Nemoto EM. Amelioration of brain damage after 12 minutes' cardiac arrest in dogs. Arch Neurol. 1976 Feb;33(2):91-5.

34. Hossmann KA. Resuscitation potentials after prolonged global cerebral ischemia in cats. Crit Care Med. 1988 Oct;16(10):964-71.

35. Shaffner DH, Eleff SM, Brambrink AM, Sugimoto H, Izuta M, Koehler RC, Traystman RJ. Effect of arrest time and cerebral perfusion pressure during cardiopulmonary resuscitation on cerebral blood flow, metabolism, adenosine triphosphate recovery, and pH in dogs. Crit Care Med. 1999 Jul;27(7):1335-42.

36. de Groot H, Rauen U. Ischemia-reperfusion injury: processes in pathogenetic networks: a review. Transplant Proc. 2007 Mar;39(2):481-4.

37. Ratych RE, Chuknyiska RS, Bulkley GB. The primary localization of free radical generation after anoxia/reoxygenation in isolated endothelial cells. Surgery. 1987 Aug;102(2):122-31.

38. Brown GC, Borutaite V. Nitric oxide inhibition of mitochondrial respiration and its role in cell death. Free Radic Biol Med. 2002 Dec 1;33(11):1440-50.

39. Wu S, Tamaki N, Nagashima T, Yamaguchi M. Reactive oxygen species in reoxygenation injury of rat brain capillary endothelial cells. Neurosurgery. 1998 Sep;43(3):577-83.

40. Lipton P. Ischemic cell death in brain neurons. Physiol Rev. 1999 Oct;79(4):1431-568.

41. Garcia JH, Liu KF, Ho KL. Neuronal necrosis after middle cerebral artery occlusion in Wistar rats progresses at different time intervals in the caudoputamen and the cortex. Stroke. 1995 Apr;26(4):636-42.

42. Konishi Y, Lindholm K, Yang LB, Li R, Shen Y. Isolation of living neurons from human elderly brains using the immunomagnetic sorting DNA-linker system. Am J Pathol. 2002 Nov;161(5):1567-76.

43. Radovsky A, Safar P, Sterz F, Leonov Y, Reich H, Kuboyama K. Regional prevalence and distribution of ischemic neurons in dog brains 96 hours after cardiac arrest of 0 to 20 minutes. Stroke. 1995 Nov;26(11):2127-33.

44. Chen J, Nagayama T, Jin K, Stetler RA, Zhu RL, Graham SH, Simon RP.
Induction of caspase-3-like protease may mediate delayed neuronal death in the
hippocampus after transient cerebral ischemia. J Neurosci. 1998 Jul 1;18(13):4914-28.

45. Stroh C, Cassens U, Samraj AK, Sibrowski W, Schulze-Osthoff K, Los M. The role of caspases in cryoinjury: caspase inhibition strongly improves the recovery of cryopreserved hematopoietic and other cells. FASEB J. 2002 Oct;16(12):1651-3.

46. Sawitzki B, Amersi F, Ritter T, Fisser M, Shen XD, Ke B, Busuttil R, Volk HD,
Kupiec-Weglinski JW. Upregulation of Bag-1 by ex vivo gene transfer protects rat livers from ischemia/reperfusion injury. Hum Gene Ther. 2002 Aug 10;13(12):1495-504.

47. Abraham WC, Robins A. Memory retention--the synaptic stability versus plasticity dilemma. Trends Neurosci. 2005 Feb;28(2):73-8.

48. Arshavsky YI. "The seven sins" of the Hebbian synapse: can the hypothesis of synaptic plasticity explain long-term memory consolidation Prog Neurobiol. 2006 Oct;80(3):99-113.

49. Rothstein TL. Recovery from near death following cerebral anoxia: A case report demonstrating superiority of median somatosensory evoked potentials over EEG in predicting a favorable outcome after cardiopulmonary resuscitation. Resuscitation. 2004 Mar;60(3):335-41.

50. Dancause N, Barbay S, Frost SB, Plautz EJ, Chen D, Zoubina EV, Stowe AM, Nudo
RJ. Extensive cortical rewiring after brain injury. J Neurosci. 2005 Nov 2;25(44):10167-79.

51. Carmichael ST. Cellular and molecular mechanisms of neural repair after stroke: making waves. Ann Neurol. 2006 May;59(5):735-42.

52. Nudo RJ. Postinfarct cortical plasticity and behavioral recovery.
Stroke. 2007 Feb;38(2 Suppl):840-5.

53. Savitz SI, Dinsmore JH, Wechsler LR, Rosenbaum DM, Caplan LR.
Cell therapy for stroke. NeuroRx. 2004 Oct;1(4):406-14.

54. Brockes JP, Kumar A.
Appendage regeneration in adult vertebrates and implications for regenerative
medicine.
Science. 2005 Dec 23;310(5756):1919-23.

55. Cooper JA Jr, Sahota JS, Gorum WJ 2nd, Carter J, Doty SB, Laurencin CT.
Biomimetic tissue-engineered anterior cruciate ligament replacement.
Proc Natl Acad Sci U S A. 2007 Feb 27;104(9):3049-54.

56. Yamamoto M, Shima T, Uozumi T, Sogabe T, Yamada K, Kawasaki T. A possible role of lipid peroxidation in cellular damages caused by cerebral ischemia and the protective effect of alpha-tocopherol administration. Stroke. 1983 Nov-Dec;14(6):977-82.

57. Christen S, Woodall AA, Shigenaga MK, Southwell-Keely PT, Duncan MW, Ames BN. Gamma-tocopherol traps mutagenic electrophiles such as NO(X) and complements alpha-tocopherol: physiological implications. Proc Natl Acad Sci U S A. 1997 Apr 1;94(7):3217-22.

58. Charnock JS, McLennan PL, Abeywardena MY. Dietary modulation of lipid metabolism and mechanical performance of the heart. Mol Cell Biochem. 1992 Oct 21;116(1-2):19-25.

59. Adembri C, Venturi L, Tani A, Chiarugi A, Gramigni E, Cozzi A, Pancani T, De Gaudio RA, Pellegrini-Giampietro DE. Neuroprotective effects of propofol in models of cerebral ischemia: inhibition of mitochondrial swelling as a possible mechanism. Anesthesiology. 2006 Jan;104(1):80-9.

60. Polster BM, Basanez G, Young M, Suzuki M, Fiskum G. Inhibition of Bax-induced cytochrome c release from neural cell and brain mitochondria by dibucaine and propranolol. J Neurosci. 2003 Apr 1;23(7):2735-43.

61. Gehlbach BK, Schmidt GA. Bench-to-bedside review: treating acid-base abnormalities in the intensive care unit - the role of buffers. Crit Care. 2004 Aug;8(4):259-65.

62. Babbs CF. CPR techniques that combine chest and abdominal compression and decompression: hemodynamic insights from a spreadsheet model. Circulation. 1999 Nov 23;100(21):2146-52.

63. Wenzel V, Lindner KH. Arginine vasopressin during cardiopulmonary resuscitation: laboratory evidence, clinical experience and recommendations, and a view to the future. Crit Care Med. 2002 Apr;30(4 Suppl):S157-61.

64. Baccino E, Cattaneo C, Jouineau C, Poudoulec J, Martrille L. Cooling rates of the ear and brain in pig heads submerged in water: implications for postmortem interval estimation of cadavers found in still water. Am J Forensic Med Pathol. 2007 Mar;28(1):80-5.

65. Nichol KL, Nordin J, Mullooly J, Lask R, Fillbrandt K, Iwane M. Influenza vaccination and reduction in hospitalizations for cardiac disease and stroke among the elderly. N Engl J Med. 2003 Apr 3;348(14):1322-32.

66. Hak E, Buskens E, van Essen GA, de Bakker DH, Grobbee DE, Tacken MA, van Hout BA, Verheij TJ. Clinical effectiveness of influenza vaccination in persons younger than 65 years with high-risk medical conditions: the PRISMA study. Arch Intern Med. 2005 Feb 14;165(3):274-80.

67. Drexler KE. Molecular engineering: An approach to the development of general capabilities for molecular manipulation. Proc Natl Acad Sci U S A. 1981 Sep;78(9):5275-5278.

68. Freitas RA Jr. What is nanomedicine? Nanomedicine. 2005 Mar;1(1):2-9.

CHAPTER THREE

Primary Institutions

Thomas O. Buford

I shall focus on two questions: [1] are any institutions primary, and [2], assuming the answer is affirmative, are they made or found. For example, we can distinguish between 'education' as a primary institution and 'education' as a secondary institution. We informally educate new persons into a society and formally school persons in a particular society. It seems clear that secondary institutions are made. But are primary institutions also made? This raises the problem of the ontological status of general social patterns of societies within which individuals formulate their own specific ones. These issues were recently raised in a work by John Searle, *The Construction of Social Reality,* where he addresses "the metaphysical burden of social reality."[1] His discussion both highlights the central metaphysical issue and allows us to focus on our two questions, which we will address shortly.

Asserting a kind of naturalism through the claim that we live in one world, the world of the sciences, Searle addresses the presence of social reality, especially institutions. He argues that we live in a world of two kinds of facts, brute and social. Brute facts have features that exist independently of our representations of them and are intrinsic to the objects themselves. A mountain's existence has no dependence on our conscious life. Social facts in contrast are rooted neither in overt behavior nor in nonconscious computational rules.[2] Rather they are rooted in consciousness, specifically intentionality and speech acts, which exist alongside and independently of brute facts.[3] Social facts have features that are not intrinsic to the natural object and are "relative to the observers and users, which are formed by collective intentionality, "we intending." That is, all social facts are created by "we intending."[4] A college degree exists through consciousness and collective intentionality; it is dependent on our conscious life.

How are institutions created? Dependent on conscious intentionality, they come into existence through the causal activity of constitutive rules, which "come in systems, and the rules individually, or, sometimes, for systems collectively, characteristically have the form 'X counts as Y or 'X counts as Y in context C.'"[5] That is, a degree from a college is "created by the application of specific rules," rules for educating and graduating students.[6] The constitutive rules concerning education create a completely new kind of entity – colleges as institutional facts; and once in existence, regulatory rules take over and govern them. The creation, maintenance, and hierarchy of institutions that occur *through* agentive functions, collective intentionality, constitutive and regulatory rules, and *within* the framework of the "basic ontology of physics, chemistry, and biology" appear to be a deliberate act or set of actions.[7]

However, he says, "the creation of institutional facts is typically a matter of natural evolution, and there need be no explicitly conscious imposition of function . . . on lower-level phenomena."[8] Though "the structure of institutions is the structure of constitutive rules," people within institutions do not appear to constitute institutions or try to follow rules, consciously or unconsciously. If we are formed by institutions, what causal roles do the constitutive rules play? To answer this problem Searle appeals to "the Background."[9] The Background is "the set of nonintentional or preintentional capacities that enable intentional states to function."[10] This resolves into the problem of the relation of the microcosm to the macrocosm, or non-intentional, neurophysiological events to intentional agentive acts, an issue beyond the confines of this paper.

Our purpose is to search for primary institutions, not to resolve all issues and/or difficulties inherent in Searle's account.

If Searle's view of collective intentionality is correct, *humans* precede all institutions and both made and make them. However, if his view of "the Background" is correct, *institutions* precede humans, who are born, live, and die in institutional structures. Humans do not make them; they are found. We face a problem, one that has been with us since the Enlightenment and a variation of the debate between the ancients and the moderns.

Let's frame the issue as an antinomy. Consider the conjunction: primary institutions are made *and* primary institutions are found. Each claim is true. Any examination of the natural world, either by a physicist or a biologist, would supply adequate evidence for the truth of the left side of the conjunct. Searle drives his stakes here, Dewey would as well, and most modern sociologists seem to agree, as does Bellah, for example.[11] However, the right side of the disjunct is true as well. Any record we have of what we would call persons indicates that they both are born into and learn in social-physical contexts. That includes forming and learning language. In that sense humans find institutions pre-existing, in some sense, their existence as humans. Thus, both claims are true, but they cannot both the true; they are contradictory. Hence, the antinomy.

I shall argue that the antinomy can best be resolved [1] by recognizing that the ontologies assumed by both claims are limited; [2] by contending that insofar as experience is anything for us it is differentiated as triadic: I-Thou-It; [3] by arguing that insofar as the triadic relation is stable and norm governed an institution is present; [4] and by claiming that as experiences, knowledge, and behavior change institutions persist. Our methodology is to scrutinize the antinomy in search of a way to resolve it. However, before we begin, we should understand the nature and place of institutions in human experience.

I.   Institutions and their Place

First, it would help us to know what we refer to by 'institution.' Various definitions have been attempted. For example, Singer in his essay, *Institutional Ethics*, says, "An institution can be thought of as (1) a relatively permanent system of social relations organized around (that is, for the protection or attainment of) some social need or value; or as (2) a recognized and organized way of meeting a social need or desire or of satisfying a social purpose."[12] Emphasizing the place of norms in institutions, Parsons says, "Then in looking for the field of empirical facts with which the theory of institutions should be concerned, I should concentrate on those uniform modes of behavior and forms of relationship which are 'sanctioned,' that stand in some kind of significant relation to normative rules to a greater or lesser degree approved by the individuals subject to them. . . . It is in the particular feature of being related to norms that the institutional aspect of these uniformity's lies."[13] Though other definitions are available, from these general characteristics emerge. Institutions are relatively permanent social relations, sanctioned by approved norms, that satisfy a social desire, need, and/or purpose. What is the place of institutions in our lives?

To answer that question, distinguish among habitat, culture, and environment. By *habitat* we mean the natural setting of human existence, "the physical features of the region inhabited by a group of people; its natural resources, actually or potentially available to the inhabitants; its climate, altitude, and other geographical features to which they have adapted themselves."[14] By *culture* we refer to "that part of the total setting that includes the material objects of human manufacture, techniques, social orientations, points of view, and sanctioned ends."[15] And by *environment* we refer to "man in his natural and cultural setting."[16] To summarize, institutions are relatively permanent social relations, sanctioned by norms that satisfy a social desire, need, and/or purpose and are situated among the ecosystem, our normative beliefs, and ourselves. With this background we can proceed to examine the antinomy: it is true *both* that institutions are constructed by humans *and* that institutions precede humans; yet that conjunction cannot be true, as it is a contradiction. We face a logical puzzle as we attempt to understand our social lives.

II.   Limited Ontologies

The antinomy is one way of articulating the central issue in the debate between social constructionists and social realists (metaphysical, in the sense that social patterns are not made by humans and have an ontological status independent of humans). Faced with the anomaly of two propositions which appear to be true but are yet contradictories, we seek a resolution to learn what is the case. Antinomies can usually be addressed by showing that one of the propositions is false. That is essentially what Searle attempts; he argues that the claim that institutions are found is false, resolving the antinomy. But there are other ways to resolve an antinomy, including the approach taken here.

The root of the anomaly in this case rests in the presuppositions underlying each proposition. That institutions are made rests on the presupposition of a certain kind of world and our relation to it. Those who agree with Searle believe a kind of naturalism. We live in the kind of world that science says it is and that humans are one system within that world, but a system that has the peculiar capacity of consciously representing that world to itself and acting intentionally with reference to those representations. For example, an institution such as education is not something that has physical properties; it is not nature, it is not brute fact. It then must be something made, rooted in human conscious intentionality, particularly speech acts; it is a social fact. However, all social facts are rooted in brute facts just as any other kind of macrocosmic action is a manifestation of microcosmic movements, neutrons, protons, particles. His is a kind of naturalism.[17] If this claim is true, that institutions are found must be false.

On the other hand, that institutions are found rests on the presupposition that humans are nestled in a Background within which they consciously, collectively form particular institutions. The Background appears to be a friendly environment which at least supports moral and social structures. Humans living in this world find this moral and social structure and live accordingly. How this Background is understood varies from thinker to thinker, from Plato to Thomas Aquinas, but it is usually interpreted within some form of idealism. The normative structure of institutions is rooted in the moral structure of the universe whether platonic forms or the ideals and purposes of God, construed in a Platonic/Augustinian or Aristotelian/Thomistic manner. And, if this claim is true, that institutions are made must be false.

Each view, however, seems to leave unrecognized the truth of the other. The naturalist denies the idealist and the idealist denies the naturalist. Each claim is exclusive, denying the other while affirming its own truth. Any view that seems to deny the truth of another calls less for acceptance than for reconsideration of the issues. These can be called limited ontologies. They are world views, yet they are exclusive; they exclude some dimension of human experience that seems to be true. The problem lies in each metaphysical world view not paying adequate attention to the richness of human experience. Each is as coherent as the other. But the antinomy remains. So, let's turn to experience and seek there a way to play fair to its multidimensional character, hoping to find primary institutions.

III.   Radical Empiricism

Experience is not simply the sensory manifold manifesting the world *through* which we grasp the world and its characteristics. Such was the view held by Descartes and the so-called British Empiricists, and today by Rorty. Rather, consider experience from another vantage point. Each of us begins doing philosophy in the context of our living in the world. All philosophies must start there, and in doing so they find common ground for discussion. Consider the words of a radical empiricist, "We find such common ground in the following postulates: -- First, the coexistence of persons. It is a personal and social world in which we live, and with which all speculation must begin. We and the neighbors are facts which cannot be questioned. Secondly, there is a law of reason valid for all and binding upon all. This is the supreme condition of any mental community. Thirdly, there is a world of common experience, actual or possible, where we meet in mutual understanding, and where the great business of life goes on."[18]

For those deeply committed to radical empiricism, the starting point for all philosophical study is *within* experience, broadly conceived. Such an empiricist claims, "We begin with experience, which is real and valid in its way. This is the world of things and persons about us, and the general order of life. Now in serious thought there can never be any question as to the validity and truth of this experience."[19]   But how is this best understood? The question is not whether things are there, but how they come to be there for us. Beginning with a radical empiricism, issues of knowledge and conduct arise as different aspects within experience conflict, requiring recognition, interpretation, and action. In this way such an empiricism avoids Cartesian methodological skepticism. This empiricism believes that the mind is not passive in its empirical life but is actively engaged in it from the outset. Something is before the mind in accordance with certain principles immanent in the imagination, that organizes the impressions of sense into connected forms and only thus does the mind reach a whole, articulated, and connected system of experience.[20] However, one must remember that the resulting understanding always occurs within experience. Now, what is implied in this kind of empiricism?

Human consciousness develops only within experience, in the midst of other persons and events, of things in the world, communication, reflection, correction, certification, and guidance. This means that as experience becomes something for us, and not simply undergone, we find that it is inherently triadic in structure composed of ourselves, other persons, and a third element, things, events, causes, norms: I, Thou, and It.[21] Further, all knowledge occurs within the triad and rests on some form of trust: *fides*, *fiducia*, and *fidelitas* (believing, trust, loyalty). It would be interesting to explore the essential element of trust in the triadic relation. But our concerns are elsewhere. It can be noted here that as rooted in the triadic character of experience, institutions are essentially, unalterably rooted in trust, without which there would be no institutions. Now we turn to address directly our two questions: (1) are any institutions primary, and (2) if so, are they made or found?

IV. Experience, Stability, and Norms

Is it possible that within the triad we find primary institutions? Let's see. Consider the members of the triadic relation, I, Thou, and It. Each individual I and Thou is numerically distinct and dyadically related both to other individual agents possessing will agency and to the It. Though we can distinguish each individual person within a triad, anything that can be said about each person, anything that each individual becomes occurs within and is deeply influenced by the other members of the triad.[22] Whatever can be said about the individuals requires a triad.

Consider next the relations within the triad. First, the triad is intrinsically situational: the It indicates both a habitat and a culture, while the I and Thou indicate psycho-physical beings related to that environment. Second, relations within the triad are structured, some of which are stable, enduring, and maintained and others are neither stable, enduring, nor maintained. For example, an Oxford college and a momentary association such as giving directions to a stranger.

Why are some relations within the triad stable, enduring, and maintained and others are not? Consider the elements within the triad. Recall not only the possibilities I, Thou, and It bring to the triad but also the needs, purposes, and agencies of I and Thou. In so far as the I-Thou-It are natural beings they operate, as best we can tell, according to the laws of physics and biology; thus, there is a kind of permanency in these relations [gravity, for example]. Undergirded but not limited by that natural context, each element within the triad both undergoes and acts on the other members of the triad. Limited by that environment as well as freed by its value possibilities, I and Thou collectively intend to achieve common purposes, satisfy common needs, and collectively live according to norms built into purposive action. Institutional structures arise, stabilize, and remain within the I-Thou-It triad on the basis of collectively intending shared needs and purposes, norms and natural law. If there were no interaction with the triad there would be no social structures, stable or otherwise. Where there is norm governed interaction within the triad to satisfy social needs or purposes there are enduring structures.

We are now approaching primary institutions. Are there stable, enduring, maintained social structures, satisfying social needs and purposes, and governed by norms and natural law that are inherent in the triadic relation itself such that without them there would be no interacting triad and with them there is an interacting triad? I suggest there are. Consider the triad again. The I and Thou in relation to each other and to the It interact on the basis of both the natural possibilities [including their own bodies] and the value [purposes and goals] possibilities within the triad. As the interaction occurs some are required for the interaction to continue occurring and others are not. For example, education is required for the interaction to continue. Without communication between the I and Thou their interaction would be severely limited, if they could interact at all without it. For I to interact with Thou, I is dependent on communication with Thou for information, knowledge, methods, procedures, and criteria. It is conceivable that an interaction could occur without communication; however, that would be only within the ecosystem (if at all). But for persons to be present, the I and Thou, communication must occur. Further, when one considers the life of the triad, particularly as it continues beyond its present members, new members must be taught the ways generated within the triad. Communication implies an It, a cause and a true value claim; it is important and we believe ourselves obligated to continue the life of the group, and communication is required to achieve that value. This communication is education, informal though it may be. In this sense education is intrinsic to and maintained within the structure of the triad. Here we find a primary institution. We are now ready to return to the second question.

V.   Made or Found?

Are institutions found or made? Now, back to the antinomy. Our short study has revealed that to attack one side of the antinomy in an attempt to show that it is false and that the other side is true is the approach of two metaphysical world-views that attempt to be exclusive, denying the truth of the other. We found them to be limited ontologies and returned to the puzzle of the antinomy. Returning to experience, not from the point of view of the methodological skepticism of a Descartes but from that of radical empiricists such as Borden Parker Bowne, we searched for structures that are implicit in experience and manifest themselves within it. We learned that experience is essentially triadic, always involving an I, a Thou, and and It. Further reflection on this structure helped us realize that three important aspects of it: [1] each of the three elements brings to the triad dispositions or potentialities, if you will, that allow for the formation of human social life, and [2] the interaction of the potentialities/dispositions are developed in terms of the natural capacities [natural law of gravity, life needs, for example] of all three and the normative life of the I and Thou in relation, and [3] the interaction necessarily involves communication among the three. Together, these provide for stability, endurance, and maintenance within the framework of the triadic structure, the condition for the continuation of the structure itself. Here we find primary institutions. That structure is required for any one individual member of the triad to come into existence and develop. [Which institutions are primary? At least education and the family. Beyond that I'm not at the moment prepared to comment.] In this sense institutions are found, and they are primary. Institutions are also made. They are *made* in the sense that they arise only within the interaction of the triad. They are potential within the triad. They come into existence, move beyond potentiality, only through the making activity of the I and Thou in relation to the It. Here we find not only primary institutions but also secondary ones as well. We can say that education as an enduring and stable social structure is creatively-found. Thus our answers to two question raised by Searle are that there are primary institutions and that they are neither found nor made; they are both found and made. The antinomy is resolved. Yet, the primary metaphysical question lingers. My claim is that they are quasi-persons. But that is for another paper.

Endnotes

1. John Searle, *The Construction of Social Reality*, p. 1.

2. Ibid., 5.

3. Ibid., 7.

4. Ibid., 10.

5. Ibid., 28.

6. Ibid., 28.

7. Ibid., 41.

8. Ibid., 125-26.

9. Ibid., 129.

10. Ibid., 129.

11. Peter L. Berger, *The Social Construction of Reality; a Treatise in the Sociology of Knowledge.* Garden City, NY, Doubleday, 1966.

12. Marcus G. Singer, "Institutional Ethics*,*" in *Ethics*, A. Phillips Griffiths (ed.), Royal Institute of Philosophy Supplement: 35. Cambridge UP, 1993*: 227-28.*

13. Talcott Parsons, "Prolegomena to a Theory of Social Institutions." *American Sociological Review* 55.3 (June 1990): 320.

14. Melville J. Herskovits, *Cultural Anthropology*, New York: Alfred A. Knopf, 1960: 95.

15. Ibid.

16. Ibid.

17. See Searle's *Mind, Brain, and Science.* Note that if Searle is correct, this implies that as something humans make, institutions can be fully grasped by the mind. They possess a structure penetrable by rational inquiry and are capable of undergoing modification. For example, consider the new institutionalism among economists, especially they heave reliance on mathematical articulation of economic institutional structures and processes.

18. Personalism 20-21.

19. Kant 131; note the presumption of trust in that view.

20. For a fuller development of this point see Thomas O. Buford, *In Search of a Calling* (Macon, Georgia: Mercer UP, 1995).

21. Our experience is inherently social. Buber, Mead, and Vico recognize this, in contrast to individualists, such as Mill and Rorty, who do not. For a fuller statement of this position see H. Richard Niebuhr, *Faith on Earth, An Inquiry into the Structure of Human Faith.* New Haven: Yale UP, 1989: 43-62.

22. Each person brings to the triad activity potentials, at least, that allow her to develop. However, the particular personality she becomes occurs in the triadic context. It is crucial to distinguish self, personality, and person: the activity potentials, the way the activity potentials develop and take form in the context of Thou and It, and the one whose these are. See Peter A. Bertocci, The Person, His Personality, and Environment." *Review of Metaphysics* 32 (1979): 605-621. See also H. Richard Niebuhr, *Faith on Earth*, and Josiah Royce, *Outlines of Psychology: An Elemental Treatise with Some Practical Applications.* New York: The Macmillan Co., 1911.

CHAPTER FOUR

Physical And Biological Aspects Of Renal Vitrification

Gregory M. Fahy, Brian Wowk, Roberto Pagotan,
Alice Chang, John Phan, Bruce Thomson, and Laura Phan*

§1. Introduction
§2. Results
§3. Materials and Methods
§4. Conclusions

§1. Introduction

The long-term banking of human organs or their engineered substitutes[1] for subsequent transplantation is a long-sought[2-4] and important[1,2,5-11] goal. Given that the full demand for vital and non-vital organ replacements may be over one million per year in the United States alone, supply chain management issues may become more and more critical as the success of laboratory construct creation increases.[1] Contemplating the possible development of emergency organ replacements with generic allografts without the availability of organ biobanking is a bit like trying to envision attempting to distribute human blood with a 24-hour shelf life limitation.

Biobanking of organ and tissue replacements has not been widely discussed perhaps in part because the technology for doing this without damage to the graft is not in hand. Although freezing can achieve limited success for some organs,[7,9,12-15] freezing of the heart, liver or kidney has not been accomplished with subsequent life support function following cooling to temperatures low enough for long-term preservation, despite work on this problem dating back to the 1950s.[3,6] Kidneys and hearts have been the most

---

widely studied organs, but neither has been reproducibly recovered after freezing to temperatures lower than about −20°C,[16–20] evidently due at least in part to mechanical damage from ice itself,[21–24] although in the case of kidneys at least, sporadic survival has sometimes been claimed after freezing to about −40 to −80°C.[25–28]

Some time ago, one of us (GMF), after witnessing transplanted dog kidneys turning deep blue and passing urine that resembled whole blood after freezing to only −30°C with 3 M glycerol (unpublished observations using the same methodology[29] used for rabbit kidney freezing), proposed a way of cooling organs to cryogenic temperatures without incurring the consequences of ice formation.[30–33] This is possible because high concentrations of cryoprotective agents reduce the likelihood and the speed of ice crystal formation, and sufficiently high concentrations can prevent ice formation completely, even at the low cooling and warming rates that are applicable to organ-sized objects.[1,34–36] Cooling an ice-free biological system to a low enough temperature eventually results in a transition from a mobile fluid state to a molecularly arrested glassy state (this transition being referred to as vitrification, or the glass transition). A glass is essentially a liquid that cannot flow over most time scales of interest to the observer,[36] and a vitrified biological system can theoretically be stored for virtually any desired length of time due to the extreme slowing of all diffusion-driven change below the glass transition temperature[37] (T_G). "Vitrification solutions"[38] are solutions of cryoprotective agents that are sufficiently concentrated to enable vitrification or virtual vitrification of a living system at the cooling rates employed for that purpose.

Major advances in vitrification technology have recently been reported,[6,39] and it is now possible to vitrify entire organs,[6] but to do so with full recovery of viability after transplantation is still difficult due in large part to devitrification. Devitrification is ice formation during rewarming, and it arises because ice nuclei, which form initially only at temperatures too low for appreciable crystal growth,[36,40] encounter temperatures during warming that maximize ice growth.[40,41] To date, small ovaries,[42–45] blood vessels,[11] heart valves,[46] corneas[47] and similar structures[48] that can all be cooled and rewarmed rather rapidly so as to avoid devitrification, are the only macroscopic structures that have been reported to recover at least in part after vitrification.

Research on vitrification of organs that require immediate vascular anastomosis upon transplantation has been carried out primarily on the rabbit kidney.[5,6,39,49-51] The rabbit kidney provides a useful illustration of the general problems of preserving both natural and laboratory-generated organ replacements. In this article, we describe the special problems of vitrifying the kidney and progress made toward their solution, including the first case of life support after vitrification and rewarming.

§2. Results

Survival of the first large solid organ after vitrification and transplantation: a case history.

In late 2002 and early 2003, several rabbit kidneys were perfused with the M22 vitrification solution,[6] vitrified and transplanted[52] back to their original donors (autografts) with immediate contralateral nephrectomy either to evaluate survival or to evaluate short-term blood reflow only for the first several minutes in vivo. No rabbit survived when perfused with M22 at 40–60 mmHg, but one of two survived after perfusion with M22 at 80 mmHg for 25 min, and the second rabbit in this small group lived for 9 days after transplantation, which was longer than any other non-surviving rabbit studied. Although anecdotal, the sole survivor proves that organ cryopreservation by vitrification can result in life-supporting function after transplantation, and a detailed examination of this case reveals many interesting aspects of the problem of successfully preserving an organ by vitrification.

The events during perfusion of the surviving kidney are shown in **Figure 1A** and, with the exception of the elevated perfusion pressure during M22 perfusion, are typical of protocols we have described for several years.[5,6,49] The venous concentration just before cooling the kidney to below T_G was 96.4% of the arterial concentration, and the absolute arteriovenous concentration difference was 330–340 mM. Under the conditions of this perfusion, this venous concentration predicts[6] an inner medullary tissue concentration that is 92.1% of the arterial concentration, which is sufficient to permit vitrification on cooling although insufficient to preclude devitrification.[6]

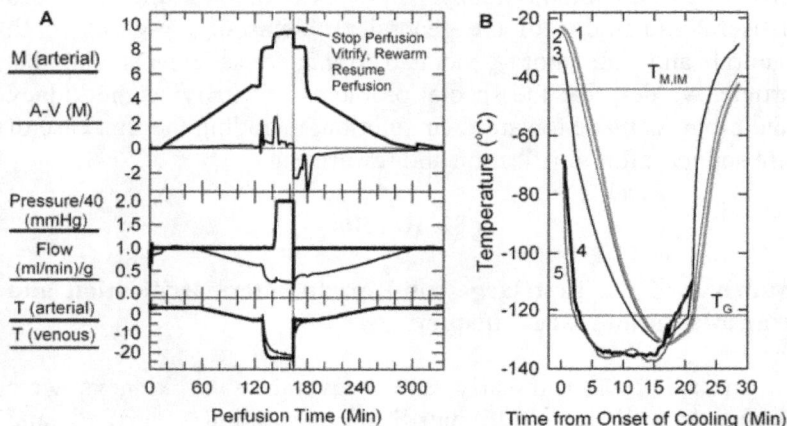

Figure 1.

A. Perfusion protocol for renal survival after vitrification and rewarming. M, molarity; A-V (M), arteriovenous difference in molarity; T, temperature in degrees Celsius. The protocol, as usual,[6, 39] employs an initial 5M plateau, a second plateau at 8.4M to allow cooling to -22°C without freezing, and a final plateau during M22 perfusion. In the experiment shown, the perfusion was interrupted at the point shown to enable the kidney to be vitrified, rewarmed, and reperfused with 8.4M cryoprotectant at -3°C.

B. Thermal history of the transplanted kidney based on invasive temperature measurements in a model rabbit kidney cooled and rewarmed by a procedure identical to that used for the vitrified-transplanted rabbit kidney. Line 1: inner medullary temperature, as documented by a thermocouple located 1.2 cm below the renal surface; line 2: outer medullary temperature, measured 7 mm below the renal surface; line 3: cortical temperature 2 mm below

Figure 1B shows the thermal history of the kidney during cooling and warming and indicates that all parts of the transplanted kidney were below T_G for about 8 min, the thermal nadir being about $-130°C$ for the cortex, outer medulla and inner medulla (approximately 7–8°C below the estimated T_G of the inner medulla[6]). The warming rate of the inner medulla from T_G to $-60°C$ was about 15°C/min, and declined to 6°C/min from $-60°C$ to the predicted inner medullary melting point ($T_{M,IM}$) of $\sim-44.2°C$. During the removal of M22, the kidney perfused normally, and during transplantation, the urine was not bloody and the kidney appearance was reasonable and seemed to be recovering at closure.

The animal became anemic on the first postoperative day and again on day 10 (**Fig. 2A**). This symptom was not previously seen after cooling to $-45°C$.[6] Fortunately, the anemia spontaneously resolved after being successfully treated, suggesting recovery of adequate renal production of erythropoietin. Acute hyperkalemia developed on days 2 and 3 but was successfully controlled. Thereafter, K^+ levels slowly rose until reaching a stable value by about day 32. Serum creatinine peaked at 14.6 mg/dl (**Fig. 2B**) on day 4 and then fell to a nadir of 3.3 mg/dl on day 24 independent of diuresis and hydration. It then slowly rose again until reaching an apparently stable value of 6.0–6.4 by day 38.

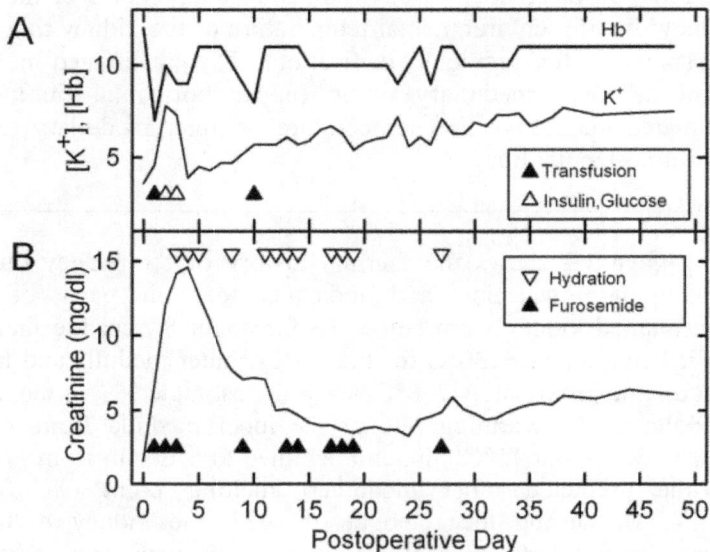

Figure 2.

A. Changes in blood levels of hemoglobin and potassium after transplantation of a previously vitrified rabbit kidney and interventions to correct both (triangles). Hyperkalemia was corrected by intravenous glucose (20 ml of 5% dextrose in 0.45% NaCl) and insulin (0.4 ml of 1 U/ml, IV). Anemia was corrected with 20 ml of whole rabbit blood (~6-8 ml/kg) on each occurrence. Blood levels were measured before corrective interventions given on the same day. [Hb], hemoglobin concentration in g/dl); [K+], potassium concentration in meq/l.

B. Postoperative creatinine levels and diuretic support history. Lower triangles indicate furosemide administration (generally 5-10 mg, IV or IM); upper triangles indicate hydration (generally 100-200 ml, consisting of equal volumes of 0.9% NaCl and 0.45% NaCl plus 5% glucose, subcutaneously). Blood levels were measured before corrective interventions given on the same day.

Clinically, the animal regained normal drinking behavior, a normal fecal output score, and a normal urine volume output score by about 1–2 weeks postoperatively, but food consumption and to a lesser extent water consumption and urine output declined on balance after day 24. The rabbit lost about 18% of its body weight by the fifth postoperative day and thereafter maintained this weight while also maintaining normal posture and behavior other than some sluggishness.

After ensuring that the animal appeared capable of living indefinitely using the vitrified kidney as the sole renal support, it was euthanized for histological follow-up on day 48. Ice formation during warming was not expected in the cortex but was expected to be equivalent to 1–2% of the total inner medullary mass,[6] so the fate of the renal medulla was of special interest. To our surprise, examination of an entire renal cross section showed that medullary damage was essentially confined to one side of the kidney, the medullary portion of the peripelvic columns on the opposite side displaying remarkably good survival (**Fig. 3**). This raises fascinating but still unresolved mechanistic questions about the origin of the observed damage, but indicates that under the experimental conditions achieved, the delivery of M22 to the medulla was sufficient to allow survival of considerable medullary mass, inspiring hope that relatively small improvements in medullary cryoprotectant delivery might enable full survival of the renal medulla.

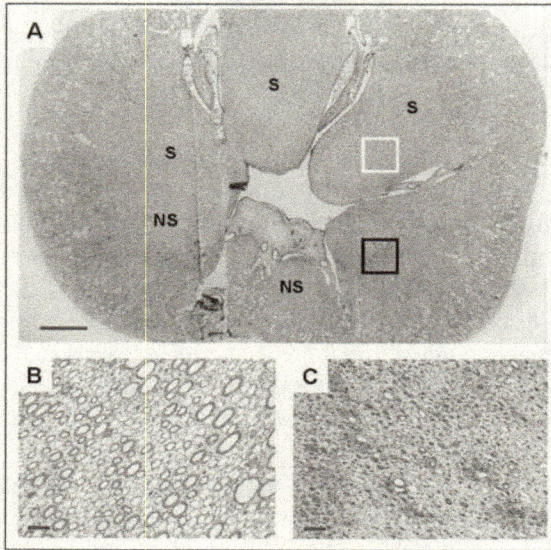

Figure 3.

A. Cross-section of the vitrified/rewarmed kidney (PAS staining) showing surviving (S) and non-surviving (NS) medullary areas; white box designates the region depicted in **Figure 3B**, and black box identifies the location of **Figure 3C**.   Non-surviving areas are confined to one side of the kidney. Scale bars: in **A**, 3 mm; in **B** and **C**, 100 microns.

Despite lack of expected freezing in the renal cortex, considerable cortical injury was observed as well. This damage ranged from reasonably mild loss of superficial cortical tubules (**Fig. 4**, top) to predominant loss of tubules in the cortex corticis with persistence of glomeruli (**Fig. 4**, middle) to loss or atrophy of both superficial tubules and associated glomeruli (**Fig. 4**, bottom). We speculate that this injury is the result of previously undiscovered stress-strain phenomena in the outer cortex caused by the establishment of large thermal gradients in relatively stiff and brittle tissue near the glass transition temperature. Lowering cooling or warming rates to avoid this form of injury is feasible in principle but will require still better distribution of M22 into the renal medulla because medullary devitrification will otherwise be exacerbated by lower cooling and warming rates, as verified by direct observation (unpublished results).

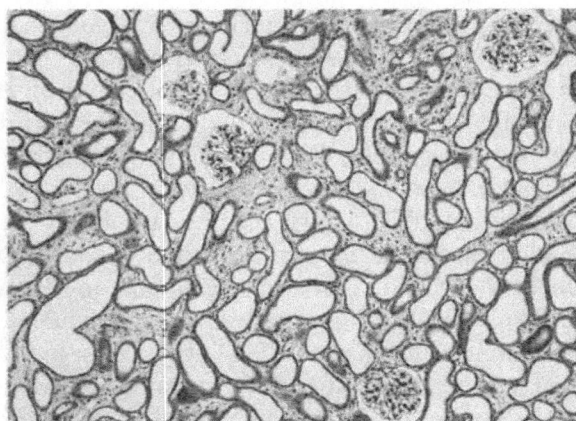

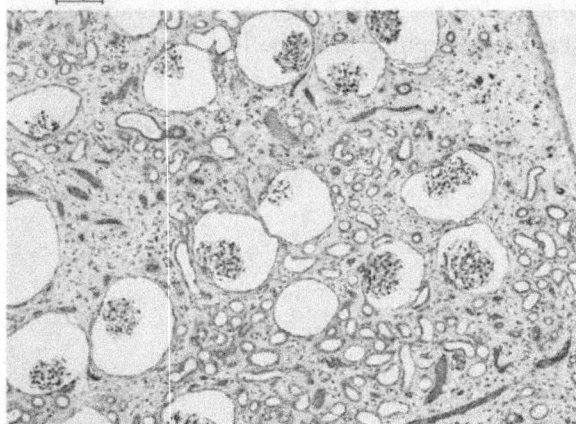

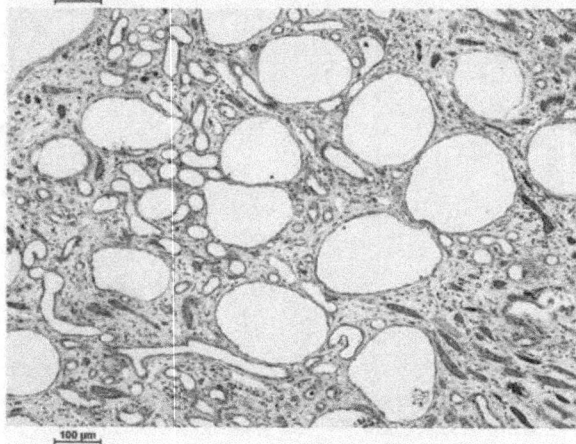

Figure 4.

The spectrum of renal cortical responses to vitrification and rewarming. **Top:** area showing predominant survival of both tubules and glomeruli. Scale bars all represent 100 microns. **Middle:** transitional zone between predominantly surviving superficial renal cortex and non-surviving cortex, showing loss of tubules but survival of glomeruli. **Bottom:** non-surviving superficial cortex, showing loss of both tubules and glomeruli, with ballooning of Bowman's capsule. PAS stain.

The problem of renal medullary water replacement.

These results identify the renal medulla as a tissue that seems to be poised at the dividing line between the success and the failure of vitrification. The survival of the medulla presumably depends on the relationship between medullary cryoprotectant delivery and medullary ice formation, and deeper insight into this relationship will be fundamental for understanding the requirements for successful vitrification and recovery of the kidney and, by extension, for the recovery of vitrified organized tissues in general.

The anatomy of the renal vasculature is organized so as to constrain medullary blood flow to a small fraction of total renal blood flow,[53] an arrangement that allows the kidney to concentrate urine but makes the task of delivering cryoprotectant to the medulla a difficult one. Anatomically, the medullary circulation is provided by the vasa recta, which originate either directly from widely-spaced points along the arcuate arteries or indirectly from efferent arterioles of juxtamedullary glomeruli, which comprise about 9% of the total number of glomeruli;[54] in either case, the originating blood vessels subdivide into many parallel vascular channels, each of which carries a small fraction of the flow that enters the originating vessel.

These anatomical limitations are a given, but medullary delivery of cryoprotectant can be influenced by factors such as perfusate viscosity, cryoprotectant delivery protocol, and the permeability and diffusivity of the cryoprotectants in the vitrification solution. In addition, the vascular system is not the only route of delivery for cryoprotectants. The medulla consists also of tubules and collecting ducts that can convey permeable cryoprotectants along their lengths, diffusing as they go. At the temperatures of our experiments[6] (−22°C to 3°C, and particularly ~−22°C or −3°C for delivery of the highest concentrations of cryoprotectants), and in the presence of more than 8 molar cryoprotectant (<~48% v/v water), no appreciable renal metabolism can be expected, and therefore tubular delivery of cryoprotectants to the medulla is presumably entirely passive and driven only by filtration at the glomerulus followed by local diffusion (no secretion, no active reabsorption, just diffusion) until delivery into the pelvis. Given that medullary blood flow amounts to only about 10% of total renal blood flow under ordinary conditions,[53] a filtration fraction in the vicinity of just 10%, which we have observed for rabbit kidneys,[49] would be sufficient to deliver enough ultrafiltrate to the medulla to match the total volume flowing through the medullary blood vessels.

The best vitrification solution known for the kidney to date is M22,[6] whose critical cooling rate (the cooling rate above which ice formation is not observed) is 0.1°C/min, and whose critical warming rate (the warming rate above which ice formation is not observed) is 0.4°C/min.[1,34] As determined from the cooling and warming curves of **Figure 1B**, the rabbit kidney can, conservatively, be cooled and warmed by conduction at about 8°C/min or more, which implies that 100% equilibration of the medulla with M22 is not required. Key questions are, what is the level of equilibration that is required, what is required to achieve it, and how can we know when we have achieved it? These questions are taken up in the next section.

Measuring and achieving adequate equilibration.

Comparisons between cryoprotectant concentrations in the urinary space and in the venous effluent revealed that the "urine" (perfusate ultrafiltrate) tends to lag far behind the venous effluent in concentration (**Fig. 5**). This is logical since the urine flow rate is a fraction of the arterial flow rate, and since the venous effluent disproportionately samples the overperfused renal cortex, which accounts for ~90% of total renal perfusate flow, and therefore under-represents poorly-equilibrated areas. In addition, the urine makes three passes through the renal medulla (descending and ascending limbs of the loop of Henle followed by passage through the collecting ducts) and therefore is in intimate osmotic/diffusive communication with the renal medulla before it is collected. For these reasons, the urine is expected to reflect medullary tissue concentrations of cryoprotectant better than is the venous effluent concentration, and experimental results described below bear out this expectation.

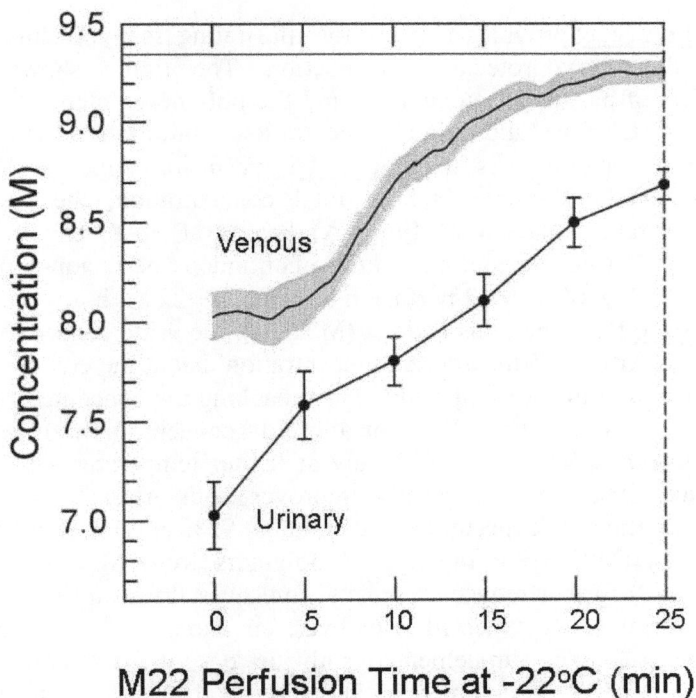

M22 Perfusion Time at -22°C (min)

Figure 5.

Difference between the venous concentration and the urinary space concentration during M22 perfusion at -22°C and 40 mmHg (n = 4 perfusions). Urine concentrations (discreet data points ± 1 SEM) determined manually; venous concentrations (line with gray "halo" consisting of ± 1 SEM) determined by computer. The time base gives time from the nominal onset of M22 perfusion, with includes a lag time as M22 makes its way through the perfusion circuit. The horizontal line near the top of the graph shows the concentration of M22, which is not fully reached even by the venous effluent by the end of M22 delivery.

Figure 6 provides a basis for illustrating many features of medullary cryoprotectant introduction. The figure shows the effects of perfusion temperature and the polymer content of M22 on arterial flow and urine concentration equilibration with the arterial perfusate. As in **Figure 1A**, perfusion with the VMP transitional solution[39] (8.4 M total concentration, the second concentration plateau of **Fig. 1A**) begins at −3°C and in the standard protocol[6] continues during continuous perfusion-cooling to −22°C to allow M22 perfusion to begin at −22°C, but we see in **Figure 6** that when this is done (M22-22), the urine concentration lags so far behind the arterial concentration that at the end of M22 perfusion, urine concentration is just reaching the concentration of VMP, or only about 90% of the full concentration of M22. Perfusion of VMP and M22 only at higher temperatures (−3°C) reduces viscosity and greatly improves both arterial flow and equilibration, as expected, allowing about 95% equilibration to be attained (M22-3). Removing all polymers from M22 at −3°C (M22NP) further reduces viscosity, improves flow, and improves equilibration, as expected. However, an anomaly is introduced when M22NP is supplemented with just one of the polymers of M22, namely, the commercial Supercool X-1000™ ice blocker[55] (X-1000). Perfusing this solution (M22NP + 2X, containing 2% w/v X-1000, or twice the usual concentration of X-1000 in M22,[6]) slows the arterial perfusion rate yet ultimately allows a degree of equilibration similar to that achieved after M22NP perfusion.

Therefore, urinary space equilibration is not proportional to the arterial flow rate.

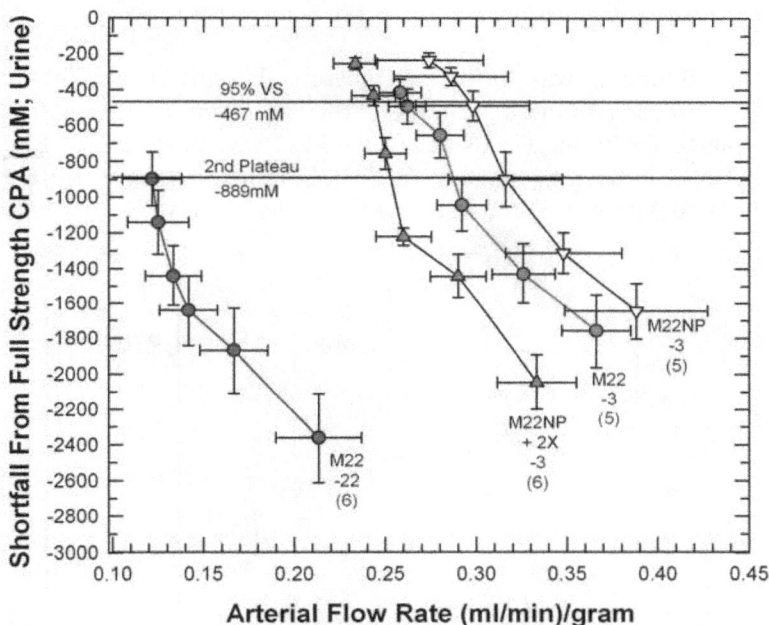

Figure 6.

Equilibration shortfalls (urine concentration minus nominal arterial concentration) in rabbit kidneys perfused with M22 at -22°C (M22 -22) or at -3°C (M22 -3) plotted as a function of arterial flow rates (which decline as higher concentrations are reached and viscosity increases). M22NP -3 refers to M22 minus all polymers, perfused at -3°C; M22NP + 2X -3 refers to M22NP containing 2% X1000 ice blocker, perfused at -3°C. Values in parentheses indicate the number of perfusions of each type. Each data point represents urine equilibration measured at 5-min intervals, beginning at VS perfusion time zero to the right and ending at VS perfusion time = 25 min to the left. The horizontal lines are "landmark" concentrations and refer to the concentrations of VMP (2nd Plateau, which falls at a shortfall of -889 mM) and 95% of full-strength vitrification solution (VS) (which, because of the negligible molarity of the polymers of M22, is essentially the

Equilibration was, however, mirrored by differences in the
urine flow rates in these groups, and the latter were in turn closely
accounted for by the viscosities of the M22 variant solutions (**Fig.
7**). Thus, it seems that urinary equilibration is more closely
correlated with urine flow rate than with arterial flow rate.

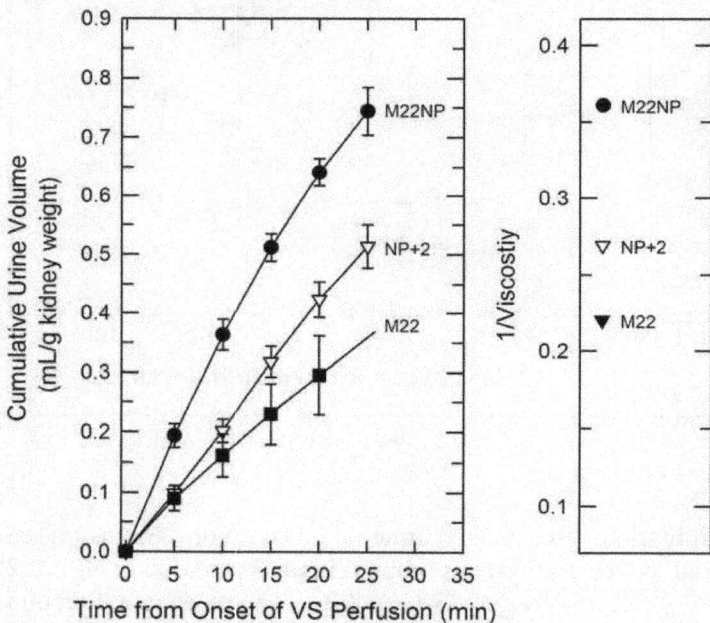

Figure 7.

Left: urine accumulation during perfusion with M22, M22NP +
2% X1000 (NP + 2), and M22NP at -3°C; right: reciprocal
viscosities of these three vitrification solutions (cP^{-1}). The total
accumulated urine volumes are inversely proportional to the total
viscosity of each VS (M22 = 4.54 cP; M22NP + 2X = 3.71 cP;
M22NP = 2.77 cP). The urine volume for M22 at 25 min was not

consistently recorded and so is indicated by extrapolation. Data points represent means ± 1 SEM.

Figure 8 answers the question of "how much equilibration is enough" and brings out a number of other important points. The left panels describe the devitrification temperatures, percent ice formed at the point of devitrification, and percent of ice melted at the tissue melting point, for urine samples collected at the end of the perfusion, and the right panels report the same information for inner medullary tissue samples (all data obtained by differential scanning calorimetry).

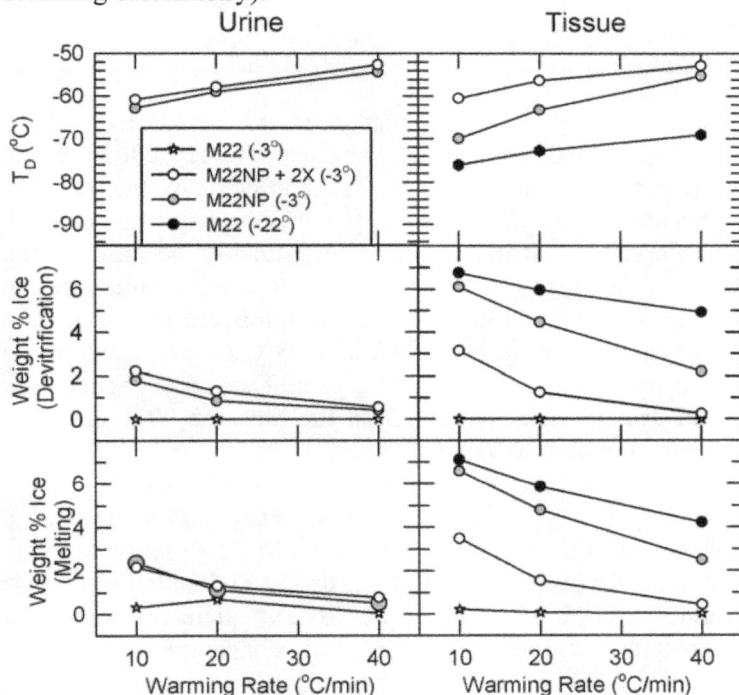

Figure 8.

Temperature, extent, and warming rate dependence of ice formation in urine (left panels) and tissue samples (right panels) obtained from kidneys subjected to the four protocols of Figure 6 and relationship between the amount of ice formed during devitrification and the amount of ice that thawed upon complete rewarming. Urine was not collected from the M22 kidneys perfused at -22°C. Upper panels: devitrification temperatures (T_D); middle panels: the percentage of sample mass that crystallizes during devitrification; lower panels: the percentage of sample mass that melts upon continued warming. Each point represents the mean of generally 5-6 independent measurements; devitrification temperatures are averaged only for those samples that devitrified. No devitrification event was observed for any specimen in the M22 -3°C group. Error bars omitted for clarity. Groups are represented as indicated in the inset. For discussion, see text.

The first thing to note is that perfusion of M22 at −22°C causes about 7% of inner medullary mass to crystallize as ice during rewarming, and this result is little affected by the warming rate. This amount of ice is substantially greater than was predicted for our surviving vitrified kidney, presumably because in the experiments of **Figure 8** we perfused at 40 mmHg rather than at 80 mmHg, which is known to make a significant difference.[6] In complete contrast, perfusion of M22 at −3°C (stars) results in no tissue ice formation at any warming rate. Therefore, the required degree of urinary space equilibration lies between 90% and 95%, and is probably close to the latter limit.

Second, perfusing M22NP at −3°C results in less ice formation than perfusing M22 at −22°C even though M22NP is a more dilute and intrinsically less stable solution; this is undoubtedly because the higher equilibration level of M22NP delivers more net cryoprotectant despite its lower total concentration. Finally, adding 2% X1000 to M22NP greatly suppresses tissue ice crystal formation, which demonstrates the ability of X1000 to usefully penetrate into and protect inner medullary tissue.

Comparing tissue results to urine results shows that tissue generally devitrifies at a lower temperature than does urine from the same kidney, and that the amount of ice formed in tissue is accordingly higher than it is in urine from the same kidney, indicating that tissue concentrations lag behind urine concentrations. Interestingly, for both urine and tissue, in most cases the percentage of sample mass that melts upon thawing is the same as the percentage that freezes during devitrification, meaning that vitrification is generally complete on cooling with the regimen used for tissue analysis.

Visual assessment of ice formation.

Although tissue biopsies allow quantitative results to be obtained as presented in **Figure 8**, we have been interested in developing methods for visualizing ice formation across entire renal cross-sections in order to be able to judge the two and three-dimensional extent of ice formation. Although these methods are still in development, we present an example of the type of information that can be obtained in **Figure 9**. In this example, the warming rate was about 1°C/min, and therefore more ice is expected than with the more rapid warming used in **Figure 8**. Nevertheless, the maximum extent of ice formation, judged by whitening of the tissue during rewarming, did not include the cortex in this example, and the ice that formed appeared to be uniformly distributed.

Maximum Extent of Devitrification

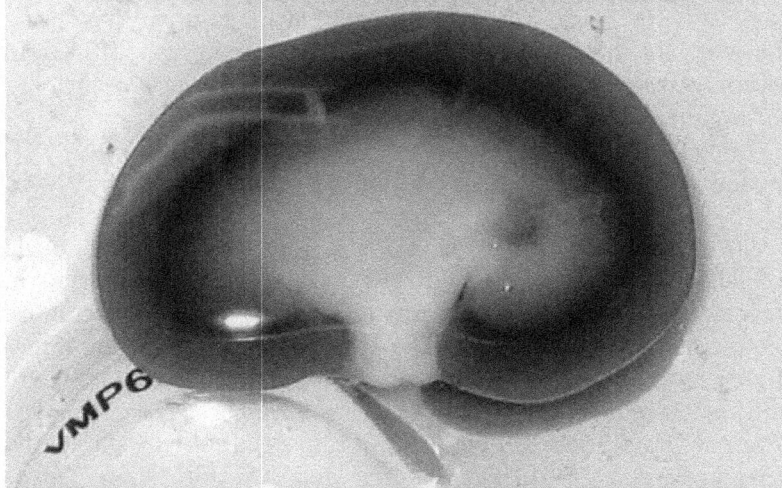

WR: 1°C/min   01:30   -40.6°

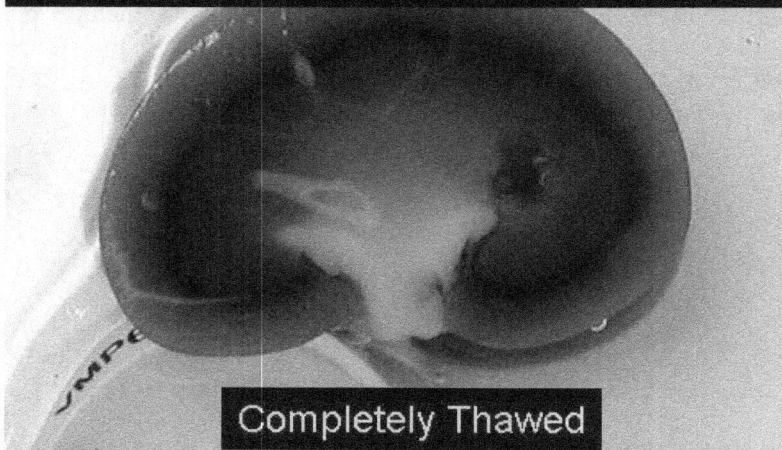

Completely Thawed

VMP6L-13   01:40   -28.4°

Figure 9.

Visual appearance of ice in an exemplary rabbit kidney cross-section during rewarming. The kidney was perfused with M22 at -22°C, cut in half, immersed in M22, vitrified in a CryoStar freezer at -135°C, and eventually rewarmed at about 1°C/min while being photographed from time to time. Rewarming was accomplished by transfer of the kidneys to an insulated box through which liquid nitrogen vapor was circulated slowly so as to allow steady warming of the contained atmosphere from just below Tg to well above the renal melting points. Times (1:30 and 1:40) represent times in hours and minutes since the onset of slow warming, and temperatures refer to ambient atmospheric temperatures near the kidney but not within the kidney itself. The upper panel shows the kidney at the point of maximum ice cross-sectional area, and the lower panel shows the kidney after complete ice melting. Both panels show the site of an inner medullary biopsy taken for differential scanning calorimetry.

Using this method and differential scanning calorimetry will eventually allow us to determine the extent to which medullary ice formation can be tolerated by the kidney. We have been able to show that medullary damage can be assessed in the acute postoperative period by removing kidneys 30 min after transplantation, flushing them to remove blood, and examining the extent of medullary blood trapping by inspection of renal cross-sections (unpublished observations). Although preliminary, such observations have identified conditions that allow blood trapping to be avoided, and as our methods improve, we should be able to use such methods to determine how much medullary ice formation, if any, is acceptable, and to select perfusion methods for evaluation by permanent transplantation.

§3. Materials and Methods

Procedure for obtaining survival after rabbit kidney vitrification.

A 12.7 gram rabbit kidney was perfused with M22, a 9.3 M vitrification solution with very low critical cooling and warming rates,[1,6,34] in an LM5 carrier solution under computer control[56] using a variation of our standard protocol[6] on December 10th, 2002 (**Fig. 1A**). Perfusion began with Renasol-14 containing 2% w/v B. Braun hydroxyethyl starch (HES) and no cryoprotectant and continued, after a pause at 5 M cryoprotectant to allow the arteriovenous (AV) concentration gradient to level, to VMP[39] in LM5 containing no HES. To distribute M22 more thoroughly than usual while minimizing damage from perfusion pressures over 40 mmHg, perfusion pressure was raised to 80 mmHg only during the 25-min period of exposure to M22 itself. The kidney was removed from the perfusion apparatus at the end of M22 perfusion and cooled in rapidly-moving nitrogen vapor[6] (**Fig. 1B**) The intra-renal thermal history was determined by inserting a three-point needle thermocouple (beads at 2, 7 and 12 mm depths; PhysiTemp, Huron, PA) into an identically-treated but non-transplanted kidney.

Rewarming was accomplished by slowly raising the environmental temperature to about −115°C in order to bring the cortical temperature to just above T_G, at which point the kidney was returned to the perfusion machine and further warmed by pouring M22 at −22°C over the renal surface for 8 min. Rewarming was completed by perfusing the kidney with VMP at −3°C, after which cryoprotectant washout was completed as usual[6] (**Fig. 1A**), transitioning from VMP in LM5 to Renasol-14 + 2% HES, and the kidney was transplanted according to our published method[52] with immediate contralateral nephrectomy.

Perfusion of kidneys with M22 and alternative vitrification solutions at 40 mmHg.

All perfusions were carried out under computer control in the general manner represented in **Figure 1A**. However, because B.

Braun discontinued the manufacturing of HES and because the use of all alternative forms of HES was associated with higher post-transplantation peak creatinine levels (unpublished results), we replaced HES with 2% w/v decaglycerol (dG) in the carrier solution at the beginning of the perfusion. We also used TransSend-4 at the beginning and end of each experiment, but retained LM5 as the carrier for VMP, M22, and their variants. Remaining protocol details other than the perfusion pressure were as reported in **Figure 1A** and elsewhere.[6]

End point measurements.

Determination of tissue freezing points (devitrification temperatures), percent ice formation and percent ice melted were all carried out by differential scanning calorimetry. The cooling and warming protocol for inner medullary samples was to cool to $-120°C$ at $10°C/min$ and to rewarm at 10, 20 or $40°C/min$ when the endpoint was devitrification. Heats of devitrification and of melting were obtained by integrating peak areas and were converted from units of joules/gram into percent ice formation by dividing by 3.34.[6] The temperatures of devitrification were taken to be the temperatures at the tops of the observed peaks. Cryoprotectant concentrations were determined from refractive index readings on the basis of appropriate calibration curves. Baseline data were freshly derived for each experiment during priming of the perfusion system. All refractive indices were recorded continuously at ~0°C using ice-immersed in-line process refractometers (AFAB Enterprises, Eutis, FL, Model PR-111) at the beginning (priming) and experimental phases of each perfusion except that urine refractive index was determined using a bench-top Bellingham Stanley RFM 330 refractometer at room temperature and converted to concentration using a separate room temperature calibration obtained using the same refractometer. Viscosities were measured using a Gilmont falling-ball viscometer (Cole Parmer) at room temperature.

§4. Conclusions

Clearly, the problem of eliminating or sufficiently limiting ice formation throughout the kidney without inducing unacceptable toxicity is a complex and many-faceted one. So far, the most promising single approach seems to be the one described in **Figure 1**, which resulted in survival after transplantation. However, the many lessons that have been learned since that experiment will undoubtedly result in methods for protecting the kidney that are more effective than those used in **Figure 1**, and that will allow better and more consistent survival to be obtained after vitrification and rewarming. Certainly, the availability of new methodologies to evaluate renal tissue resistance to ice formation will be helpful, and the use of microwave rewarming to reduce the likelihood of damage from devitrification could also be highly beneficial for our efforts to solve the very complex problem of fully successful renal vitrification.

Because of its unique vascularization, the kidney may be the most challenging organ of them all to vitrify and rewarm successfully. If so, continued progress with the kidney should be encouraging for the future vitrification and recovery of other complex living systems, including laboratory-produced organ and tissue replacements, whose accessibility to cryoprotectant may be significantly greater than that of the renal medulla.

Acknowledgements

Transplantation of the surviving kidney was carried out by Dr. Jun Wu, who is now in private dental practice. The authors wish to thank Ms. Perlie Tam for expert surgical assistance. Supported by 21st Century Medicine, Inc., All procedures involving animal use were done according to USDA standards and with IACUC approval.

References

1. Fahy GM, Wowk B, Wu J. Cryopreservation of complex systems: the missing link in the regenerative medicine supply chain. *Rejuvenation Res.* 2006;9:279–291. [PubMed]

2. Starzl TE. A look ahead at transplantation. *J Surg Res.* 1970;10:291–297. [PubMed]

3. Smith AU. Problems in the resuscitation of mammals from body temperatures below 0°C. *Proc R Soc Lond B Biol Sci.* 1957;147:533–544. [PubMed]

4. Karow AM., Jr . The organ bank concept. In: Karow AM Jr, Abouna GJM, Humphries AL Jr, editors. *Organ Preservation for Transplantation.* Boston: Little, Brown and Company; 1974. pp. 3–8.

5. Khirabadi B, Fahy GM. Permanent life support by kidneys perfused with a vitrifiable (7.5 molar) cryoprotectant solution. *Transplantation.* 2000;70:51–57. [PubMed]

6. Fahy GM, Wowk B, Wu J, Phan J, Rasch C, Chang A, et al. Cryopreservation of organs by vitrification: perspectives and recent advances. *Cryobiology.* 2004;48:157–178. [PubMed]

7. Wang X, Chen H, Yin H, Kim S, Lin Tan S, Gosden R. Fertility after intact ovary transplantation. *Nature.* 2002;415:385. [PubMed]

8. Karlsson JO, Toner M. Cryopreservation. In: Lanza RP, Langer R, Vacanti J, editors. *Principles of Tissue Engineering.* Second Edition. San Diego: Academic Press; 2000. pp. 293–307.

9. Arav A, Revel A, Nathan Y, Bor A, Gacitua H, Yavin S, et al. Oocyte recovery, embryo development and ovarian function after cryopreservation and transplantation of whole sheep ovary. *Hum Reprod.* 2005;20:3554–3559. [PubMed]

10. Kaiser J. New prospects for putting organs on ice. *Science.* 2002;295:1015. [PubMed]

11. Song YC, Khirabadi BS, Lightfoot F, Brockbank KG, Taylor MJ. Vitreous cryopreservation maintains the function of vascular grafts. *Nat Biotechnol.* 2000;18:296–299. [PubMed]

12. Bedaiwy MA, Jeremias E, Gurunluoglu R, Hussein MR, Siemianow M, Biscotti C, et al. Restoration of ovarian function after autotransplantation of intact frozen-thawed sheep ovaries

with microvascular anastomosis. *Fertil Steril.* 2003;79:594–602. [PubMed]

13. Martinez-Madrid B, Dolmans M-M, van Langendonckt A, Defrere S, Donnez J. Freeze-thawing intact human ovary with its vascular pedicle with a passive cooling device. *Fertil Steril.* 2004;82:1390–1394. [PubMed]

14. Hamilton R, Holst HI, Lehr HB. Successful preservation of canine small intestine by freezing. *J Surg Res.* 1973;14:313–318. [PubMed]

15. Fahy GM. Analysis of "solution effects" injury: rabbit renal cortex frozen in the presence of dimethyl sulfoxide. *Cryobiology.* 1980;17:371–388. [PubMed]

16. Elami A, Gavish Z, Korach A, Houminer E, Schneider A, Schwalb H, et al. Successful restoration of function of frozen and thawed isolated rat hearts. *J Thorac Cardiovasc Surg.* 2008;135:666–672. [PubMed]

17. Toledo-Pereyra LH. Organ freezing. *J Surg Res.* 1982;32:75–84. [PubMed]

18. Pegg DE, Green CJ, Walter CA. Attempted canine renal cryopreservation using dimethyl sulphoxide, helium perfusion and microwave thawing. *Cryobiology.* 1978;15:618–626. [PubMed]

19. Smith AU. The effects of glycerol and of freezing on mammalian organs. In: Smith AU, editor. *Biological Effects of Freezing and Supercooling.* London: Edward Arnold, Ltd; 1961. pp. 247–269.

20. Kubota S, Lillehei RC. Some of the problems associated with kidneys frozen to −50°C or below. *Low Temp Med.* 1976;2:95–105.

21. Pegg DE, Diaper MP. The mechanism of cryoinjury in glycerol-treated rabbit kidneys. In: Pegg DE, Jacobsen IA, Halasz

NA, editors. *Organ Preservation, Basic and Applied Aspects.* Lancaster: MTP Press, Ltd; 1982. pp. 389–393.

22. Karow AM, Jr, Shlafer M. Ultrastructure-function correlative studies for cardiac cryopreservation. IV. Prethaw ultrastructure of myocardium cooled slowly (<=2°C/min) or rapidly (>=70°C/sec) with or without dimethyl sulfoxide (DMSO) *Cryobiology.* 1975;12:130–143. [PubMed]

23. Hunt CJ. Studies on cellular structure and ice location in frozen organs and tissues: the use of freeze-substitution and related techniques. *Cryobiology.* 1984;21:385–402. [PubMed]

24. Pollack GA, Pegg DE, Hardie IR. An isolated perfused rat mesentery model for direct observation of the vasculature during cryopreservation. *Cryobiology.* 1986;23:500–511. [PubMed]

25. Halasz NA, Rosenfield HA, Orloff MJ, Seifert LN. Whole organ preservation II. Freezing studies. *Surgery.* 1967;61:417–421. [PubMed]

26. Halasz NA, Miller S. Rewarming methods for whole organ freezing. In: Norman JC, editor. *Organ Perfusion and Preservation.* New York: Appleton-Century-Crofts; 1968. pp. 731–737.

27. Guttman FM, Lizin J, Robitaille P, Blanchard H, Turgeon-Knaack C. Survival of canine kidneys after treatment with dimethylsulfoxide, freezing at −80°C, and thawing by microwave illumination. *Cryobiology.* 1977;14:559–567. [PubMed]

28. Lehr H. Progress in long-term organ freezing. *Transplant Proc.* 1971;3:1565.

29. Fahy GM. Activation of alpha adrenergic vasoconstrictor response in kidneys stored at −30°C for up to 8 days. *Cryo Letters.* 1980;1:312–317.

30. Fahy GM. Prospects for vitrification of whole organs. *Cryobiology.* 1981;18:617.

31. Fahy GM, Hirsh A. Prospects for organ preservation by vitrification. In: Pegg DE, Jacobsen IA, Halasz NA, editors. *Organ Preservation, Basic and Applied Aspects.* Lancaster: MTP Press; 1982. pp. 399–404.

32. Fahy GM, MacFarlane DR, Angell CA, Meryman HT. Vitrification as an approach to cryopreservation. *Cryobiology.* 1984;21:407–426. [PubMed]

33. Fahy GM, MacFarlane DR, Angell CA. Recent progress toward vitrification of kidneys. *Cryobiology.* 1982;19:668–669.

34. Wowk B, Fahy GM. Toward large organ vitrification: extremely low critical cooling and warming rates of M22 vitrification solution. *Cryobiology.* 2005;51:362.

35. Wowk B, Fahy GM. Ice nucleation and growth in concentrated vitrification solutions. *Cryobiology.* 2007:330.

36. Wowk B. Thermodynamic aspects of vitrification. *Cryobiology.* 2009:59. (in press) [PubMed]

37. Fahy GM. Vitrification: An overview. In: Liebermann J, Tucker MJ, editors. *Vitrification in Assisted Reproduction: A User's Manual and Troubleshooting Guide.* London: Informa Healthcare; 2007. (in press)

38. Rall WF, Fahy GM. Ice-free cryopreservation of mouse embryos at −196°C by vitrification. *Nature.* 1985;313:573–575. [PubMed]

39. Fahy GM, Wowk B, Wu J, Paynter S. Improved vitrification solutions based on predictability of vitrification solution toxicity. *Cryobiology.* 2004;48:22–35. [PubMed]

40. Fahy GM. The role of nucleation in cryopreservation. In: Lee REJ, Warren GJ, Gusta LV, editors. *Biological ice nucleation and its applications.* St. Paul: APS Press; 1995. pp. 315–336.

41. Fahy GM. Vitrification. In: McGrath JJ, Diller KR, editors. *Low Temperature Biotechnology: Emerging Applications and Engineering Contributions.* New York: American Society of Mechanical Engineers; 1988. pp. 113–146.

42. Courbiere B, Massardier J, Salle B, Mazoyer C, Guerin J-F, Lornage J. Follicular viability and histological assessment after cryopreservation of whole sheep ovaries with vascular pedicle by vitrification. *Fertil Steril.* 2005;84:1065–1071. [PubMed]

43. Sugimoto M, Maeda S, Manabe N, Miyamoto H. Development of infantile rat ovaries autotransplanted after cryopreservation by vitrification. *Theriogenology.* 2000;53:1093–1103. [PubMed]

44. Salehnia M. Autograft of vitrified mouse ovaries using ethylene glycol as cryoprotectant. *Exp Anim.* 2002;5:509–512. [PubMed]

45. Migishima F, Suzuki-Migishima R, Song S-Y, Kuramochi T, Azuma S, Nishijima M, et al. Successful cryopreservation of mouse ovaries by vitrification. *Biol Reprod.* 2003;68:881–887. [PubMed]

46. Brockbank KG, Song YC. Morphological analyses of ice-free and frozen cryopreserved heart valve explants. *J Heart Valve Dis.* 2004;13:297–301. [PubMed]

47. Armitage WJ, Hall SC, Routledge C. Recovery of endothelial function after vitrification of cornea at −110°C. *Invest Ophthalmol Vis Sci.* 2002;43:2160–2164. [PubMed]

48. Taylor MJ, Song YC, Brockbank KG. Vitrification in tissue preservation: new developments. In: Fuller BJ, Lane N, Benson EE, editors. *Life in the frozen state.* Boca Raton: CRC Press; 2004. pp. 603–641.

49. Fahy GM, Ali SE. Cryopreservation of the mammalian kidney II. Demonstration of immediate ex vivo function after introduction and removal of 7.5 M cryoprotectant. *Cryobiology.* 1997;35:114–131. [PubMed]

50. Fahy GM, da Mouta C, Tsonev L, Khirabadi BS, Mehl P, Meryman HT. Cellular injury associated with organ cryopreservation: chemical toxicity and cooling injury. In: Lemasters JJ, Oliver C, editors. *Cell Biology of Trauma.* Boca Raton: CRC Press; 1995.

51. Khirabadi BS, Fahy GM, Ewing L, Saur J, Meryman HT. 100% survival of rabbit kidneys chilled to −32°C after perfusion with 8 M cryoprotectant at −22°C. *Cryobiology.* 1994;31:597.

52. Wu J, Ge X, Fahy GM. Ultrarapid nonsuture mated cuff technique for renal transplantation in rabbits. *Microsurgery.* 2003;23:1–5.

53. Ofstad J, Aukland K. Renal circulation. In: Seldin DW, Giebisch G, editors. *The kidney, physiology and pathophysiology.* New York: Raven Press; 1985. pp. 471–496.

54. Kaissling B, Kritz W. Structural analysis of the rabbit kidney. *Adv Anat Embryol Cell Biol.* 1979;56:1–123. [PubMed]

55. Wowk B, Leitl E, Rasch CM, Mesbah-Karimi N, Harris SB, Fahy GM. Vitrification enhancement by synthetic ice blocking agents. *Cryobiology.* 2000;40:228–236. [PubMed]

56. Fahy GM. Organ perfusion equipment for the introduction and removal of cryoprotectants. *Biomed Instrum Technol.* 1994;28:87–100. [PubMed]

CHAPTER FIVE

Latest Advances In Antiaging Medicine

Terry Grossman*

People are very interested in learning what they can do to live longer as well as to remain healthy during their later years. Numerous books have written about this topic, and, in fact, I have contributed to the growing number of titles by writing two books specifically addressing this topic, The Baby Boomers' Guide to Living Forever (Hubristic Press, 2000) and Fantastic Voyage: Live Long Enough to Live Forever (Rodale Press, October 2004), which I coauthored with Ray Kurzweil.

Changing Views of Aging

Views about aging are changing rapidly. The old view of aging was that much of one's latter years were basically spent in a state of disability. Conventional wisdom was that seniors of advanced age (the "oldest of the old") would spend their final years unable to contribute to society, with several of these years often requiring skilled care such as a nursing home. The new view of aging is that people will remain physically and mentally healthy until very old age. This is what people today are looking for in their later years: to remain vibrant and in very good health, as contributing members of society. This paradigm shift is possible for many people today with present-day strategies, so this is a realistic goal for many people.

We have the technology and medical knowledge to enable a significant portion of the population to live to more than 100 years of age already. Several factors are intersecting to create a scenario where this will become increasingly common in the years ahead.

* Reprinted by permission from *The Keio Journal of Medicine* [2005 June; 54(2): 85–94].

It is even possible that within the next few decades, life spans of 100 years may appear quite short. Today, human life expectancy has been steadily increasing by about 2.2 months per year. [1] This progress has been steady for at least the past 60 years (see Figure 1 [next page]). Due to the acceleration in the rate of technological change, within 10-15 years, some authors feel that life expectancy will continue to increase and at an increasing rate in the years ahead. [2] Others researchers put the maximum life expectancy for the human population at 85 - 90 years, although individuals within the population may live longer. [3]

Evolution Is Not on Our Side

The problem is that, historically, from a biological point of view, evolution was not on our side to live so long. We need to recall that biological evolution took place in an era of scarcity. In primordial times, there were a limited number of calories to go around for our ancient predecessors. Since older individuals in the community were simply competing for the limited food and other resources available, it was not in the best interest for the survival of the tribe or community to allow its elderly to remain alive too long. There was a survival advantage to a community whose members remained alive and healthy (just) long enough to reproduce and raise their children; but then it would be best if "aging genes" were to express themselves and cause adults to age rapidly, essentially to shrivel up and blow away. A community whose individuals didn't survive much beyond the time their offspring were self-sufficient would possess a natural selection advantage over a community where the members lived into ripe-old age (and decrepitude). In the first community, it is more likely that the younger members would have a greater chance of surviving because of less competition for scarce calories and these would be genes increasing survival advantage. It is important to realize that our DNA really hasn't changed very much in the past 100,000 years or so. These "aging genes" remains key letters in the genetic alphabet soup in which we still swim today. But, in modern times, the conditions that selected for these genes in millennia past no longer apply. Today we live in an era of abundance. Granted, many people still live under conditions where there isn't enough food to go around, and a significant portion of

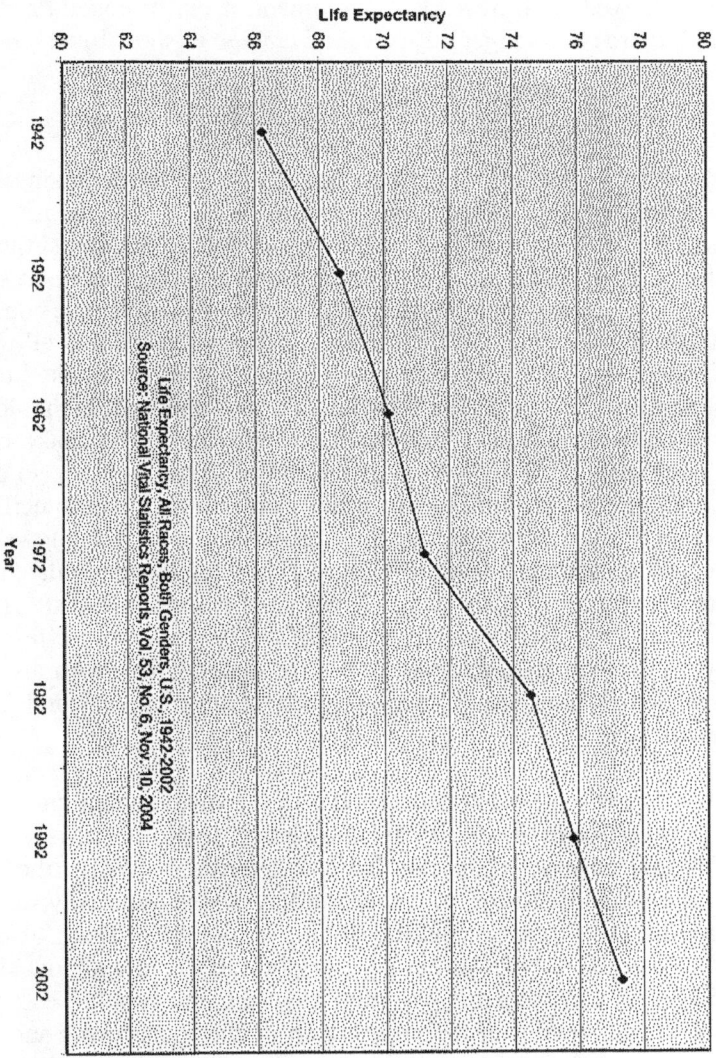

Fig. 1 Life expectancy U.S. 1942–2002.

human kind still lives precariously close to starvation, but this reflects political and social issues more than lack of technological capability. If allowed, technology could easily solve the problem of inadequate food for all peoples of the world. [4] So, technology has allowed us to override this remnant of our biological heritage, and, there is no longer a survival advantage to short-lived humans.

The Grandmother Hypothesis

Keeping the elderly around longer is also good for business: the business of the human race. Seniors represent a repository of wisdom. It is a shame and tragedy for people to die at just the time they are at the apogee of their accumulated knowledge. It seems ironic to allow people to accumulate knowledge over the course of a lifetime and then cut them down at their pinnacle and allow this wisdom be lost to the cemetery plot or "go up in smoke" on the funeral pyre. It is in the best interest of a modern, technological society to keep the elderly alive and vibrant and to make use of their vast collective knowledge base. This is the so-called "grandmother hypothesis" in which "older individuals specifically retain the ability to preserve their knowledge base that allows them to focus on transmitting to children knowledge accumulated over generations." [5] This is modern evolution at its finest, and by utilizing the intellectual resources of our seniors more effectively, we will be able to evolve even more quickly. But we must keep our bodies (and minds) in our senior years in good condition to accomplish this.

In most cases, people can achieve this goal of remaining healthy until very late in life. The knowledge to do so exists today. Chronic disease does not happen overnight. While the overt clinical manifestation of a disease may appear suddenly, most of the disease processes that constitute the leading causes of death, such as cancer, heart disease, kidney failure, etc. are almost always decades in the making. We now have the ability to assess where we are personally in the progression of these processes and, with this knowledge, take sensible and effective and preventive action to halt and reverse the lethal march towards disease, disability and death. Effective testing is the key; uncovering incubating disease processes and curing them before symptoms appear and it is too late.

The leading cause of death in the United States is coronary heart disease. Yet, 50 per cent of men and 64 per cent of women discover they have heart disease by suffering a fatal heart attack. This means that half of men and almost two-thirds of women with heart disease receive no warning symptoms such as chest pain people before succumbing to a fatal myocardial infarction. [6] By aggressively utilizing non-invasive diagnostic testing available today, many people could take effective, preventive action before suffering a myocardial infarction.

The second leading cause of death is cancer, but, all too often, cancer is not detected until after it has metastasized and then is very difficult to treat. The same applies to strokes, the third leading cause of death. Strokes typically come on without warning, all too often resulting in permanent damage.

But, it does not need to be this way. By aggressively applying the medical knowledge we have today, people can assess where they are in the progression of disease and take effective action to halt and reverse the disease process before irreversible damage occurs.

The Aging Process: No One's Friend

We need to begin by taking aging for what it is. Aging is no one's friend. Aging is our enemy. A fundamental principle of antiaging medicine is, in fact, that aging can be regarded as a disease, an enemy. And, as with any enemy, we must fight it with all the tools and weapons at our disposal: with multiple independent methodologies. We don't go to war with one soldier or one weapon. If we want to win, we attack our enemy with everything we have. Since aging is our enemy, we should use everything we have to fight it.

But, on the road to longevity, we have other enemies as well. We need to acknowledge that we are our own worst enemies. We sabotage our health by making poor lifestyle choices. We know we shouldn't eat doughnuts or potato chips, but we do. We know we should exercise regularly and control the stress in our lives, but we don't. We ignore our genetics. We know that lung cancer runs in our family, but we smoke anyway. We know alcohol-related liver disease and cirrhosis runs in our family, but we still drink alcohol.

In addition, the disease processes themselves are our enemy. Cardiovascular disease, cancer, pneumonia, and Alzheimer's disease are nobody's friend. In addition, there is a vocal body of opposition to longevity research. For instance, some researchers feel any efforts to extend human longevity beyond its current 80 or so years are actually immoral. [7] This was epitomized by an enormously insensitive statement made by former governor of Colorado, Richard Lamm, who stated in 1984 that the elderly "have a duty to die." [8] Some two decades later, Mr. Lamm is now approaching his 70th birthday. One wonders if he still feels the same way (and if he thinks it applies to him).

The Three Bridges to Longevity

One way to look at our path to longevity is to regard it as a journey over three sequential bridges. Bridge One is based on therapies that exist today. Bridge One therapies consist of the best of present-day medicine, including biotechnology breakthroughs that are occurring every day. Bridge One will take us to Bridge Two, which consists of the full blossoming of the biotechnology revolution. Then Bridge Two will take us to Bridge Three, the nanotechnology/ artificial intelligence revolution, which will lead to life spans that are currently incomprehensible, but which will soon be commonplace, measuring in the hundreds of years.

There are several Bridge One technologies that people can do themselves without the aid of medical professionals. These include things like proper diet, caloric restriction, adequate exercise, stress management, aggressive nutritional supplementation, detoxification and proper care of your brain.

There are also things that can be done with the help of medical professionals. For example, you can have your genomics tested; you can have tests done to determine specifics of your metabolism such as methylation, inflammation and glycation defects. You can undergo testing to detect heart disease early and then take measures before it has a chance to strike. You can detect and prevent cancer before it has a chance to spread. You can avail yourself of bio identical hormone replacement therapies.

Bridge One – What People Can Do Themselves

Diet

There are 4 sources of calories: carbohydrates, proteins, fats and alcohol. People can make good and bad choices for each of these types of calories. Carbohydrates, for example, are not inherently bad; rather there are good and bad carbohydrates. Similarly, proteins, fats and alcohol are not inherently bad, but there are good and bad choices within each of these categories.

Whole grains like brown rice, legumes such as lentils and beans, and above-ground green vegetables are examples of good carbohydrates. They have a low glycemic index, meaning they do not turn into sugar quickly in the body. There are "bad" carbohydrates, however. Bad carbohydrates are ones which tend to turn into sugar quickly, causing excessive weight. Refined sugar, refined flour products, breads, root vegetables that grow underground, such as potatoes and beets, have a high glycemic index and are not optimal sources of energy. [9] Consumption of these should be limited.

Proteins are available in good and bad varieties. Fish and seafood, vegetable-based protein sources such as tofu and other soy products are good proteins. Less beneficial proteins are fatty red meats, which are high in saturated fat, and egg yoke, high in inflammatory arachidonic acid.

There are good and bad fats. Fish and fish oils are good as are avocado and olive oils; raw, unsalted nuts and seeds contain beneficial fats. Bad fats include deep-fried foods, processed vegetable oils like corn oil, and "trans" fats such as are found in margarine. [10, 11]

People are often surprised to learn that alcohol is also available in good and bad varieties. Red wine has benefit, as it is rich in phytonutrients such as resveratrol, an anticancer agent. [12, 13, 14] Beer is an example of a somewhat bad alcohol, since beer is rich in amylose, which has a high glycemic index. Although this is the conventional wisdom, this is controversial. [15]

Caloric Restriction
Caloric restriction has been proven to extend lifespan of laboratory animals. [16, 17] Lifespan is how long the longest living individual in a species can live. Currently, the maximum human lifespan seems to be about 120 years. No human being has ever been scientifically documented as living more than 124 years. But, like Olympic records, we expect this soon to be broken and rebroken again and again as Bridge Two and Three therapies come into their own.

For now, however, caloric restriction is the only thing proven to help lifespan. For example, fruit flies have a lifespan measured in days or weeks; laboratory rodents have a lifespan of two or three years. In animal experiments involving caloric restriction, researchers were able to extend the healthy lifespan of laboratory animals and insects drastically. Increases in lifespan of 50 - 100 percent or more were seen. We know that consuming excess calories reduce life expectancy; conversely, restricting calories can extend lifespan, at least in laboratory animals. [18]

In yeast experiments done by David Sinclair, caloric restriction was shown to activate the SIRT1gene that expresses SirT1 deacetylase that stabilizes DNA, extending lifespan. The authors postulate that caloric restriction "could extend life-span by inducing SIRT1 expression and promoting the long-term survival of irreplaceable cells". [19] Polyphenols are a dietary substance that also activates this gene and may extend lifespan by 70 percent in yeast. A potent polyphenol known as resveratrol is found in red wine.

Rodent experiments have been done by Richard Weindrich at the University of Wisconsin, and he has done gene profiling to find out which genes are turned on and which genes are turned off during the aging process. It seems only a handful of genes are involved. By caloric restriction, he was able to show 30-50% increase in lifespan in rodents, fish, spiders and other animals. We know that in youth certain genes are active and certain genes are not, and in aging, these often change. We know that caloric restriction can move those genetic switches in older animals to more closely resemble the gene expression of a younger animal. [20]

For humans we have the results of real life experiments, not laboratory data. For example, on the Japanese island of Okinawa, a common phrase is "hara hachi bu," which means "belly 80 percent full." This concept means you should try to eat approximately 80 percent of what you would need to feel full. You try to leave the table a little hungry. The caloric intake of the average adult Okinawan is 1800 calories per day compared to Americans who eat 2500 calories or more each day. This seems clearly related to their lifestyle choices and not genetics, because as Okinawans move away from their homes on Okinawa and adopt different lifestyles, they live shorter lifespans like their neighbors. So there is nothing genetic about their long lifespans. They live longer because of their lifestyle choices, particularly caloric restriction. [21, 22]

Other Lifestyle Choices

Japan is a land of other healthy lifestyle choices – at least as far as diet is concerned. Even though many Japanese smoke and live high stress lifestyles, the number of Japanese over 100 years of age has doubled in the past five years. This is a new phenomenon felt to be due to improved health-care coupled with their underlying healthy diet. In 1998 there were 10,000 Japanese over 100 years of age. In 2003, five years later, the number had doubled over 20,000. Women, by the way, constitute 85 percent of this centenarian population in Japan.

It is interesting to compare this to the United States, a land characterized by bad lifestyle choices, certainly as regards diet. A leading cause of death in the United States is, in fact, poor lifestyle choices. Half of the deaths in the U.S. are the result of easily modifiable lifestyle choices. The number one cause of death in the U.S. is tobacco, which causes 18.1 percent of deaths. The second is bad diet and lack of exercise, which causes 16.6 percent of deaths. [23]

Another healthy lifestyle choice that people can make is regular aerobic exercise. Proper aerobic exercise consists of getting your heart rate up to 60-80 percent of your maximum predicted heart rate (which is equal to 220 less your age) and keeping it there for a sustained period of time. Some type of aerobic exercise should be done continuously for 30 minutes at least every other day. In addition to aerobic exercise, weight training and stretching are valuable.

In addition we can take nutritional supplements. Modern farming methods unfortunately have led to significant decreases in the vitamin and mineral content of our food. [24] Also, almost no one eats enough fruits and vegetables to get adequate nutrition without taking supplementation. Chemical reactions in the body occur because of enzymes and many enzymes in the body need vitamins and minerals as cofactors in order to function properly. By taking nutritional supplements, you can ensure that the raw minerals are always available to ensure that your youth-sustaining enzymes function optimally at all times. By taking supplements we can help neutralize the now outdated genetic time bombs programmed within our genes designed to take us out of competition for scarce calories as we age. We can help accomplish some of the same goals by caloric restriction, namely turning genes of youth on and genes of aging off at least for a while. [25] In effect, by aggressive supplementation and other lifestyle choices we're trying to "reprogram our biochemistry."

Bridge One – What People Can Do With Physician Assistance

Genomics testing

One thing doctors can do is to help people test their genes. Genomics testing tells you what genes you have. The Human Genome Project was recently completed and all 35,000 human genes have been identified. But, the key to genomics is the realization that genes just represent *tendencies* in most cases. Most people do not have a "breast cancer gene" or an "Alzheimer's gene." You may have a gene that increases your *risk* of breast cancer or Alzheimer's disease, particularly if you make the wrong lifestyle choices. But you can often change this outcome by making *proper* lifestyle choices. But you won't know what specific lifestyle choices to make unless you know what specific genes you have, and this is where genomics can provide critical information.

Here are some practical examples of what genomics testing can tell you. The cytochrome P 450 (CYP 450) genes code for proteins that help the body detoxify toxins in the liver. One cytochrome P 450 gene is known as C17. Women who have a mutated C17 gene can have an increased risk of breast cancer or osteoporosis. [26, 27] If a woman performs genomics testing and finds she has mutated C17, she should avoid taking estrogen, since she would be at higher risk of developing breast cancer by doing so. Another CYP 450 gene is 1A1. Some mutations of this gene may render people much more susceptible to lung cancer. Individuals who have mutated CYP 1A1 should be especially careful not to smoke because they are at much higher risk of developing lung cancer. A mutated CYP 450 2E1 gene increases susceptibility to alcohol-related liver disease and cirrhosis. [28] People with this mutation should be very careful with regard to alcohol consumption. So these are examples of how genomics testing can help you make specific lifestyle choices.

Early Detection of Cardiovascular Disease and Cancer
Cardiovascular disease is the leading cause of death in many developed countries. Medical professionals can help their patients avoid heart disease with early detection. The usual blood tests that are used to determine risk of heart disease (cholesterol subfractions, triglycerides, etc.) are not adequate. Highly sensitive-CRP, homocysteine and lipoprotein (a) levels need to be checked as well. In addition, the "acceptable limits" given by many reference laboratories are also often inappropriate. As one example, the reference laboratory I use in my medical practice considers levels of homocysteine less than 15 as acceptable; yet risk of heart disease increases with levels greater than 7.0. Cholesterol levels under 200 are considered acceptable, but levels between 130-160 may be optimal. Specific cardiovascular-related genomics tests are available as well and can provide useful patient-specific information.

Recently, non-invasive test for delineation of atherosclerosis have become available. Ultrafast CT scans can quantitate the amount of calcified plaque in the coronary arteries. Virtual angiograms and MRIs are becoming available in several metropolitan areas, obviating the need for invasive angiography.

Cancer is the second leading cause of death. Medical professionals can help with early detection and prevention of cancer. They can also suggest supplements like vitamin C, selenium, curcumin, fish oil, and folic acid. A diet rich in soy products and green tea is cancer-protective as well. It is important to make lifestyle choices including regular aerobic exercise and strict avoidance of smoking and exposure to passive cigarette smoke as well. These are proven methods of reducing cancer risk. Physicians can help detect cancer in the very earliest stages even before a detectable lesion is present by utilizing the DR- 70 test which is undergoing FDA testing as a possible screening test for colorectal cancer. [29] This blood test only requires a few cancer cells to be present in the body for detection. This can help detect cancer in a very early stage before it has had a chance to metastasize.

Hormone replacement
High hormone levels are a hallmark of youth, while declining hormone levels are characteristic of advancing age. At around 30 years of age an individual's hormones begin to decrease, which then begins to accelerate the aging process. By restoring hormone levels to more youthful levels, people can often experience the benefits of higher hormone levels similar to what they had when they were younger. Antiaging hormones include estrogen, progesterone, testosterone, thyroid, melatonin and human growth hormone (hGH). The cornerstone of hormone replacement therapy, however, is to only replace hormones whose levels are low and for which there is a demonstrable deficiency. The indiscriminate prescribing of anabolic steroid hormones, such as human growth hormone for bodybuilders who merely wish to enhance the appearance of their physique, yet have adequate levels of hGH, is an inappropriate use of hormone replacement therapy and is not recommended.

Optimally one should use bio identical hormones, hormones which are chemically identical in their structure to those found naturally in the human body. For example when prescribing estrogen replacement for women, we want to use estriol, estrone and estradiol, which are the same estrogens found naturally in the body. Similarly, we prefer to use bio identical progesterone, not a progestin like medroxprogesterone (which is not progesterone at all, but a synthetic drug, which possesses progesterone-like activity). We use testosterone itself (not medroxytestosterone), which can improve muscle mass and sex drive in both men and women. Melatonin is useful in promoting deep sleep and also fights some cancers [30] and, in certain cases of documented growth hormone deficiency, we have found the use of human growth injections of value.

Bridge Two (The Biotechnology Revolution)

Bridge Two consists largely of the biotechnology revolution, which has already begun and which will find its full expression over the course on the next 15 years. Among the most important items of Bridge Two will include stem cell therapy, therapeutic cloning, recombinant gene technology or "pharming" (genetically modifying bacteria, plants and farm animals to produce desired proteins) and developing a deeper understanding of the human genome with the ultimate goal of creating designer proteins ("proteomics"). All of these therapies are very exciting and will change medicine as we know it, so that is completely unrecognizable from the medicine of today. By optimizing care of our present biological bodies and eliminating the expression of undesirable genes, the life expectancy for the most people should easily exceed 100 years. [31]

Stem Cell Therapies
Stem cells occur naturally in the human body and have the ability to differentiate themselves into many different cell types; for example, a stem cell found in a hair follicle can be coaxed to turn into a heart muscle cell or a nerve cell depending on the chemical environment in which it finds itself. Recently, there has been considerable political and ethical debate about the use of a specific type of stem cell that is found early in fetal development -- the

embryonic stem cell. Embryonic stem cells are characterized by an extreme degree of "plasticity," namely, the ability to change or differentiate into any type of tissue. Since the production of embryonic stem cells involves the destruction of human embryos, this has created a moral dilemma for many political and religious leaders, particularly in the United States.

Fetal stem cells are found in two main varieties: totipotent and pluripotent stem cells. Totipotent stem cells are found in the embryo immediately after fertilization. Because some people feel that as soon as the embryo starts to divide it constitutes a human being, it is sacrosanct and must not be destroyed. A little further along the line of cell division come the pluripotent stem cells. These cells are not as plastic as the totipotent stem cells and cannot form every cell type, at least not with today's technologies, so they are not as valuable to scientists as the embryonic stem cells. However, even the totipotent stem cells can be encouraged to differentiate into many different cell types under the influence of specific growth factors. In the near future, it is not unrealistic to envision a scenario in which a patient suffers, say, a myocardial infarction and receives a transplant of cloned heart muscle cells generated from his own stem cells to replace the region of infarcted myocardium. Patients with spinal cord injuries or a history of cerebrovascular accident may soon be able to receive stem cell implants to regenerate damaged tissues and restore function.

A small number of stem cells persist into adult life. Most adult tissues have a few stem cells, but they are rare. Their function remains incompletely understood, but we may soon be able to transform these adult stem cells into any cell type in the body. As we perfect this technology the debate over embryonic stem cells will end, and we will be able to harness the full potential of this therapy. A group in South Korea recently was able to create a pure clone of embryonic stem cells, obviating the need for actual embryos. Woo Suk Hwang and Shin Yong Moon of Seoul National University have successfully cloned a line of human pluripotent stem cells. This research paves the way for production of human-replacement tissues and organs from a cloned stem cell line. [32]

134

Pharming
Another very powerful technique is known as "pharming." It involves recombinant technology. This refers to modifying or inserting desired genes into animals, plants and bacteria. Then these so-called "pharm" animals or plants create the desired proteins. An interesting variant of this therapy involves genetically modifying bananas or tomatoes to create a vaccine against hepatitis B. In order to get vaccinated, the patient would simply eat the banana or tomato. It is estimated that these plant derived vaccines could be produced for less than 2 cents per dose, a 99 per cent saving from conventional vaccines. So, in effect, there would be bananas and tomatoes growing in the field producing hepatitis vaccine. When we give patients human insulin, it is a bioengineered product that is made by recombinant bacteria; similarly, human growth hormone for injection is fabricated in the laboratory by recombinant bacteria. There now exist "pharms" comprised entirely of different "pharm" animals, in which the milk of the goats is used to create silk threads stronger than those created by spiders, [33] and, in the fields are grown plants that are particularly high in protein such as tobacco and corn, which have had their genetic structure modified to create other desirable proteins.

Proteomics
Proteomics, in many ways, is an even more exciting field of development. Proteomics involves systematically creating desired proteins in the laboratory de novo. This may be the most important medical technological development to occur within the next two decades. In a typical scenario, doctors will determine what protein a patient needs to recover from a given illness and then the proteomics engineers in the laboratory will fabricate the needed protein molecules. The full expression of this therapy still lies 10-15 years away, but when this therapy is mature, it will offer incredibly powerful therapies. Proteomics will also be utilized for diagnostic tests, and medical diagnoses will be able to be made much more quickly.

A major problem with proteomics relates to the three-dimensional configuration of proteins. While the chemical structure of a protein molecule can be written as a linear chain of amino acids, determining in advance the sequence of amino acids that will be needed to create a desired three-dimensional protein molecule with folding and cross-linking of the amino acids is an incredibly difficult task. Determining what chain of amino acids is required to create a desired three-dimensional structure remains a problem that exceeds the computational capacities of even the world's fastest computers. There are some new supercomputers designed to work on the protein folding problem. IBM has just introduced the Blue Gene/L, a supercomputer that operates at a speed out of 360 trillion operations per second and one of its main goals is solving the protein folding problem. It is anticipated within the next 10 years some early proteomics therapies will be available and significantly affect our abilities to diagnose and treat many diseases.

Cloning
There are two types of cloning: reproductive cloning and therapeutic cloning. In reproductive cloning, you recreate an entirely genetically identical organism. Reproductive cloning has already been used in animals, but due to the thorny moral issues involved, has not been used in humans. "Dolly the sheep" was the first cloned animal, and there have been many additional animal species that had been cloned. Genetics Savings & Clone, Inc., for example, is a private company that is in business of cloning beloved pets for their owners.

This is to be differentiated from therapeutic cloning in which you create individual tissues, not entire organisms. Therapeutic cloning uses germ cell lines before they are implanted and causes them to differentiate and develop into specific tissues or organs. Currently, researchers are growing simpler organs such as corneas and urinary bladders, and soon should be able to grow skin, blood vessels and other complex tissues and organs by cloning them from germ cells.

Gene therapies
Gene therapies are also very powerful aspects of the Bridge Two biochemical revolution. Some of the current therapies available include RNA interference (RNAi), and antisense RNA, which are available now, but are in a very primitive stage of development.

RNA interference therapy involves taking short double-stranded segments of RNA and inserting them into the cytoplasm of the cell. These match and lock onto portions of the native messenger RNA (mRNA) that is transcribed from mutated genes. This causes the native RNA segment to be cut apart and destroyed, silencing expression of the undesirable gene. Antisense RNA also blocks the mRNA created by the defective genes so they are unable to make undesired proteins. Antisense therapy uses mirror-image sequences of RNA (antisense RNA), which stick to the abnormal protein-encoding RNA, preventing it from being expressed. [34]

It is hoped that these therapies which effect gene expression indirectly, by interfering with the mRNA coded for by the defective genes, will eventually lead to "somatic gene therapy," which is the holy grail of gene therapy. In somatic gene therapy the goal is to insert a desired gene directly into the patient's genome and delete or turn off undesired or defective genes. Somatic gene therapy is most important gene therapy of all, but is probably 25 years away. But when this therapy is mature, it will be possible take any type of adult or somatic cell in the body and turn it into any other cell type because all somatic cells contain all the genes needed to create every cell type. We will, for example, be able to turn hair follicle or fat cells into heart muscle cells simply by manipulating their genes.

Gains are already being made in this arena by using adult stem cells. Adult liver stem cells have been transformed into pancreatic cells [35] and adult muscle stem cells have been transformed into heart muscle cells, neural tissue, and blood vessels. [36]

Yet, as exciting as these Bridge Two therapies will be, they will pale in comparison to the Bridge Three therapies to which they will lead, the full flowering of which will occur 25 to 30 years from now

Bridge Three (The Nanotechnology/AI Revolution)

The hallmarks of Bridge Three include development of artificial intelligence (AI) and molecular nanotechnology (MNT). Artificial Intelligence refers to nonbiological (computer) intelligence, while nanotechnology is a term coined by Dr. K. Eric Drexler at MIT in 1978 and which he elaborated in his seminal book, "Engines of Creation." [37] MNT refers to machines that operate on the scale of nanometers (10^{-9} m) and which will be able to create molecules one atom at a time much like our cellular machinery does today. As AI and MNT technologies combine with and are incorporated into our biological bodies, serious increases in human lifespan well beyond 120 years will be not only possible, but highly probable. [38]

Artificial Intelligence
Artificial Intelligence refers to forms of nonbiological intelligence that will soon rival the capacity of the human brain and then rapidly exceed human intelligence by a billion-fold or more within the next few decades. Limited forms of artificial intelligence exist today such as pattern recognition software programs which allow computers to "understand" human language (and which I am using right now to type this article). Other AI programs write poetry and create original works of art. But the type of advanced AI that we will see unfold during Bridge Three will center on the "reverse engineering" of the human brain, a task which experts in the field estimate will be completed within 20 years. Reverse engineering (understand the inner workings) of the brain will not only provide key insights into how the brain performs its many functions, but will lead to a radical increase in human intelligence resulting from the merging of our current biological brain power with nonbiological (computer) intelligence. As we ultimately merge our biological thinking with artificial intelligence, we will vastly expand both the scope and speed of the thinking experience.

This development will provide key insights into how the human brain performs its pattern recognition and cognitive functions. These insights in turn will greatly accelerate the development of artificial intelligence in nonbiological systems such as nanobots, a key to the biological application of nanotechnology. Microscopic, but intelligent nanobots will course through our blood stream, bodily organs, and brain, downloading software either from the Internet or from our own personal "Wi-Fi" network that will enable us to overcome virtually any obstacle to keeping us healthy. This will ultimately lead to dramatic increases in human longevity.

Nanotechnology

Nanotechnology refers to engineering done on the scale of nanometers and will enable scientists to manipulate atoms, moving them from place to place, one at a time. The textbooks for the practical medical application of nanotechnology are already being written, and Robert Frietas, M.D. has already completed 2 of the 4 volumes of his "Nanomedicine" series of books.

According to Dr. Freitas, "The comprehensive knowledge of human molecular structure so painstakingly acquired during the 20th and early 21st centuries will be used in the 21st century to design medically-active microscopic machines. These machines, rather than being tasked primarily with voyages of pure discovery, will instead most often be sent on missions of cellular inspection, repair, and reconstruction." He goes on to say "If the idea of placing millions of autonomous nanorobots inside one's body might seem odd, even alarming, the fact is that the body already teems with a vast number of mobile nanodevices." [39]

Nanobiotic Red Blood Cells – Respirocytes

Programmable red blood cells are an example of nanotechnological devices on Freitas' drawing board. Like most of our biological systems, red blood cells perform their function of oxygenating the body very inefficiently. So, Freitas has redesigned them for optimal performance.

With an ounce or two of Freitas' "respirocytes" (nanobiotic red blood cells) circulating along with your normal blood, you could go hours without oxygen. You would be able to go "scuba diving" without scuba gear. An ounce or two or respirocytes in your blood stream could provide several hours worth of oxygen. An untrained teenager with respirocyte-enhanced blood would be easily able to outrun a highly trained, but unenhanced Olympic athlete. Astronauts could explore other planets without oxygen tanks. In a less frivolous application, in the event of a cardiac arrest, an injection of respirocytes administered by emergency medical personnel would allow the patient's brain and other vital organs to be perfused with oxygen for several hours while they awaited definitive repair of the underlying process that led to the cardiac arrest. Analyses have shown that these respirocytes could be hundreds or thousands of times more efficient than our own biological blood.

In a television documentary entitled "Beyond Human", which was originally broadcast nationwide in the U.S. on 15 May 2001, respirocytes were shown being injected into the bloodstream of a smoke inhalation victim who was suffering from carbon monoxide poisoning. The video depicted the respirocytes passing through the arteries on their way to the capillary bed, where they then released their store of oxygen in the patient's tissues to help keep the victim alive while awaiting definitive care.

Nanobiotic White Blood Cells – Microbivores
Nanotechnology will also create programmable white blood cells. These nanobiotic white blood cells ("microbivores") will patrol the bloodstream, seeking out and destroying undesirable bacteria, viruses and other pathogens.

These nanobots will download software from the Internet for particular problems, and could be programmed to recognize and destroy cancer cells before they would have a chance to grow and spread. [40]

Won't it be nice to have some programmable white blood cells in your blood stream to defend against the possibility of biological warfare attack? It would be a simple matter to realize the organism that had been released and your microbivores would immediately download the appropriate program from the Internet to destroy that pathogen. So programmable microbivores will be the ultimate defense against biological weapons of mass distraction or biological warfare agents, as well as against any influenza pandemic or other potential pathogen.

Conclusion

According to Harvey Mackey, 61 percent of famous people didn't achieve their most noble achievements until after 60 years of age.

So, there is a great deal of advantage to remaining in good health until your later decades of life. In the 19th century, Victor Hugo said "Forty is the old age of youth, fifty the youth of old age." At the beginning of the 21st century, we can now safely say that 50 or 60 is only the beginning of the second half of life. On December 31, 2004, there were 41 documented "supercentenarians, individuals over 110 years of age. [41] By the middle of this century, many longevity researchers believe these numbers will increase dramatically.

We have discussed some of the Bridge One strategies we can use to keep our bodies in good health until the Biotechnology (Bridge Two) and Nanotechnology/AI (Bridge Three) revolutions find full expression. There is no doubt that the 21st century will be worth living to experience, By following the Bridge One strategies that exist today, many people alive today will have the opportunity to see it in its entirety.

Acknowledgements

I wish to acknowledge Raymond Kurzweil, my coauthor of Fantastic Voyage: Live Long Enough to Live Forever (Rodale, 2004) who originated many of the original concepts presented herein.

This paper is based on a lecture given at Keio University on 22 September 2004.

References

1. National Vital Life Statistics Reports, Vol. 53, No. 6, Nov. 10, 2004. http://www.cdc.gov/nchs/data/dvs/nvsr53_06t12.pdf

2. Kurzweil R, Grossman T: Fantastic Voyage: Live Long Enough to Live Forever. Emmaus PA, Rodale, 2004, pp. 4-5.

3. Olshanski SJ., Carnes BA. & Cassel C: In search of Methuselah: estimating the upper limits to human longevity. Science 1990; 250:634-640

4. Prakash CS, Conko G.: "Technology That Will Save Billions From Starvation". www.agbioworld.org/biotech_info/articles/prakash/prakashart/save-billions.html

5. Lewis K: Human longevity: an evolutionary approach. Mech Ageing Dev 1999 Jun 1;109(1):43-51.

6. http://www.americanheart.org/downloadable/heart/10797367 29696HDSStats2004UpdateREV3-19-04.pdf.

7. Fukuyama F: Our Post Human Future: Consequences of the Biotechnology Revolution; New York, Farrar Straus, 2002.

8. https://www.cato.org/dailys/7-28-96.html

9. Revised International Table of Glycemic Index (GI) and Glycemic Load (GL) Values—2002. http://diabetes.about.com/library/mendosagi/ngilists.htm

10. Khor GL: Dietary fat quality: a nutritional epidemiologist's view. Asia Pac J Clin Nutr 2004 Aug; 13(Suppl):S22.

11. Hu FB, Willett WC: Optimal diets for prevention of coronary heart disease. JAMA 2002 Nov 27;288(20): 2569-2578.

12. Wollin SD, Jones PJ: Alcohol, red wine and cardiovascular disease. J Nutr 2001 May; 131(5):1401-1404.

13. Wu JM, Wang ZR, Hsieh TC, Bruder JL, Zou JG, Huang YZ: Mechanism of cardioprotection by resveratrol, a phenolic anti-oxidant present in red wine. Int J Mol Med. 2001 Jul; 8(1):3-17.

14. Renaud SC, Gueguen R, Conard P, Lanzmann-Petithory D, Orgogozo JM, Henry O: Moderate wine drinkers have lower hypertension-related mortality: a prospective cohort study in French men. Am J Clin Nutr 2004 Sep; 80(3):621-5.

15. "Beer Carbohydrates - The Real Story," http://www.anheuser-busch.com

16. Li D, Sun F, Wang K: Protein profile of aging and its retardation by caloric restriction in neural retina. Biochem Biophys Res Commun 2004 May 21;318(1):253-258.

17. Bergamini E, Cavallini G, Donati A, Gori Z: The role of macroautophagy in the ageing process, anti-ageing intervention and age-associated diseases. Int J Biochem Cell Biol 2004 Dec; 36(12):2392-2404.

18. Ingram DK, Anson RM, de Cabo R, Mamczarz J, Zhu M, Mattison J, Lane MA, Roth GS: Development of calorie restriction mimetics as a prolongevity strategy. Ann N Y Acad Sci 2004; Jun;1019:412-423.

19. Cohen HY, Sinclair D: Calorie restriction promotes mammalian cell survival by inducing the SIRT1 deacetylase. Science 2004 Jul 16;305(5682):390-392.

20. Weindruch R, Kayo T, Lee CK, Prolla TA : Microarray profiling of gene expression in aging and its alteration by caloric restriction in mice. J Nutr. 2001 Mar;131(3):918S-923S.

21. Yamori Y, Miura A, Taira K: Implications from and for food cultures for cardiovascular diseases: Japanese food, particularly Okinawan diets. Asia Pac J Clin Nutr. 2001;10(2):144-145.

22. Willcox BJ, Yano K, Chen R, Willcox DC, Rodriguez BL, Masaki KH, Donlon T, Tanaka B, Curb JD: How much should we eat? The association between energy intake and mortality in a 36-year follow-up study of Japanese-American men. J Gerontol A Biol Sci Med Sci 2004; Aug;59(8):B789-795.

23. Mokdad AH, Marks JS, Stroup DF, Gerberding JL: Actual causes of death in the United States, 2000. JAMA. 2004 Mar 10;291(10):1238-1245

24. Worthington V: Effect of agricultural methods on nutritional quality: a comparison of organic with conventional crops. Altern Ther Health Med 1998 Jan;4(1):58-69.

25. Fenech M: Micronutrients and genomic stability: a new paradigm for recommended dietary allowances (RDAs). Food Chem Toxicol. 2002; Aug;40(8):1113-7.

26. Tworoger SS, Chubak J, Aiello EJ, Ulrich CM, Atkinson C, Potter JD, Yasui Y, Stapleton PL, Lampe JW, Farin FM, Stanczyk FZ, McTiernan A: Association of CYP17, CYP19, CYP1B1, and COMT polymorphisms with serum and urinary sex hormone concentrations in postmenopausal women. Cancer Epidemiol Biomarkers Prev 2004 Jan;13(1):94-101.

27. Somner J, McLellan S, Cheung J, Mak YT, Frost ML, Knapp KM, Wierzbicki AS, Wheeler M, Fogelman I, Ralston SH, Hampson GN: Polymorphisms in the P450 c17 (17-hydroxylase/17,20-Lyase) and P450 c19 (aromatase) genes: association with serum sex steroid concentrations and bone mineral density in postmenopausal women. J Clin Endocrinol Metab. 2004 Jan;89(1):344-51.

28. Villeneuve JP, Pichette V: Cytochrome p450 and liver diseases. Curr Drug Metab. 2004 Jun;5(3):273-282.

29. Kerber A, Trojan J, Herrlinger K, Zgouras D, Caspary WF, Braden B: The new DR-70 immunoassay detects cancer of the gastrointestinal tract: a validation study. Aliment Pharmacol Ther 2004 Nov 1;20(9):983-987.

30. Fraschini F, Demartini G, Esposti D, Scaglione F: Melatonin involvement in immunity and cancer. Biol Signals Recept 1998 Jan-Feb;7(1):61-72.

31. de Grey AD: An engineer's approach to the development of real antiaging medicine. Sci Aging Knowledge Environ 2003 Jan 08;2003(1):VP1.

32. Hwang WS, Ryu YJ, Park JH, Park ES, Lee EG, Koo JM, Jeon HY, Lee BC, Kang SK, Kim SJ, Ahn C, Hwang JH, Park KY, Cibelli JB, Moon SY: 2004. Evidence of a Pluripotent Human Embryonic Stem Cell Line Derived from a Cloned Blastocyst. Science Mar 12; 303(5664): 1669-1674

33. http://fluid.stanford.edu/~mbrennan/interests/insects/spider_si lk.html

34. Biroccio A, Leonetti C, Zupi G: The future of antisense therapy: combination with anticancer treatments. Oncogene 2003 Sep 29;22(42):6579-6588

35. Yang L, Li S, Hatch H, Ahrens K, Cornelius JG, Petersen BE, Peck AB: In vitro trans-differentiation of adult hepatic stem cells into pancreatic endocrine hormone-producing cells. Proc Natl Acad Sci U S A 2002 Jun 11;99(12):8078-8083.

36. Qu-Petersen Z, Deasy B, Jankowski R, Ikezawa M, Cummins J, Pruchnic R, Mytinger J, Cao B, Gates C, Wernig A, Huard J: Identification of a novel population of muscle stem cells in mice: potential for muscle regeneration. J Cell Biol 2002 May;157:851-864.

37. Drexler, KE. Engines of Creation; New York, Doubleday, 1986.

38. Kurzweil R: The Singularity is Near: When Humans Transcend Biology. New York, Viking, 2005 (prepublication manuscript).

39. Freitas, RA. Nanomedicine, Volume I: Basic Capabilities. Austin TX, Landes Pub, 1999.

40. Freitas R. "Robots in the bloodstream: the promise of nanomedicine" as reported on www.kurzweilai.net/ meme/frame.html?main=/articles/art0410.html.

41. Table of World-Wide Living Supercentenarians for the Year 2003. Rejuvenation Research 2004, May; 7:1, 83-85. See http:// www.liebertonline.com/doi/pdf/10.1089/154916804323105143.

CHAPTER SIX

The Will To Believe

William James

Note From The Editor – 2010

This chapter, based on an 1896 address by William James to the Philosophical Clubs of Yale and Brown Universities, is reprinted from *The Will to Believe and Other Essays in Popular Philosophy* by William James (1897).

Following a first reading or initial skimming of this classic essay by William James, it is sometimes concluded that its primary purpose is meant to be a defense of religion. As I read it, however, James is really emphasizing that there are particular conditions for particular kinds of beliefs properly held, but that the details will vary from individual to individual. Since, according to James, there are many kinds of truth: "a rule of thinking which would absolutely prevent me from acknowledging certain kinds of truth if those kinds of truth were really there, would be an irrational rule." Indeed, contrary to religious or philosophical absolutism, James is arguing that we should view our beliefs as tentative or uncertain, always open to revision.

In the recently published *Life* by Leslie Stephen of his brother, Fitzjames, there is an account of a school to which the latter went when he was a boy. The teacher, a certain Mr. Guest, used to converse with his pupils in this wise: "Gurney, what is the difference between justification and sanctification? Stephen, prove the omnipotence of God!" etc. In the midst of our Harvard freethinking and indifference we are prone to imagine that here at your good old orthodox College conversation continues to be somewhat upon this order; and to show you that we at Harvard have not lost all interest in these vital subjects, I have brought with me tonight something like a sermon on justification by faith to read to you – I mean an essay in justification *of* faith, a defense of our right to adopt a believing attitude in religious matters, in spite of the fact that our merely logical intellect may not have been coerced. "The Will to Believe," accordingly, is the title of my paper.

I have long defended to my own students the lawfulness of voluntarily adopted faith; but as soon as they have got well imbued with the logical spirit, they have as a rule refused to admit my contention to be lawful philosophically, even though in point of fact they were personally all the time chock-full of some faith or other themselves. I am all the while, however, so profoundly convinced that my own position is correct, that your invitation has seemed to me a good occasion to make my statements more clear. Perhaps your minds will be more open than those with which I have hitherto had to deal. I will be as little technical as I can, though I must begin by setting up some technical distinctions that will help us in the end.

♦

Let us give the name of *hypothesis* to anything that may be proposed to our belief; and just as the electricians speak of live and dead wires, let us speak of any hypothesis as either *live* or *dead*. A live hypothesis is one which appeals as a real possibility to him to whom it is proposed. If I ask you to believe in the Mahdi, the notion makes no electric connection with your nature – it refuses to scintillate with any credibility at all. As an hypothesis it is completely dead. To an Arab, however (even if he be not one of the Madhi's followers), the hypothesis is among the mind's possibilities: it is alive. This shows that deadness and liveness in an hypothesis are not intrinsic properties, but relations to the individual thinker. They are measured by his willingness to act. The maximum of liveness in hypothesis means willingness to act irrevocably. Practically, that means belief; but there is some believing tendency wherever there is willingness to act at all.

Next, let us call the decision between two hypotheses an *option*. Options may be of several kinds. They may be:

- 1, *living* or *dead*;

- 2, *forced* or *avoidable*;

- 3, *momentous* or *trivial*;

and for our purpose we may call an option a *genuine* option when it is of the forced, living, and momentous kind.

1. A living option is one in which both hypotheses are live ones. If I say to you: "Be a theosophist or be a Mohammedan," it is probably a dead option, because for you neither hypothesis is likely to be alive. But if I say: "Be an agnostic or be Christian," it is otherwise: trained as you are, each hypothesis makes some appeal, however small, to your belief.

2. Next, if I say to you: "Choose between going out with your umbrella or without it," I do not offer you a genuine option, for it is not forced. You can easily avoid it by not going out at all. Similarly, if I say, "Either love me or hate me," "Either call my theory true or call it false," your option is avoidable. You may remain indifferent to me, neither loving nor hating, and you may decline to offer any judgment as to my theory. But if I say, "Either accept this truth or go without it," I put on you a forced option, for there is no standing place outside of the alternative. Every dilemma based on a complete logical disjunction, with no possibility of not choosing, is an option of this forced kind.

3. Finally, if I were Dr. Nansen and proposed to you to join my North Pole expedition, your option would be momentous; for this would probably be your only similar opportunity, and your choice now would either exclude you from the North Pole sort of immortality altogether or put at least the chance of it into your hands. He who refuses to embrace a unique opportunity loses the prize as

surely as if he tried and failed. Per contra, the option is trivial when the opportunity is not unique, when the stake is insignificant, or when the decision is reversible if it later prove unwise. Such trivial options abound in the scientific life. A chemist finds an hypothesis live enough to spend a year in its verification: he believes in it to that extent. But if his experiments prove inconclusive either way, he is quit for his loss of time, no vital harm being done.

It will facilitate our discussion if we keep all these distinctions well in mind.

◆◆

The next matter to consider is the actual psychology of human opinion. When we look at certain facts, it seems as if our passional and volitional nature lay at the root of all our convictions. When we look at others, it seems as if they could do nothing when the intellect had once said its say. Let us take the latter facts up first.

Does it not seem preposterous on the very face of it to talk of our opinions being modifiable at will? Can our will either help or hinder our intellect in its perceptions of truth? Can we, by just willing it, believe that Abraham Lincoln's existence is a myth, and that the portraits of him in McClure's Magazine are all of some one else? Can we, by any effort of our will, or by any strength of wish that it were true, believe ourselves well and about when we are roaring with rheumatism in bed, or feel certain that the sum of the two one-dollar bills in our pocket must be a hundred dollars? We can *say* any of these things, but we are absolutely impotent to believe them; and of just such things is the whole fabric of the truths that we do believe in made up – matters of fact, immediate or remote, as Hume said, and relations between ideas, which are either there or not there for us if we see them so, and which if not there cannot be put there by any action of our own.

In Pascal's *Thoughts* there is a celebrated passage known in literature as Pascal's wager. In it he tries to force us into Christianity by reasoning as if our concern with truth resembled our concern with the stakes in a game of chance. Translated freely his words are these: You must either believe or not believe that

God is – which will you do? Your human reason cannot say. A game is going on between you and the nature of things which at the day of judgment will bring out either heads or tails. Weigh what your gains and your losses would be if you should stake all you have on heads, or God's existence: if you win in such case, you gain eternal beatitude; if you lose, you lose nothing at all. If there were an infinity of chances, and only one for God in this wager, still you ought to stake your all on God; for though you surely risk a finite loss by this procedure, any finite loss is reasonable, even a certain one is reasonable, if there is but the possibility of infinite gain. Go, then, and take holy water, and have masses said; belief will come and stupefy your scruples – Cela vous fera croire et vous abetira. Why should you not? At bottom, what have you to lose?

You probably feel that when religious faith expresses itself thus, in the language of the gaming-table, it is put to its last trumps. Surely Pascal's own personal belief in masses and holy water had far other springs; and this celebrated page of his is but an argument for others, a last desperate snatch at a weapon against the hardness of the unbelieving heart. We feel that a faith in masses and holy water adopted willfully after such a mechanical calculation would lack the inner soul of faith's reality; and if we were ourselves in the place of the Deity, we should probably take particular pleasure in cutting off believers from their infinite reward. It is evident that unless there be some pre-existing tendency to believe in masses and holy water, the option offered to the will by Pascal is not a living option. Certainly no Turk ever took to masses and holy water on its account; and even to us Protestants these seem such foregone impossibilities that Pascal's logic, invoked for them specifically, leaves us unmoved. As well might the Mahdi write to us, saying, "I am the Expected One whom God has created in his effulgence. You shall be infinitely happy if you confess me; otherwise you shall be cut off from the light of the sun. Weigh, then, your infinite gain if I am genuine against your finite sacrifice if I am not!" His logic would be that of Pascal; but he would vainly use it on us, for the hypothesis he offers us is dead. No tendency to act on it exists in us to any degree.

The talk of believing by our volition seems, then, from one point of view, simply silly. From another point of view it is worse than silly, it is vile. When one turns to the magnificent edifice of the physical sciences, and sees how it was reared; what thousands of disinterested moral lives of men lie buried in its mere foundations; what patience and postponement, what choking down of preference, what submission to the icy laws of outer fact are wrought into its very stones and mortar; how absolutely impersonal it stands in its vast augustness – then how besotted and contemptible seems every little sentimentalist who comes blowing his voluntary smoke-wreaths, and pretending to decide things from out of his private dream! Can we wonder if those bred in the rugged and manly school of science should feel like spewing such subjectivism out of their mouths? The whole system of loyalties which grow up in the schools of science go dead against its toleration; so that it is only natural that those who have caught the scientific fever should pass over to the opposite extreme, and write sometimes as if the incorruptibly truthful intellect ought positively to prefer bitterness and unacceptableness to the heart in its cup.

> It fortifies my soul to know
> That, though I perish, Truth is so –

sings Clough, while Huxley exclaims: "My only consolation lies in the reflection that, however bad our posterity may become, so far as they hold by the plain rule of not pretending to believe what they have no reason to believe, because it may be to their advantage so to pretend [the word 'pretend' is surely here redundant], they will not have reached the lowest depth of immorality." And that delicious *enfant terrible* Clifford writes: "Belief is desecrated when given to unproved and unquestioned statements for the solace and private pleasure of the believer.... Whoso would deserve well of his fellows in this matter will guard the purity of his belief with a very fanaticism of jealous care, lest at any time it should rest on an unworthy object, and catch a stain which can never be wiped away.... If [a] belief has been accepted on insufficient evidence [even though the belief be true, as Clifford on the same page explains] the pleasure is a stolen one.... It is sinful because it is stolen in defiance of our duty to mankind. That duty is to guard ourselves from such beliefs as from a pestilence which may shortly master our own body and then spread to the rest of the town.... It is wrong always, everywhere, and for every one, to believe anything upon insufficient evidence."

♦♦♦

All this strikes one as healthy, even when expressed, as by Clifford, with somewhat too much of robustious pathos in the voice. Free-will and simple wishing do seem, in the matter of our credences, to be only fifth wheels to the coach. Yet if any one should thereupon assume that intellectual insight is what remains after wish and will and sentimental preference have taken wing, or that pure reason is what then settles our opinions, he would fly quite as directly in the teeth of the facts.

It is only our already dead hypotheses that our willing nature is unable to bring to life again. But what has made them dead for us is for the most part a previous action of our willing nature of an antagonistic kind. When I say "willing nature," I do not mean only such deliberate volitions as may have set up habits of belief that we cannot now escape from – I mean all such factors of belief as fear and hope, prejudice and passion, imitation and partisanship, the circumpressure of our caste and set. As a matter of fact we find ourselves believing, we hardly know how or why. Mr. Balfour gives the name of "authority" to all those influences, born of the intellectual climate, that make hypotheses possible or impossible for us, alive or dead. Here in this room, we all of us believe in molecules and the conservation of energy, in democracy and necessary progress, in Protestant Christianity and the duty of fighting for "the doctrine of the immortal Monroe," all for no reasons worthy of the name. We see into these matters with no more inner clearness, and probably with much less, than any disbeliever in them might possess. His unconventionality would probably have some grounds to show for its conclusions; but for us, not insight, but the prestige of the opinions, is what makes the spark shoot from them and light up our sleeping magazines of faith. Our reason is quite satisfied, in nine hundred and ninety-nine cases out of every thousand of us, if it can find a few arguments that will do to recite in case our credulity is criticized by some one else. Our faith is faith in some one else's faith, and in the greatest matters this is most the case. Our belief in truth itself, for instance, that there is a truth, and that our minds and it are made for each other – what is it but a passionate affirmation of desire, in which our social system backs us up? We want to have a truth; we want to believe that our experiments and studies and discussions must put us in a continually better and better position towards it; and on this line we agree to fight out our thinking lives. But if a Pyrrhonistic sceptic asks us *how we know* all this, can our logic find a reply? No! certainly it cannot. It is just one volition against another – we willing to go in for life upon a trust or assumption which he, for his part, does not care to make.

As a rule we disbelieve all facts and theories for which we have no use. Clifford's cosmic emotions find no use for Christian feelings. Huxley belabors the bishops because there is no use for sacerdotalism in his scheme of life. Newman, on the contrary, goes over to Romanism, and finds all sorts of reasons good for staying there, because a priestly system is for him an organic need and delight. Why do so few "scientists" even look at the evidence for telepathy, so called? Because they think, as a leading biologist, now dead, once said to me, that even if such a thing were true, scientists ought to band together to keep it suppressed and concealed. It would undo the uniformity of Nature and all sorts of other things without which scientists cannot carry on their pursuits. But if this very man had been shown something which as a scientist he might do with telepathy, he might not only have examined the evidence, but even have found it good enough. This very law which the logicians would impose upon us – if I may give the name of logicians to those who would rule out our willing nature here – is based on nothing but their own natural wish to exclude all elements for which they, in their professional quality of logicians, can find no use.

Evidently, then, our non-intellectual nature does influence our convictions. There are passional tendencies and volitions which run before and others which come after belief, and it is only the latter that are too late for the fair; and they are not too late when the previous passional work has been already in their own direction. Pascal's argument, instead of being powerless, then seems a regular clincher, and is the last stroke needed to make our faith in masses and holy water complete. The state of things is evidently far from simple; and pure insight and logic, whatever they might do ideally, are not the only things that really do produce our creeds.

♦♦♦♦

Our next duty, having recognized this mixed-up state of affairs, is to ask whether it be simply reprehensible and pathological, or whether, on the contrary, we must treat it as a normal element in making up our minds. The thesis I defend is, briefly stated, this: *Our passional nature not only lawfully may, but must, decide an option between propositions, whenever it is a genuine option that cannot by its nature be decided on intellectual grounds; for to say, under such circumstances, "Do not decide, but leave the question open," is itself a passional decision – just like deciding yes or no – and is attended with the same risk of losing the truth.* The thesis thus abstractly expressed will, I trust, soon become quite clear. But I must first indulge in a bit more of preliminary work.

♦♦♦♦♦

It will be observed that for the purposes of this discussion we are on "dogmatic" ground – ground, I mean, which leaves systematic philosophical skepticism altogether out of account. The postulate that there is truth, and that it is the destiny of our minds to attain it, we are deliberately resolving to make, though the skeptic will not make it. We part company with him, therefore, absolutely, at this point. But the faith that truth exists, and that our minds can find it, may be held in two ways. We may talk of the *empiricist* way and of the *absolutist* way of believing in truth. The absolutists in this matter say that we not only can attain to knowing truth, but we can *know when* we have attained to knowing it; while the empiricists think that although we may attain it, we cannot infallibly know when. To *know* is one thing, and to know for certain *that* we know is another. One may hold to the first being possible without the second; hence the empiricists and the absolutists, although neither of them is a skeptic in the usual philosophic sense of the term, show very different degrees of dogmatism in their lives.

If we look at the history of opinions, we see that the empiricist tendency has largely prevailed in science, while in philosophy the absolutist tendency has had everything its own way. The characteristic sort of happiness, indeed, which philosophies yield has mainly consisted in the conviction felt by each successive school or system that by it bottom-certitude had been attained. "Other philosophies are collections of opinions, mostly false; *my* philosophy gives standing-ground forever" – who does not recognize in this the key-note of every system worthy of the name? A system, to be a system at all, must come as a *closed* system, reversible in this or that detail, perchance, but in its essential features never!

Scholastic orthodoxy, to which one must always go when one wishes to find perfectly clear statement, has beautifully elaborated this absolutist conviction in a doctrine which it calls that of "objective evidence." If, for example, I am unable to doubt that I now exist before you, that two is less than three, or that if all men are mortal then I am mortal too, it is because these things illumine my intellect irresistibly. The final ground of this objective evidence possessed by certain propositions is the *adequatio intellectus nostri cum re* [the equating of our minds with the thing]. The certitude it brings involves an *aptitudinem ad extorquendam certum assensum* [aptitude for extracting clear and positive assent] on the part of the truth envisaged, and on the side of the subject *a quietem in cognitione* [repose in apprehension], when once the object is mentally received, that leaves no possibility of doubt behind; and in the whole transaction nothing operates but the *entitas ipsa* [existence itself] of the object and the *entitas ipsa* of the mind. We slouchy modern thinkers dislike to talk in Latin – indeed, we dislike to talk in set terms at all; but at bottom our own state of mind is very much like this whenever we uncritically abandon ourselves: You believe in objective evidence, and I do. Of some things we feel that we are certain: we know, and we know that we do know. There is something that gives a click inside of us, a bell that strikes twelve, when the hands of our mental clock have swept the dial and meet over the meridian hour. The greatest empiricists among us are only empiricists on reflection: when left to their instincts, they dogmatize like infallible popes. When the Cliffords tell us how sinful it is to be

Christians on such "insufficient evidence," insufficiency is really the last thing they have in mind. For them the evidence is absolutely sufficient, only it makes the other way. They believe so completely in an anti-Christian order of the universe that there is no living option: Christianity is a dead hypothesis from the start.

◆◆◆◆◆◆

But now, since we are all such absolutists by instinct, what in our quality of students of philosophy ought we to do about the fact? Shall we espouse and indorse it? Or shall we treat it as a weakness of our nature from which we must free ourselves, if we can?

I sincerely believe that the latter course is the only one we can follow as reflective men. Objective evidence and certitude are doubtless very fine ideals to play with, but where on this moonlit and dream-visited planet are they found? I am, therefore, myself a complete empiricist so far as my theory of human knowledge goes. I live, to be sure, by the practical faith that we must go on experiencing and thinking over our experience, for only thus can our opinions grow more true; but to hold any one of them – I absolutely do not care which – as if it never could be reinterpretable or corrigible, I believe to be a tremendously mistaken attitude, and I think that the whole history of philosophy will bear me out. There is but one indefectibly certain truth, and that is the truth that Pyrrhonistic skepticism itself leaves standing – the truth that the present phenomenon of consciousness exists. That, however, is the bare starting-point of knowledge, the mere admission of a stuff to be philosophized about. The various philosophies are but so many attempts at expressing what this stuff really is. And if we repair to our libraries what disagreement do we discover! Where is a certainly true answer found? Apart from abstract propositions of comparison (such as two and two are the same as four), propositions which tell us nothing by themselves about concrete reality, we find no proposition ever regarded by any one as evidently certain that has not either been called a falsehood, or at least had its truth sincerely questioned by some one else. The transcending of the axioms of geometry, not in play but in earnest, by certain of our contemporaries (as Zöllner and

Charles H. Hinton), and the rejection of the whole Aristotelian logic by the Hegelians, are striking instances in point.

No concrete test of what is really true has ever been agreed upon. Some make the criterion external to the moment of perception, putting it either in revelation, the *consensus gentium*, the instincts of the heart, or the systematized experience of the race. Others make the perceptive moment its own test – Descartes, for instance, with his clear and distinct ideas guaranteed by the veracity of God; Reid with his "common-sense"; and Kant with his forms of synthetic judgment a priori. The inconceivability of the opposite; the capacity to be verified by sense; the possession of complete organic unity or self-relation, realized when a thing is its own other, are standards which, in turn, have been used. The much lauded objective evidence is never triumphantly there; it is a mere aspiration or *Grenzbegriff*, marking the infinitely remote ideal of our thinking life. To claim that certain truths now possess it is simply to say that when you think them true and they *are* true, then their evidence is objective, otherwise it is not. But practically one's conviction that the evidence one goes by is of the real objective brand, is only one more subjective opinion added to the lot. For what a contradictory array of opinions have objective evidence and absolute certitude been claimed! The world is rational through and through – its existence is an ultimate brute fact; there is a personal God – a personal God is inconceivable; there is an extramental physical world immediately known – the mind can only know its own ideas; a moral imperative exists – obligation is only the resultant of desires; a permanent spiritual principle is in every one – there are only shifting states of mind; there is an endless chain of causes – there is an absolute first cause; an eternal necessity – a freedom; a purpose – no purpose; a primal One – a primal Many; a universal continuity – an essential discontinuity in things; an infinity – no infinity. There is this – there is that; there is indeed nothing which some one has not thought absolutely true, while his neighbor deemed it absolutely false; and not an absolutist among them seems ever to have considered that the trouble may all the time be essential, and that the intellect, even with truth directly in its grasp, may have no infallible signal for knowing whether it be truth or no. When, indeed, one remembers that the most striking practical application to life of the doctrine of objective certitude has been the conscientious labors of the Holy Office of the Inquisition, one feels less tempted than ever to lend the doctrine a respectful ear.

But please observe, now, that when as empiricists we give up the doctrine of objective certitude, we do not thereby give up the quest or hope of truth itself. We still pin our faith on its existence, and still believe that we gain an ever better position towards it by systematically continuing to roll up experiences and think. Our great difference from the scholastic lies in the way we face. The strength of his system lies in the principles, the origin, the *terminus a quo* of his thought; for us the strength is in the outcome, the upshot, the *terminus ad quem*. Not where it comes from but what it leads to is to decide. It matters not to an empiricist from what quarter an hypothesis may come to him: he may have acquired it by fair means or by foul; passion may have whispered or accident suggested it; but if the total drift of thinking continues to confirm it, that is what he means by its being true.

◆◆◆◆◆◆◆

One more point, small but important, and our preliminaries are done. There are two ways of looking at our duty in the matter of opinion – ways entirely different, and yet ways about whose difference the theory of knowledge seems hitherto to have shown very little concern. *We must know the truth;* and *we must avoid error* – these are our first and great commandments as would-be knowers; but they are not two ways of stating an identical commandment, they are two separable laws. Although it may indeed happen that when we believe the truth A, we escape as an incidental consequence from believing the falsehood B, it hardly ever happens that by merely disbelieving B we necessarily believe A. We may in escaping B fall into believing other falsehoods, C or D, just as bad as B; or we may escape B by not believing anything at all, not even A.

Believe truth! Shun error! These, we see, are two materially different laws; and by choosing between them we may end by coloring differently our whole intellectual life. We may regard the chase for truth as paramount, and the avoidance of error as secondary; or we may, on the other hand, treat the avoidance of error as more imperative, and let truth take its chance. Clifford, in the instructive passage which I have quoted, exhorts us to the latter course. Believe nothing, he tells us, keep your mind in suspense forever, rather than by closing it on insufficient evidence incur the awful risk of believing lies. You, on the other hand, may think that the risk of being in error is a very small matter when compared with the blessings of real knowledge, and be ready to be duped many times in your investigation rather than postpone indefinitely the chance of guessing true. I myself find it impossible to go with Clifford. We must remember that these feelings of our duty about either truth or error are in any case only expressions of our passional life. Biologically considered, our minds are as ready to grind out falsehood as veracity, and he who says, "Better go without belief forever than believe a lie!" merely shows his own preponderant private horror of becoming a dupe. He may be critical of many of his desires and fears, but this fear he slavishly obeys. He cannot imagine any one questioning its binding force. For my own part, I have also a horror of being duped; but I can believe that worse things than being doped may happen to a man in this world: so Clifford's exhortation has to my ears a thoroughly fantastic sound. It is like a general informing his soldiers that it is better to keep out of battle forever than to risk a single wound. Not so are victories either over enemies or over nature gained. Our errors are surely not such awfully solemn things. In a world where we are so certain to incur them in spite of all our caution, a certain lightness of heart seems healthier than this excessive nervousness on their behalf. At any rate, it seems the fittest thing for the empiricist philosopher.

◆◆◆◆◆◆◆◆

And now, after all this introduction, let us go straight at our question. I have said, and now repeat it, that not only as a matter of fact do we find our passional nature influencing us in our opinions, but that there are some options between opinions in which this influence must be regarded both as an inevitable and as a lawful determinant of our choice.

I fear here that some of you my hearers will begin to scent danger, and lend an inhospitable ear. Two first steps of passion you have indeed had to admit as necessary – we must think so as to avoid dupery, and we must think so as to gain truth; but the surest path to those ideal consummations, you will probably consider, is from now onwards to take no further passional step.

Well, of course, I agree as far as the facts will allow. Wherever the option between losing truth and gaining it is not momentous, we can throw the chance of *gaining truth* away, and at any rate save ourselves from any chance of *believing falsehood*, by not making up our minds at all till objective evidence has come. In scientific questions, this is almost always the case; and even in human affairs in general, the need of acting is seldom so urgent that a false belief to act on is better than no belief at all. Law courts, indeed, have to decide on the best evidence attainable for the moment, because a judge's duty is to make law as well as to ascertain it, and (as a learned judge once said to me) few cases are worth spending much time over: the great thing is to have them decided on any acceptable principle, and got out of the way. But in our dealings with objective nature we obviously are recorders, not makers, of the truth; and decisions for the mere sake of deciding promptly and getting on to the next business would be wholly out of place. Throughout the breadth of physical nature facts are what they are quite independently of us, and seldom is there any such hurry about them that the risks of being duped by believing a premature theory need be faced. The questions here are always trivial options, the hypotheses are hardly living (at any rate not living for us spectators), the choice between believing truth or falsehood is seldom forced. The attitude of skeptical balance is therefore the absolutely wise one if we would escape mistakes. What difference, indeed, does it make to most of us whether we have or have not a theory of the Röntgen rays, whether we believe or not in mind stuff, or have a conviction about the causality of conscious states? It makes no difference. Such options are not forced on us. On every account it is better not to make them, but still keep weighing reasons *pro et contra* with an indifferent hand.

I speak, of course, here of the purely judging mind. For purposes of discovery such indifference is to be less highly recommended, and science would be far less advanced than she is if the passionate desires of individuals to get their own faiths confirmed had been kept out of the game. See for example the sagacity which Spencer and Weismann now display. On the other hand, if you want an absolute duffer in an investigation, you must, after all, take the man who has no interest whatever in its results: he is the warranted incapable, the positive fool. The most useful investigator, because the most sensitive observer, is always he whose eager interest in one side of the question is balanced by an equally keen nervousness lest he become deceived. [1] Science has organized this nervousness into a regular technique, her so-called method of verification; and she has fallen so deeply in love with the method that one may even say she has ceased to care for truth by itself at all. It is only truth as technically verified that interests her. The truth of truths might come in merely affirmative form, and she would decline to touch it. Such truth as that, she might repeat with Clifford, would be stolen in defiance of her duty to mankind. Human passions, however, are stronger than technical rules. "*Le coeur a ses raisons*," as Pascal says, "*que la raison ne connait pas* [The heart has its reasons which the reason does not understand]*"; and however indifferent to all but the bare rules of the game the umpire, the abstract intellect, may be, the concrete players who furnish him the materials to judge of are usually, each one of them, in love with some pet "live hypothesis'" of his own. Let us agree, however, that wherever there is no forced option, the dispassionately judicial intellect with no pet hypothesis, saving us, as it does, from dupery at any rate, ought to be our ideal.

The question next arises: Are there not somewhere forced options in our speculative questions, and can we (as men who may be interested at least as much in positively gaining truth as in merely escaping dupery) always wait with impunity till the coercive evidence shall have arrived? It seems a priori improbable that the truth should be so nicely adjusted to our needs and powers as that. In the great boarding house of nature, the cakes and the butter and the sirup seldom come out so even and leave the plates so clean. Indeed, we should view them with scientific suspicion if they did.

♦♦♦♦♦♦♦♦♦

Moral questions immediately present themselves as questions whose solution cannot wait for sensible proof. A moral question is a question not of what sensibly exists, but of what is good, or would be good if it did exist. Science can tell us what exists; but to compare the *worths*, both of what exists and of what does not exist, we must consult not science, but what Pascal calls our heart. Science herself consults her heart when she lays it down that the infinite ascertainment of fact and correction of false belief are the supreme goods for man. Challenge the statement, and science can only repeat it oracularly, or else prove it by showing that such ascertainment and correction bring man all sorts of other goods which man's heart in turn declares. The question of having moral beliefs at all or not having them is decided by our will. Are our moral preferences true or false, or are they only odd biological phenomena, making things good or bad for *us*, but in themselves indifferent? How can your pure intellect decide? If your heart does not *want* a world of moral reality, your head will assuredly never make you believe in one. Mephistophelian skepticism, indeed, will satisfy the head's play instincts much better than any rigorous idealism can. Some men (even at the student age) are so naturally coolhearted that the moralistic hypothesis never has for them any pungent life, and in their supercilious presence the hot young moralist always feels strangely ill at ease. The appearance of knowingness is on their side, of naïveté and gullibility on his. Yet, in the inarticulate heart of him, he clings to it that he is not a dupe, and that there is a realm in which (as Emerson says) all their wit and intellectual superiority is no better than the cunning of a fox. Moral skepticism can no more be refuted or proved by logic than intellectual skepticism can. When we stick to it that there is truth (be it of either kind), we do so with our whole nature, and resolve to stand or fall by the results. The skeptic with his whole nature adopts the doubting attitude; but which of us is the wiser, Omniscience only knows.

Turn now from these wide questions of good to a certain class of questions of fact, questions concerning personal relations, states of mind between one man and another. *Do you like me or not?* for example. Whether you do or not depends, in countless instances, on whether I meet you half-way, am willing to assume that you must like me, and show you trust and expectation. The previous faith on my part in your liking's existence is in such cases what makes your liking come. But if I stand aloof, and refuse to budge an inch until I have objective evidence, until you shall have done something apt, as the absolutists say, *ad extorquendum assensum meum* [at extorting my assent], ten to one your liking never comes. How many women's hearts are vanquished by the mere sanguine insistence of some man that they *must* love him! He will not consent to the hypothesis that they cannot. The desire for a certain kind of truth here brings about that special truth's existence; and so it is in innumerable cases of other sorts. Who gains promotions, boons, appointments, but the man in whose life they are seen to play the part of live hypotheses, who discounts them, sacrifices other things for their sake before they have come, and takes risks for them in advance? His faith acts on the powers above him as a claim, and creates its own verification.

A social organism of any sort whatever, large or small, is what it is because each member proceeds to his own duty with a trust that the other members will simultaneously do theirs. Wherever a desired result is achieved by the co-operation of many independent persons, its existence as a fact is a pure consequence of the precursive faith in one another of those immediately concerned. A government, an army, a commercial system, a ship, a college, an athletic team, all exist on this condition, without which not only is nothing achieved, but nothing is even attempted. A whole train of passengers (individually brave enough) will be looted by a few highwaymen, simply because the latter can count on one another, while each passenger fears that if he makes a movement of resistance, he will be shot before any one else backs him up. If we believed that the whole carful would rise at once with us, we should each severally rise, and train robbing would never even be attempted. There are, then, cases where a fact cannot come at all unless a preliminary faith exists in its coming. *And where faith in a fact can help create the fact*, that would be an insane logic which should say that faith running ahead of scientific evidence is the "lowest kind of immorality" into which a thinking being can fall. Yet such is the logic by which our scientific absolutists pretend to regulate our lives!

◆◆◆◆◆◆◆◆◆◆

In truths dependent on our personal action, then, faith based on desire is certainly a lawful and possibly an indispensable thing.

But now, it will be said, these are all childish human cases, and have nothing to do with great cosmic matters, like the question of religious faith. Let us then pass on to that. Religions differ so much in their accidents that in discussing the religious question we must make it very generic and broad. What then do we now mean by the religious hypothesis? Science says things are; morality says some things are better than other things; and religion says essentially two things.

First, she says that the best things are the more eternal things, the overlapping things, the things in the universe that throw the last stone, so to speak, and say the final word. "Perfection is eternal" – this phrase of Charles Secrétan seems a good way of putting this first affirmation of religion, an affirmation which obviously cannot yet be verified scientifically at all.

The second affirmation of religion is that we are better off even now if we believe her first affirmation to be true.

Now, let us consider what the logical elements of this situation are *in case the religious hypothesis in both its branches be really true.* (Of course, we must admit that possibility at the outset. If we are to discuss the question at all, it must involve a living option. If for any of you religion be a hypothesis that cannot, by any living possibility be true, then you need go no farther. I speak to the "saving remnant" alone.) So proceeding, we see, first that religion offers itself as a *momentous* option. We are supposed to gain, even now, by our belief, and to lose by our nonbelief, a certain vital good. Secondly, religion is a *forced* option, so far as that good goes. We cannot escape the issue by remaining skeptical and waiting for more light, because, although we do avoid error in that way *if religion be untrue,* we lose the good, *if it be true,* just as certainly as if we positively chose to disbelieve. It is as if a man should hesitate indefinitely to ask a certain woman to marry him because he was not perfectly sure that she would prove an angel after he brought her home. Would he not cut himself off from that particular angel possibility as decisively as if he went and married some one else? Skepticism, then, is not avoidance of option; it is option of a certain particular kind of risk. *Better risk loss of truth than chance of error* – that is your faith vetoer's exact position. He is actively playing his stake as much as the believer is; he is backing the field against the religious hypothesis, just as the believer is backing the religious hypothesis against the field. To preach skepticism to us as a duty until "sufficient evidence" for religion be found, is tantamount therefore to telling us, when in presence of the religious hypothesis, that to yield to our fear of its being error is wiser and better than to yield to our hope that it may be true. It is not intellect against all passions, then; it is only intellect with one passion laying down its law. And by what, forsooth, is the supreme wisdom of this passion warranted? Dupery for dupery, what proof is there that dupery through hope is so much worse than dupery through fear? I, for one, can see no proof; and I simply refuse obedience to the scientist's command to imitate his kind of option, in a case where my own stake is important enough to give me the right to choose my own form of risk. If religion be true and the evidence for it be still insufficient, I do not wish, by putting your extinguisher upon my nature (which feels to me as if it had after all some business in this matter), to forfeit my sole chance in life of getting upon the winning side – that chance depending, of course, on my willingness to run the risk of acting as if my passional need of taking the world religiously might be prophetic and right.

All this is on the supposition that it really may be prophetic and right, and that, even to us who are discussing the matter, religion is a live hypothesis which may be true. Now, to most of us religion comes in a still further way that makes a veto on our active faith even more illogical. The more perfect and more eternal aspect of the universe is represented in our religions as having personal form. The universe is no longer a mere *It* to us, but a *Thou*, if we are religious; and any relation that may be possible from person to person might be possible here. For instance, although in one sense we are passive portions of the universe, in another we show a curious autonomy, as if we were small active centers on our own account. We feel, too, as if the appeal of religion to us were made to our own active good will, as if evidence might be forever withheld from us unless we met the hypothesis halfway. To take a trivial illustration: just as a man who in a company of gentlemen made no advances, asked a warrant for every concession, and believed no one's word without proof, would cut himself off by such churlishness from all the social rewards that a more trusting spirit would earn, so here, one who should shut himself up in snarling logicality and try to make the gods extort his recognition willy-nilly, or not get it at all, might cut himself off forever from his only opportunity of making the gods' acquaintance. This feeling, forced on us we know not whence, that by obstinately believing that there are gods (although not to do so would be so easy both for our logic and our life) we are doing the universe the deepest service we can, seems part of the living essence of the religious hypothesis. If the hypothesis *were* true in all its parts, including this one, then pure intellectualism, with its veto on our making willing advances, would be an absurdity; and some participation of our sympathetic nature would be logically required. I, therefore, for one, cannot see my way to accepting the agnostic rules for truth seeking, or willfully agree to keep my willing nature out of the game. I cannot do so for this plain reason, that *a rule of thinking which would absolutely prevent me from acknowledging certain kinds of truth if those kinds of truth were really there, would be an irrational rule.* That for me is the long and short of the formal logic of the situation, no matter what the kinds of truth might materially be.

I confess I do not see how this logic can be escaped. But sad experience makes me fear that some of you may still shrink from radically saying with me, *in abstracto*, that we have the right to believe at our own risk any hypothesis that is live enough to tempt our will. I suspect, however, that if this is so, it is because you have got away from the abstract logical point of view altogether, and are thinking (perhaps without realizing it) of some particular religious hypothesis which for you is dead. The freedom to "believe what we will" you apply to the case of some patent superstition; and the faith you think of is the faith defined by the schoolboy when he said, "Faith is when you believe something that you know ain't true." I can only repeat that this is misapprehension. *In concreto*, the freedom to believe can only cover living options which the intellect of the individual cannot by itself resolve; and living options never seem absurdities to him who has them to consider. When I look at the religious question as it really puts itself to concrete men, and when I think of all the possibilities which both practically and theoretically it involves, then this command that we shall put a stopper on our heart, instincts, and courage, and *wait* – acting of course meanwhile more or less as if religion were *not* true [2] till doomsday, or till such time as our intellect and senses working together may have raked in evidence enough – this command, I say, seems to me the queerest idol ever manufactured in the philosophic cave. Were we scholastic absolutists, there might be more excuse. If we had an infallible intellect with its objective certitudes, we might feel ourselves disloyal to such a perfect organ of knowledge in not trusting to it exclusively, in not waiting for its releasing word. But if we are empiricists, if we believe that no bell in us tolls to let us know for certain when truth is in our grasp, then it seems a piece of idle fantasticality to preach so solemnly our duty of waiting for the bell. Indeed we *may* wait if we will – I hope you do not think that I am denying that – but if we do so, we do so at our peril as much as if we believed. In either case we *act*, taking our life in our hands. No one of us ought to issue vetoes to the other, nor should we bandy words of abuse. We ought, on the contrary, delicately and profoundly to respect one another's mental freedom: then only shall we bring about the intellectual republic; then only shall we have that spirit of inner tolerance without which all our outer tolerance is soulless, and which is empiricism's glory; then only shall we live and let live, in speculative as well as in practical things.

I began by a reference to Fitzjames Stephen; let me end by a quotation from him. "What do you think of yourself? What do you think of the world? ... These are questions with which all must deal as it seems good to them. They are riddles of the Sphinx, and in some way or other we must deal with them.... In all important transactions of life we have to take a leap in the dark.... If we decide to leave the riddles unanswered, that is a choice; if we waver in our answer, that, too, is a choice: but whatever choice we make, we make it at our peril. If a man chooses to turn his back altogether on God and the future, no one can prevent him; no one can show beyond reasonable doubt that he is mistaken. If a man thinks otherwise and acts as he thinks, I do not see that any one can prove that *he* is mistaken. Each must act as he thinks best; and if he is wrong, so much the worse for him. We stand on a mountain pass in the midst of whirling snow and blinding mist through which we get glimpses now and then of paths which may be deceptive. If we stand still we shall be frozen to death. If we take the wrong road we shall be dashed to pieces. We do not certainly know whether there is any right one. What must we do?

'Be strong and of a good courage.' Act for the best, hope for the best, and take what comes.... If death ends all, we cannot meet death better." [3]

Notes

1. Cf. Wilfrid Ward's essay, "The Wish to Believe," in his *Witnesses to the Unseen*, Macmillan & Co., 1893.

2. Since belief is measured by action, he who forbids us to believe religion to be true, necessarily also forbids us to act as we should if we did believe it to be true. The whole defence of religious faith hinges upon action. If the action required or inspired by the religious hypothesis is in no way different from that dictated by the naturalistic hypothesis, then religious faith is a pure superfluity, better pruned away, and controversy about its legitimacy is a piece of idle trifling, unworthy of serious minds. I myself believe, of course, that the religious hypothesis gives to the world an expression which specifically determines our reactions, and makes them in a large part unlike what they might be on a purely naturalistic scheme of belief.

3. *Liberty, Equality, Fraternity*, p. 353, 2d edition, London, 1874.

CHAPTER SEVEN

Politics, Death, And Camus's Late Anarchic Style

John Randolph LeBlanc [1]

I. Death and Politics

Albert Camus's circumstances made him a political man. After all, how could a man, born of a father killed in the First World War, coming of age in the interwar years, an activist journalist, a member in the Resistance movement against the German occupiers of France, a brief proponent of the post-Liberation purges in France, and, finally, a witness to the bloody decolonization of his native Algeria, not be "political"? Yet, in the midst of the war for Algerian decolonization, there was Camus, four years before his premature death, giving a lecture in Algiers, struggling to articulate a commitment to something larger than the claims of either side and, not coincidentally, his own uneasy sense of himself: "I am not a political man, and my passions and inclinations do not lead me to public platforms. I step onto the podium only when forced to by the pressure of circumstances and by my conception of my function as a writer" (*Resistance* 132).

Why this reticence despite a biography and, indeed, a corpus that suggests otherwise? One answer for Camus, I wish to argue, is his recognition of the troubling, proximate, and inevitable presence of death in any form of politics, a recognition that sent Camus to some unlikely places looking for a different way of political being. Many scholars concerned with the question argue that Camus's politics, while never simple, were of the left and sought a social democratic order. Camus said as much in an October 1, 1944 editorial in *Combat*:

> What we want is to make justice compatible with freedom. It seems this is not clear enough. We shall therefore call justice a social state in which each individual starts with equal opportunity, and in which the country's majority cannot be held in abject conditions

by a privileged few. And we shall call freedom a political climate in which the human being is respected both for what he is and for what he says. (*Between Hell and Reason* 57)

Camus called this hybrid political order "true popular democracy." It was the kind of order he believed had to be realized in France and the only legitimate object of any postwar political struggle. This comprehensive vision would prove short-lived but despite the postwar political complexities Camus would face, or appear not to, the core of these commitments—protecting individual freedom in a just environment—persisted in Camus's postwar work.

As he tried to confront the exigencies of the purges, the Cold War, and the Algerian War of decolonization, Camus struggled to find a political home. Stephen Eric Bronner characterizes Camus as a "bohemian and a non-conformist" who also was "a socialist with syndicalist sympathies, who called on people to make use of civil liberties in deepening democracy" and should be read as "bringing some heart and a sense of moral urgency into a socialist democratic movement grown increasingly stale and technocratic" (152-53). Indeed, in the aftermath of the Liberation, Camus's voice is strong, confident, and self-assured. His mission is to exorcise of the demons of collaboration and to help complete the social democratic revolution in France. But this commitment catches him up in the temptation to justify killing. As a participant in the Resistance, Camus had worked against the Nazi occupiers in France up to and beyond the point of justifying killing. In the fourth of his propagandistic "Letters to a German Friend," Camus makes clear that confronting injustice means destroying the enemy: "I am fighting you because your logic is as criminal as your heart... [A]t the very moment when we are going to destroy you without pity, we still feel no hatred for you" (*Resistance* 30-31). Here we find Camus the erstwhile humanist drawing a sharp distinction between friend and enemy, articulating a nearly messianic justification of violence, and advocating the destruction of the enemy in order to save France. While it is true that Camus expresses doubt, reflected in his tone here is the certainty required to justify killing another human being. What's more, this certainty carries over into the immediate aftermath of the Liberation. Defending the purges in a six month long correspondence with Francois Mauriac, Camus again argued that "there are times when we must silence our feelings and renounce our peace of mind. Ours is such a time, and it's terrible law, with which it is futile to argue, forces us to destroy a living part of this country in order that we may save its very soul" (*Between Hell and Reason* 66). The "terrible law" of his time had emerged fully with the coming of the Second World War. In another response to Mauriac, Camus announces his disdain for murder, but insists that "since 1939 we have truly learned that not to destroy certain men would be to betray the good of this country" (*Between Hell and Reason* 72). Camus's hope that the purges would be finished quickly and efficiently was unfulfilled. Even in its failure, however, he held to the efficacy of the idea: "The word purge is painful enough in itself. That which it describes has become hateful. *It could have succeeded only if undertaken without vengeance or frivolity*" (*Between Hell and Reason* 112, my emphasis).

It is tempting to label "Letters to a German Friend" as mere propaganda and his post-Liberation work in *Combat* as so much "heat of the moment" zeal. But Camus's certainty that a new democratic socialist France was about to be born and could therefore justify the breaking of a few eggs in the name of "human justice" is too important to be so quickly overlooked. It is formative of postwar Camus. During the purges, people were killed, exiled or, ruined and Camus had advocated it in his exchanges with Mauriac. While circumstance surely plays a part, I would argue that the importance of this work is that during this period, Camus slips into and, indeed, embraces all that is wrong with the modern form of politics, that is, the politics he refuses before the war, defies during his time in the Resistance albeit using its methods against it) and rejects in his later years. In collaborators, Camus found a readily identifiable evil that could and had to be rooted out in the name of preserving the possibility of a collective "good," that is, the realization of the "relative utopia" of socialist-democratic France. Consequently, this part of his work is critical to understanding his postwar political positioning. His experience at this crucial time teaches him once and for all that modern political life is built upon a foundation of death and dying. On the other side of his support for the purges,

Camus realizes that to participate in political life is to affiliate oneself with death, especially the killing of others. Camus's temporary collaboration with what he might later have called the modern politics of death—and his subsequent rejection of it— forces him to turn away from this politics and attempt to conceive a new ethical version of politics. In an excellent essay on Camus's use of the terms "man" and "justice," Martin Crowley notes that "Murder" is Camus's "synecdochic figure for political violence"(Crowley 101). For Camus, violence is part and parcel of politics, but the concern with murder transcends "mere" violence in Camus's political consciousness. Camus learns from his support of the purges that participation in politics implicates one in death—certainly the death of others, but also the death of any certainty regarding the content of "human justice." Thereafter, for Camus the term "political" must be reclaimed for a politics that defies death through creativity rather than one that is necessarily the bringer of death.

His refusal to justify killing after the purges go awry leads to what David Carroll aptly characterizes as Camus's "political agnosticism" (Carroll 103). Camus's postwar political commitments—his crusade against capital punishment ("Reflections on the Guillotine"), his refusal to embrace either side in the Cold War ("Neither Victims Nor Executioners"), his unwillingness to embrace the terroristic practices of either side in the Algerian Civil War, and, finally, his implied sympathies with what can be called nonviolent anarchism—are each grounded in an adamant denial of death as an acceptable political practice or outcome. Carroll rightly argues that during the Algerian War we see Camus struggle "to reconcile his refusal to legitimize murder with an activist, socially responsible politics." What Camus seeks, Carroll says, is "to find a 'third way' that would be able to effectively oppose injustices without supporting the deadly means used in revolutionary struggles for national independence—or the counter-revolutionary and counterterrorist means used against them" (Carroll 104). This chapter is an attempt to flesh out a possible reading of that "third way" and its sources in Camus's experience.

Camus's committed political agnosticism is a function of participating in and reflecting upon the implications of modern politics. Though he resisted it, often turning back to ancient Greece, Camus was a modern—in place if not in spirit. Whatever its philosophical shortcomings—and they were well documented beginning with the critique of Jean-Paul Sartre—his reading of the character and implications of modernity in *The Rebel* is both nuanced and powerful. For Camus, himself an advocate of reason and the reasonable, modernity is about our dependence on rationalized order and structure, either their confining presence or, conversely, what we become in their absence. In *The Rebel*, we find Camus chronicling the march of modern human beings away from their humanity. Human reason harnesses some of the powers of the natural world, but assumes a dangerous standpoint outside its natural home. A dependence on rational structures influences political organization, artistic productions, philosophical explanations and justifications for them. In short, there is a self-destructive disconnect between the way modern human beings use their reason and human being itself. Camus's Nietzschean inheritance is a desire to reject the critical binaries that he deploys in his postwar work, rebellion and revolution in philosophy and politics, realism and formalism in art. That he struggles to do so, I would argue, makes him a modern.

Chief among Camus's concerns is how the modern conceives of its relationship to the natural world. Here, science poses a particular problem for Camus. The day after the Americans dropped the atomic bomb on Hiroshima, Camus wrote "The world is what it is, which is to say, nothing much" (*Between Hell and Reason* 110). Weary of years of war and destruction Camus knew that our capacity for technological development had ceased (if it had ever done so) consulting the requirements of the human, that is, the experience of two world wars amply demonstrated the degree to which our technology had outstripped our ability to know what it meant. "We can," he wrote, "sum it up in one sentence. Our technical civilization has just reached its greatest level of savagery. We will have to choose, in the more or less near future, between collective suicide and the intelligent use of our scientific conquests" (*Between Hell and Reason* 110). The bomb here becomes the quintessential image of the modern: a remarkable achievement of the modern mind deployed to the end of mass murder by remote control. That the bomb was deployed to "political" ends was particularly poignant for Camus. It represented the cross-fertilization between technological development and political power, that is, between our desire to control the world and our desire to escape responsibility. Quite literally, Camus suggests, what we seek is "remote control" of which the bomb is the best illustration. On the one hand, the more we wish to control things, the less control we have. The science that created the bomb had also discovered penicillin in order to cure disease. On the other hand, we try to remove ourselves from the world by relinquishing agency and responsibility to rational institutions or mere rationalizations like the aspiration to objectivity. This was an experience with which Camus was quite familiar. In a notebook entry in the Spring of 1949 as he sketched out a preface to his *Actuelles* Camus mused: 'One of my regrets is having sacrificed too much to objectivity. Objectivity at times is a self-indulgence. Today things are clear and what belongs to the concentration camp, even socialism, must be called a concentration camp. In a sense, I shall never again be polite.... I strove toward objectivity, contrary to my nature. This is because I distrusted freedom." (*Notebooks 1942-1951* 211). The implications for both science and politics are clear. Science seeks "objectivity" or detachment but it is the product of human being and understanding. A politics basing itself on this view of science—and Camus's certainty during the purges, for instance--falls prey to the same difficulty. Wrapping oneself in "objectivity" is an attempt to escape responsibility for our free actions. While Camus took responsibility for what he believed had to be done in the aftermath of the liberation, he nonetheless discovered that the twin desires to control the world and escape responsibility drives us to destroy that which we seem finally unable to control—namely, ourselves.

Another way we relate to our world, and one with which Camus was intimately familiar, was art. If Camus knew himself to be no scientist, he did consider himself to be an artist and he found that the tension between knowledge and control had no less colored the art of modernity. Artistic manifestations of the coupling of the modern desire for order with the disjointed, lost soul of the modern person were no accident. For instance, the many forms taken by the artistic movement known as Modernism reflected the fractured and self- and other-destructive rebellions against the conditions of modernity that Camus describes in his work. Tyrus Miller in *Late Modernism: Politics, Fiction, and the Arts Between the Wars* surveys some of the diverse forms that modernism has taken in scholarly accounts. "Modernism," Miller writes, "is the liberation of formal innovation; the destruction of tradition; the renewal of decadent conventions or habit-encrusted perceptions; the depersonalization of art; the radical subjectivization of art. And so on" (4). If we look for a coherent aesthetic in this congeries of tendencies we will find it difficult. Taking a step back Camus, again in *The Rebel*, was able to discern two nihilistic trends in modern art: the realist and the formalist. "The realist artist and the formal artist try to find unity where it does not exist," he writes, "in reality and its crudest state, or in the imaginative creation which wants to abolish all reality" (*The Rebel* 269). The one is art in which the form overwhelms the content, while in the other the content overflows the form. In the latter instance, the event—the represented--enslaves the creator while in the former the creator tries to turn the tables on reality. In both cases, what we have is the destruction of the artistic process by removing the human capacity to judge between and among the elements of the artistic endeavor. Here the artist abdicates responsibility for ordering his or her experience of the world. In modern art, we have the contradiction at the heart of modernity and the reason Camus cannot sign off on any comprehensive view of it. The movement of the modern seems to imply a rejection—of imposed rules and forms that no longer speak to the lived conditions of human being—but it responds to the corresponding sense of homelessness by imposing totalizing restrictions and, as Foucault's work reminds us, suffocating disciplines. The modernist approaches to art may be seen as responses to the larger modern project--to impose absolute unity on the world—that also characterize modern politics, particularly, but not exclusively, its totalitarian forms. The assertion or imposition of a false unity renders reality simpler and more manageable, that is, all is a single articulable unity *or* all is chaotic and incoherent and not worth trying to figure out. The assertion may be comforting in the short term, but it is disastrous in the long term. In the meantime, the modern either clings desperately to the script or wanders aimlessly in pre-apocalyptic chaos. In either instance, the modern finds what Camus finds: a world barely habitable by human beings.

Camus's own creative productions reveal that the brutal, dehumanizing facts of modernity were his constant companions. Camus began with the term "absurdity," not to ally himself with existentialism as such, but rather to signify his sense of human homelessness in the midst of what Simone Weil called the "icy pandemonium of modernity." Camus's novels--the art form with which he is chiefly concerned in *The Rebel*--are filled with characters who find themselves out of place in their environments, often quite ordinary surroundings. Meursault is an alien to his environment in *The Stranger*, yet that environment is filled with people who make themselves feel better by executing a man for not crying at his mother's funeral. All the while, the nameless murdered "Arab" on the beach is mostly ignored at Meursault's trial. If characters in works like *The Plague* (e.g., the vaguely criminal Cottard) and *The Fall* (e.g., Clamence, the ex-patriate Frenchman and "judge-penitent") appear poorly adapted to their circumstances, Camus's depictions challenge us to ask after the role the environments of modernity play in developing those characters. Cottard, for instance, is transformed completely as the plague's grip on Oran tightens. As the environment becomes more hostile to the busy but meaningless pursuits the Oranais call human life, Cottard becomes more recognizably social.

Cottard's transformation from an attempted suicide to someone nearly comfortable in his own skin in the difficult circumstances of *The Plague* represents another tendency in the development of Camus's characters: the all-too-frequent disconnect between the requirements of circumstances and the character of his characters. Circumstances consistently, if not inevitably, undermine certainties we have about our own agency. In Camus's work, those who appear to be comfortable, either in their stations or in their own skins, are revealed as less than they seem by changes in circumstance. As the pestilence wanes, Cottard returns to his former self. The smug certitude of the magistrate Othon in the same novel is undone by the illness of his son—his office cannot protect him from the effects of the plague after all. Government officials and their medical advisors hesitate (Dr. Rieux is a notable exception to which I'll return) at the outbreak of the plague, haggling over what to name the outbreak. In Camus's straightforward depiction, bureaucratic concerns outweigh the health and safety of those the officials are charged with governing. In *The Stranger*, Meursault's prosecutor and interrogators refer to him as "Monsieur Anti-Christ" as they turn his trial into a prosecution for refusing to live up to societal expectations rather than for his real crime--the killing of a human being on the beach. Their official motivation seems to be keeping up appearances through a prosecution and execution, not any ongoing commitment to justice or any kind of human life (Meursault's or the Arab's). Nor do Camus's depictions of religious characters offer relief. Meursault's priest in *The* Stranger and Father Paneloux in *The Plague* find their worldviews utterly inadequate to deal with the modern world that confronts them and the persons created by that world.

The failure of these latter "institutional agents" points up another characteristic of the modern world for Camus: its preoccupation with form and its codependence upon it. In social and political terms, this preoccupation with form means a faith in institutional arrangements. As I suggested above, Camus's direct experience with institutions was less than heartening. Institutionalizing collective action (or action in the service of some utopian ideal) tempts one to abrogate responsibility for one's actions. Agency and accountability can be lost in collective service of some greater good—either well-defined and utopian (extremely dangerous, thus murder) or ill-defined and irresponsible (desperate, thus suicide *cum* murder). The postmodern thinker Jean-Francois Lyotard observed that politics always about institutionalizing something. In Camus's work, institutionalization (cum, bureaucratization) has two critical dimensions: (1) we create some "thing" to do our killing for us; (2) reliance/dependence on institutions and their categories "kills" spontaneity, flexibility, originality, responsiveness, humanity. I will argue that Camus's "anarchic" turn to the creative individual as a model of the responsible political actor is a direct response to these difficulties.

Another dimension of Camus's political critique was his disdain for ideologies. Camus's work suggests a distinction be drawn between the reasonable and the rational. Ideological constructions, in which the universe is defined and the answers provided because the questions have been anticipated, were dangerous fictions but also very human temptations for Camus. Instead of having to puzzle out "what to do" in full knowledge that "we don't understand," we are more comfortable with having ready-made answers provided for us. Ideologies remake a kind of "divine ideal" for modern politics, fostering a discourse which becomes a focal point of political action to the detriment of acting when action is necessary. The operative example is again the debate among local officials in Oran over what to call the plague when, as Rieux reminds them, what to do about it is more important than what to call it. In that novel, Joseph Grand serves as counter to this tendency, recognizing an unspoken duty to resist the plague while adhering to his ideal. Heroic in his own minimalist way, he works on his perfect novel (or, the first line of it) after doing his bit for the sanitation squads. Modern politics and its epistemological dependence upon ideologies engender a way of thinking—inherited from modern science-- that makes life formulaic. Embracing the formulae frees us from engaging the contradictions that characterize human existence and being responsible for the consequences. Camus works within this framework in order to work beyond it, to reveal its vacuity. His focus is on the human beings generated by modernity because he was unable, at least for long, to posit an alternative set of arrangements. In other words, Camus implicitly rejects the adequacy of institutional arrangements to the task of humanizing our environment. Politicians, church officials, members of professional organizations all fail to create adequate responses to extraordinary or, indeed ordinary, conditions. The modern's reliance upon rational institutions is at best futile in Camus's work and at worst tragic. His work points to breaking the habit of counting on institutions to set rules and to name and settle our problems. This difficult willingness to admit our solitude and come to terms with it through creativity is what he identifies as "The Wager of Our Generation" in *Resistance, Rebellion, and Death* and is at the core of his anarchist bent.

II. Art and the Style of Anarchy

In a notebook entry in April 1950, Camus wrote "After *The Rebel*, free creation" (*Notebooks 1942-1951* 254). Ten months later, in February 1951, Camus revisited the idea: "After *The Rebel*. Agreessive, obstinate rejection of the system. The aphorism henceforth" (*Notebooks 1942-1951* 269). Taken together, these two entries suggest the direction that Camus sought after the publication of *The Rebel* in late 1951. "Free creation" marks the aspiration to a kind of creativity unhindered by the expectations of his contemporaries and, therefore, of himself. Camus knew *The Rebel* signaled a break with many of his former colleagues, even if he could not anticipate the level of scorn that would characterize the resulting very public split with Sartre. At the same time, his will to an "aggressive, obstinate rejection of the system" suggests a patent distrust of systems of any sort, particularly those that govern politics on the one hand and literary endeavors on the other. "The aphorism henceforth" clearly refers to his literary efforts, the aphorism being a Nietzschean form giving much freedom to the author and placing the onus of interpretation and understanding on the reader. One can with some justification suggest that Camus is also thinking about an aphoristic form of political action. In politics, the aphorism suggests actions that are creative, effective, if not readily predictable, and that have implications and subtler shades of meaning than are apparent on their surface. Because such action precludes a final reading, an aphoristic politics remains open and forces the participant to keep a vigilant critical eye on his or her political environment. There is art to the aphorism and Camus's consideration of the work of the artist as an apt analogy for political agency toward the end of *The Rebel* suggested a turn that would manifest itself in his postwar political writing. Free creation beyond systems in an aphoristic medium bears the earmarks of anarchism. In the balance of this chapter, I will argue that Camus's late work often suggests a desire for a political agency that resembles a non-violent form of anarchism.

It is not necessary to argue that Camus is an anarchist or that he allied himself with anarchist groups in any serious way. Anarchists were part of the intellectual and political environment that he inhabited, but so were Marxists, Communists, Liberals, Capitalists, etc. What I am arguing here is that Camus's late, that is, postwar political thought bears the mark of an anarchist approach. Transcending death means creating a new politics and it is, I will argue, a politics resembling non-violent anarchism. His turn to the artistic process bears the hallmarks of the anarchist response to the terror by chaos or excessive formalism characteristic of modern politics. Before turning to the content of what I will call his anarchist style, we must seek what connections there are between Camus and the anarchists. Camus wrote for anarchist periodicals, wrote favorably of trade unions and syndicalist movements, and was even sympathetic to the Russian anarchists he depicts in "The Just Assassins" (*Caligula and Three Other Plays*) and to whom he refers as "The Fastidious Assassins" in *The Rebel*. This connection of Camus with anarchism has been underplayed in the literature because nowhere does Camus claim to *be* an anarchist. Camus's absurdist starting point, his relationship (or not) to existentialism, his critique of modernity, and the humanist aspirations of his politics and political thought are all familiar staples of Camus scholarship. In what follows, I argue that we should bear these familiar approaches in mind as we think about Camus, but we should supplement our understanding of Camus—both as critic and as engaged intellectual—with the anarchist tendencies that inform all of his late work. We will find that it is from the spontaneous, improvisational—if ultimately principled--possibilities of creative being rather than the absurdist starting point that Camus offers the possibility of humanizing our relations with one another.

Scholars have noted the connection that I wish to explore here. Lou Marin demonstrated that Camus the journalist had many affiliations with anarchists and anarchism. In an occasional paper written for the *Dag Hammarskjold Foundation*, the pseudonymous Marin documents Camus's "contributions to the pacifist, anarchist and syndicalist press in France and Algeria during the 1940s and the 1950s" and shows "the wide field of friendly personal relations Camus had with French anarchists and non-conformist syndicalists during this time" (9). Marin finds that "Camus not only contributed to anarchist journals; he also defended anarchists" (9). Marin argues that these periodicals and personal connections gave

Camus the opportunity to work out "a real alternative both to the capitalist West and to the state-socialist East. This perspective," Marin says, "which goes beyond capitalism and Marxism, is the one that has always been held by anarchism" (20). For Marin, the content of this anarchism seems largely to be one of critique; it consists in a refusal to participate in the destructive and partisan politics of the period and a defense of those who dared to so act.

Jeffrey Isaac makes a bit more of the lurking anarchist tendencies in Camus in his comparative analysis of the political thought of Camus and Hannah Arendt. In working through the notions of democracy that influenced the two authors, Isaac identifies a critical anarchist form: "both Arendt and Camus can be considered anarchists of a sort--democratic anarchists interested in opening up spaces for political agency that tend to be closed by the modern state, nurturing forms of public identity that defy the boundaries of sovereign nation-states" (148). Critical in Isaac's account is his careful distinction between anarchism and its reputation. The "anarchist advocacy of direct action," he argues, "is often confused with a violent politics of the deed, and its opposition to authoritarianism is confused with a principled refusal of all authority (148). At its heart, Isaac insists, anarchism only refuses modern forms of authority—particularly those bureaucratizing, rule-based, and oppressive forms found in the modern state. What the apparent "antipolitics" of anarchism advocates, Isaac writes invoking the anarchist Voline, is

mutualism, solidarity, and voluntary cooperation rather than command and obedience. Such an anarchism celebrates spontaneity but not mindless activism, political participation but not organicism. It is compatible with political authority, but only when such authority is proximate, responsive, and above all provisional. (148)

The emphases on spontaneity and the provisional nature of human authority are crucial. One response to the absurdity of modernity we find in Camus's work is a confidence in mindful improvisation and it is in this improvisation that we find Camus's latent anarchism. Improvisation—what Camus calls "style"--becomes a way to discover, act upon, and reinterpret what we are--not just individually, but collectively. The idea is played out most pointedly in his novel *The Plague* where circumstance draws characters like Dr. Rieux, the stranded Jean Tarrou, and Joseph Grand out of themselves and into collective action with others. The members of the sanitation squads create out of necessity, but their responses are generated out of a humane rejection of the chief of necessity's requirements, that is, the inevitability of death. Nonetheless, their actions must also reckon with the natural world. Camus holds out the possibility that a due attention to the guidance of an admittedly benign version of human nature could guide ethical action in community. Camus frequently referred to the Greeks as a model from which we should learn this lesson. David Sprintzen notes that Camus's support of anarcho-syndicalism is best understood when we appreciate his desire to construct human community on the basis of what that movement indentified as "natural conditions of existence" (258). These conditions included particularly the relationship of human beings to their labor and to the natural world. As we will see, Camus grounded his understanding of these relationships n a broadly humane notion of creativity, a creative way of being fully compatible with improvisation and spontaneity.

I think this aspiration to a creative, consciously improvised and more spontaneous human (and political) existence is the primary reason that Camus, in *The Rebel* and for the rest of his abbreviated life, looked to art as a kind of ethical model. In his essay "Create Dangerously," we find Camus calling the artist home, away from mere production and back to what he termed an "authentic" form of creation:

> Civilization is only possible if, by renouncing the nihilism of formal principles and the nihilism without principles, the world rediscovers the road to a creative synthesis. In the same way, in art the time of perpetual commentary and factual reporting is at the point of death; it announces the advent of creative artists. (*Resistance* 273)

In his diagnosis of the perils of relying upon institutions to order our lives and embracing nihilistic responses, I think Camus opens himself to the possibilities of anarchist forms of creativity. The difference between nihilism and anarchism has been elegantly captured by George Woodcock. "The nihilist," he writes, "using the term in a general sense, believes in no moral principle and no natural law; the anarchist believes in a moral urge powerful enough to survive the destruction of authority and still to hold society together in the free and natural bonds of fraternity" (15; see also Wood). It is certain that Camus would not sign off on the physical destruction of sites of authority, but he doesn't have to. On his reckoning, "civilization" was destroying itself. Withdrawing our consent to this destruction is much more what Camus has in mind and the artist can do just that. The time of the artist as producer in a world that values production, he argues, is at an end. For Camus, the "society based on production is only productive, not creative" (*Resistance* 273). By the end of the Second World War and with the advent of the Cold War, Camus saw modern society excelled in producing shortcuts to self-destruction. Camus develops the necessity for creation and an understanding of what it entails in his essays on art and in *The Rebel*. Artistic creation, for Camus, is the antidote to the life of mindless acquiescence and the futile destructiveness of the modern. His anarchist tendencies are closest to those of Tolstoy who wrote that "the efforts of those who wish to improve our social life should be directed towards the liberation of themselves from national governments, whose evil, and above all whose futility is in our time becoming more and more apparent" (qtd. in Woodcock: 225).

The kinship of ideas between Camus and anarchists is critical to my analysis and we can find an anarchic base to Camus's emphasis on creation. In a notebook entry from the summer of 1947 Camus reminds himself to look at the journal "*Crapouillot*: issue on Anarchy. Tailhade: Recollections of a Prosecutor. Stirner: *The Ego and His Own*" (*Notebooks 1942-1951* 163). The reference to Max Stirner is interesting because when we think about Camus's notions of human community, Stirner might, in fact, come to mind. In *The Ego and His Own*, Stirner distinguishes what he calls "association" from other collective forms. "The goal to be achieved is not another State (the 'people's State' let's say)," wrote Stirner in 1843, "but rather association, *the ever-fluid, constantly renewed association of all that exists*" (qtd. in Guerin: 21; my emphasis). Stirner makes clear that by the time an association hardens into a political party or a society or a community it is only an "association in an arrested state" (qtd. in Guerin: 22). Without overstating what Camus could have derived from Stirner, I think that Stirner's sense of human living together as a perpetual re-association that has to be constantly recreated would have great appeal for the French-Algerian. It was clear to Camus that the façade of civilization, that it is settled and permanent, was not only false but deadly. The human being must be creatively engaged in his or her life, his or her environment but he or she must also recognize limits—the most notable of these being the life of another. Here too Stirner's conception would have appealed; his view of association and practice was not one of license, but one in which "some limitation upon freedom is inevitable everywhere" (qtd. in Guerin: 23).

Except for vague affiliations with the Left, postwar Camus lacked confidence in any particular political program. The very notion of a "political program" suggests the hardening of ideas into settled dogma and here enters Camus's work in *The Rebel*. The "anarchic" tendencies in his political thinking arise particularly in the aftermath of his publication of *The Rebel* for two reasons. First, the clarity of purpose and mission he experienced during the Second World War, particularly during his time with the Resistance, had been obliterated by the end of the 1940s. The purges, the emerging conflict between East and West, colonialism, and the split with Sartre all suggested to Camus that current political conflicts had become two sides of the same coin and that political engagement in its current form meant choosing death over against an idea he had of human being. Second, these same experiences revealed to Camus that politics, seen as the province of certainties embodied in collectivities and ideologies, was exceedingly dangerous. As Camus experienced during the purges, collectivities and ready-made political and philosophical programs take us away from ourselves and the complexities of human living together. Camus's response is to turn to our responsibilities as individual human selves by refashioning politics in terms of ethics, that is, in terms of individuals risking interaction with one another and with the world in which they move around. There are decided shortcomings to the resulting approach; not the least of these is that the refusal to choose sides after *The Rebel* on for instance, the Algerian question, also implicates Camus in seemingly needless deaths. But death awaited either choice and, for all the shortcomings in Camus's political positionings, as Tony Judt has argued, Camus may well have been right not to choose and to demand better of human beings. Whatever his contemporaries and critics thought of his positions, it is worthwhile to consider how this might have been so.

We can make some generalizations about Camus's postwar politics and see that his attempt to get beyond what his contemporaries—either Left or Right--viewed as acceptable orthodoxy put him in a rather anarchist frame. In 1949, the pacifist and anarchist Alex Comfort identified Camus as "on the verge of something that is already filtering into psychiatry—the identity of the ruler in modern urban cultures with the criminal, the fact that power and authority as we know them, with their most important activity, killing, whether by atom bomb, gallows, or assassination, are themselves the enemies of society and that power itself is the Plague, with which we all, through acquiescence, are infected" (118). I think Comfort's insight here is telling of Camus's affiliation with a non-violent form of anarchism. While we have already seen Camus's critique of modern politics, I now want to place his thought in a more particularly anarchist space. Camus's political thought takes two quasi-anarchist tacks. The first is the one with which he spends the most time: his rejection of modern politics in most of its contemporary forms. This rejection includes a rejection of the State, a Tolstoyan rejection of the use of violence and terror as means to the ends of politics, a rejection of the false conceptions of freedom that littered his political environment, and, as I have suggested, his rejection of all ready-made political programs in which the thinking is done for you and the responsibility lies with others or in abstract principles. The second tack is the more positive embrace of the possibility of an alternative form of political life. This move depends upon his faith in the human creative capacity and translates into a willingness to risk spontaneity in defense of freedom and justice. The last two terms, Camus argues, along with other critical terms like democracy, must be reclaimed from their abusive keepers on the Left and the Right. The risk Camus is willing to take is the *differentia specifica* of the anarchist that Woodcock articulated above: the belief "in a moral urge powerful enough to survive the destruction of authority and still to hold society together in the free and natural bonds of fraternity" (15).

For Camus, like anarchists from Stirner to Kropotkin to Comfort, the State had become the focal point of political society. Contemporary political society is "despicable," Camus wrote to Gabriel Marcel in 1948, because it values "abstractions and terrors" more highly than "human love" (*Resistance* 78). The law, revered in the West as protector of the individual against the mechanism of the State had long since ceased to perform this function. A product of the State, the law had come to serve only its creator. Camus's early postwar hopes that socialism would move human political community beyond the need for the apparatus of the State proved disastrously shortsighted. "There is no need for anyone to show us that socialism can, quite as well as capitalism, foment wars," Camus told an interviewer in 1957. "All it takes is a little will to power and there is scarcely any nation without that" (*Resistance* 167). Furthermore, Camus was convinced that neither Left nor Right could wrest itself free from the imperious presence of the State. Both were overly reliant on its mechanisms and on its promises. Camus's disgust assumes particular critical power in his 1957 essay "Reflections on the Guillotine." Government officials had taken over the role of executioner from the priests and with similarly amorphous justifications. "Hence," Camus wrote, "our society must now defend herself not so much against the individual as against the State" (*Resistance* 227). The death penalty in general and the guillotine in particular were, for Camus, the perfect image of modern politics as death. By exchanging murder for murder, the State could not argue it protected its citizens from death. "There will be no lasting peace either in the heart of individuals or in social customs," Camus held, "until death is outlawed" (*Resistance* 234).

Nor could the willingness to use death as an instrument of political power be separated from the fundamental weakness in any politics, particularly a politics based on the existence of the State. Violence, Camus had come to see, is the failure of politics.

No cause, Camus lectured the two sides in the Algerian Civil War, "justifies the death of the innocent" (*Resistance* 134). To be sure, Camus had to live with the purge experience even as he denied the efficacy of death in politics, but Camus was analytically prepared for the task. Comfort argued that Camus's work expressed keen "clinical insight" which "hands us our own experience with the reasons which make it what it is" (118). Camus's political experience and his understanding of human psychology kept him from being naïve enough to believe in the possibility of a political world—or any other kind of world—totally free of violence or even murder. In "Reflections," Camus was clear that the "death instinct" was as fundamental in human beings as the instinct to live and that the two were rarely in equilibrium. Rather, the two instincts were "variable forces constantly waxing and waning" (*Resistance* 190-91). It was the role of the law as reason to protect its people from the death instinct and nourish the instinct to life. But the law that serves the State is interested only in ridding itself of its complicity in its own failures manifest in violence and murder. On Camus's argument, murder, even legalized murder, feeds the death instinct, offering not protection, but quite the wrong object lessons. In the preface to his Algerian Reports, Camus argued that

> When violence answers violence in a growing frenzy that makes the simple language of reason impossible, the role of the intellectuals cannot be, as we read every day, to excuse from a distance one of the violences and condemn the other. This has the double result of enraging the violent group that is condemned and encouraging to greater violence the violent group that is exonerated." (*Resistance* 116)

An environment so informed with violence ceases to function as anything worthy of the mantle "human." The responsible observer, let alone participant, has an obligation to refuse to participate in this spiraling of violence, but this refusal, as Camus discovered, carries with it a significant burden, including implication in death itself.

Along with the perversion of political power and institutions went perversions in the way human beings think of themselves as political actors and the language they use to explain themselves.

In his "Homage to an Exile," Camus diagnosed what he identified as a "decrease in liberal energies, the prostituting of words, the slandered justification of oppression, the insane admiration of force" prevalent in his political environment. The crippling of intelligence, the pitting of labor against intellect and vice versa, and the abuse of terms like freedom all served to divide his contemporaries from one another and to distract them from their real object: "the unconditional defense of freedom" (*Resistance* 105). Instead, they embrace political parties and programs uncritically:

> In short, all flee responsibility, the effort of being consistent or of having an opinion of one's own, in order to take refuge in the parties or groups that will think for them, express their anger for them, and make their plans for them. Contemporary intelligence seems to measure the truth of doctrines and causes solely by the number of armored divisions that each can put in the field.
> (*Resistance* 100-01)

The implications here are obvious. Group think is no think and rather than concern themselves with the conditions of human beings and their fates, political collectives rest content in justifying themselves through doctrines. The emphasis on doctrines or on discourse—and Camus has some of his former friends in mind-- allows for the embrace of abstractions over actions. When actions are taken, they are of an "all or nothing" bent, in the name of some State or ideology or both, and critical consideration of their implications is dismissed out of hand. "Thenceforth," Camus writes, "everything is good that justifies the slaughter of freedom, whether it be the nation, the people, or the grandeur of the State" (*Resistance* 101).

Despite this bleak picture and this trenchant set of criticisms which were echoed not only in the anarchists but also in the work of people like Hannah Arendt, Camus still could not find it in himself to abandon human being. Consequently, his critical work embraces the possibility of an alternative, that is to say, "creative" form of political life. On October 26, 1954 we find Camus rejecting the destructive dimensions of any reactionary political effort while we also hear the faint resonances of the anarchic in his tone and aspirations:

The opposite of reaction is not revolution, but creation. The world is in an unending state of reaction and thus unendingly in danger of revolution. What defines progress, if it is such, is that without compromise, creators of all kinds triumph over the mind, over reaction, and over inactivity without revolution being necessary. When there are no more of these creators, revolution in inevitable. (*Notebooks 1951-1959* 111)

Creating without compromise is marks Camus's refusal to be hemmed in by the absence of imagination characterizing his world (and ours). For this alternative model of creative politics, he turns to the role and function of the artist and writer who, Camus believes, "becomes unreal if he remains in the ivory tower or sterilized if he spends his time galloping around the political arena" (*Resistance* 238). The artist, Camus told an interviewer in 1957, must share the misfortune of his time *and* tear himself away "in order to consider the misfortune and give to it a form. (*Resistance* 238). Camus had shared in the misfortune of his time and in many ways did so until his death. He had resisted discernible evil, had blood on his hands which may or may not have been innocent, had turned away from violence and murder as political practices, and was now being pilloried because he attempted to give that misfortune a form from outside the warp and woof of ideological European politics. It was not the life of the artist that Camus appeals to here. What interests him is the artist who creatively articulates his or her encounter with the world in all its beauty and madness. It is the artist who is not only willing but compelled to rearticulate his or her encounters with that changing reality. In other words, the artists imposes a unity on the world, but it is the unity of a moment and of a particular experience. It is the unity of the aphorism which, once uttered, is abandoned by its author in all its beauty and mystery. Creativity without finality is the object of Camus's idea of the artist's role and, given that his political environment was run through with efforts at finalizing human existence, we can see easily how this view would appeal to Camus in his rethinking of political life.

The attentive reader is entitled to ask why a creative approach to political existence would not simply recreate the difficulties of the era Camus had lived and was living through. What keeps creativity from fashioning the concentration camps in new and more efficient forms? The only answer, and it is as unsatisfactory as it is challenging, is that there is a faith at work in Camus. This faith in individual human beings if not their creations resembles the challenge that anarchists of all stripes make of the world in which they find themselves. "Our faith," Camus said in a lecture on Hungary in 1957, "is that throughout the world, beside the impulse toward compulsion and death that is darkening history, there is a growing impulse toward persuasion and life, a vast emancipator movement called culture that is made up both of free creation and free work" (*Resistance* 164). The free creation Camus sought after *The Rebel* persisted as an aspiration not only for himself but for the oppressed. Tapping into the creator within human beings, Camus believed, would render not only the tools and resolve of liberation, but engender the necessary respect for the limit represented by the human other.

In Camus, creative action does not take us out of this world. If anything, it grounds and recommits us to reshaping this life by overcoming the forms in which it is imposed upon us. The properly creative artist can be a model of free, just human life. The seeds of this position had been sown in *The Rebel*, where Camus argued that artistic creation was the site of a "living transcendence, of which beauty carries the promise, which can make this mortal and limited world preferable to and more appealing than any other" (*The Rebel* 258) The creative artist's work, then, is *in this world*. What that work transcends is a condition in which the choices we make are not of our own making. We are called away both from ready answers and evasions, to engage difficulties of expression and articulation. In the artistic process as Camus finds it, the artist expresses his humanity in the work of choosing what to create, how to create it, the material to be used and the experience to be articulated. In other words, the transcendence of which Camus speaks is immanent. Some may see the work as a means of communication (ala Tolstoy), but for Camus's artist the product has the ethereal reality of the spoken word, the conciseness and mystery of the aphorism. Once articulated, the experience of the artist demands rearticulation. It is the process, then, and not the product *per se*, that makes artistic creation a possible site of Camus's version of authentic humanity.

We can, then, find Camus's anarchist tendencies in his notion of *style*. In the act of creation, it is "style" that limits the artistic reconstruction of the world. "The unity in art," Camus writes in *The Rebel*, "appears at the limit of the transformation that the artist imposes on reality" (*The Rebel* 268). Style is this limit imposed by the artist "by his language and by a redistribution of elements derived from reality" (*The Rebel* 271). Style is the means through which the artist gives form to reality. "In this domain," Camus argues, "as in others, any unity that is not a unity of style is a mutilation. Whatever may be the chosen point of view of an artist, one principle remains common to all creators: stylization, which supposes the simultaneous existence of reality and of mind which gives reality its form" (*The Rebel* 271). Style is the consistent application of the creative sensibility to the problem of disordered reality. The essential element of the act of creation is maintaining the interaction of form and content. This same tension must be articulated and maintained by the actor who would transform the social world: "It is the same thing with creation as with civilization," Camus writes, "it presumes uninterrupted tension between form and matter, between evolution and the mind, and between history and values" (*The Rebel* 270-71)

How does style translate into social action—and what makes it anarchistic? Camus's novel *The Plague* offers the best illustration. Following his regretted support of the purges in post-occupation France, Camus wants no part of a social or political program—ala some forms of anarchism--focused on the destruction of the existing order. On his view, the existing order is eating itself. As I've noted, Camus's work—literature, philosophy, journalism--is run through with descriptions and discussions of the self-destructive tendencies of modernity. The task, the real challenge, is to preserve the human in the context of that cultural self-immolation. Camus's faith is not in human institutions, but in the human capacity for self-creation (style).

Camus's anarchistic commitment is that self-creation does not lead to nihilism, but to a common community of creative human beings. Artistic creation he sees as a solitary endeavor that finally points toward collective human existence. It is the same with civilization as it is with art for Camus. The obligation to respond to crisis in the name of human being is most clearly articulated in *The Plague*. In the novel, we find Dr. Rieux and the other members of the improvised sanitation squads—who diagnose, treat, and generally provide for the sick and dying--fighting a seemingly impossible battle against a seemingly intractable enemy. Yet each member of the group maintains a commitment to a human ideal—life in the face of death. Dr. Rieux, who organizes the groups, remains committed to the work of the doctor in service of humanity rather than God or state. Jean Tarrou commits himself to the tiny details that make up the mosaic of human existence, in revolt against any conception of the need for a larger meaning of things. Tarrou is of particular interest to Alex Comfort who writes that in Tarrou we find the remedy for the plague of "irresponsibility." Comfort reads Tarrou's contention "'that on this earth there are pestilences and there are victims, and it's up to us, so far as possible, not to join forces with the pestilences'" (Comfort 118-119) as a call for resisting the plague without participating in it. Comfort argues that in Camus's work, we must ignore the promises of the pestilences, "whether they offer us democracy, peace, freedom, national greatness, justice, righteous punishment, or anything else as cover for the filth of their sores and the stink of their murders" (Comfort 119). But saying "no" to the plague asserts only part of our humanity in the novel. One must also carry on and this means not destroying one's environment just for the sake of its destruction. Thus Joseph Grand is the nominal hero of the novel. He works his job as a bureaucrat, makes his contribution to the sanitation squad, and maintains his commitment to love—of the novel whose first line he cannot get just right and, we finally learn, of his former wife to whom he finally writes.

Camus never claims to be an anarchist, but there is in his late work an appreciation for the improvisation of anarchic style and a grudging if very real faith in human being as such. In *Anarchist Modernism*, Allan Antliff argues that in the early twentieth century "the liberated creativity fueling anarchist modernism anticipated similarly creative—and liberating—rebellions against capitalism

(215).   Though anarchist modernism "withered" its impulses remained in the work of later artists like Camus. In an effort to break free of the constraints of ideology, of –isms, of the fetish for progress and the certainty of answers that had led to the murder of 50 million human beings in twenty five years, Camus argued that the impulse to create had to be reclaimed, not according to an established orthodoxy, but, in one of Camus's many Nietzschean moments, according to the needs of the human creator.

> The free artist is the one who, with great effort, creates his own order. The more undisciplined what he must put in order, the stricter will be his rule and the more he will assert his freedom... Art lives only on the constraints it imposes on itself; it dies of all others. Conversely, if it does not constrain itself, it indulges in ravings and becomes a slave to mere shadows. (*Resistance* 268)

As in the relationship between the political actor and the world, there is an interplay here between the artist and his world. Freedom entails discipline of self—but discipline the creative self imposes on itself.   This ascetic discipline, along with a commitment to one's own authenticity and a mindfulness of the limits the world imposes upon one, is the difference between anarchism and nihilism. We should stop short of labeling Camus an anarchist as he never so labels himself.   But at the end of "Create Dangerously," we find what might easily be taken for the coda of a nonviolent anarchist:

> Some will say that...hope lies in a nation; others, in a man. I believe rather that it is awakened, revived and nourished by millions of solitary individuals whose deeds and works every day negate frontiers and the crudest implications of history. As a result, there shines forth fleetingly the ever threatened truth that each and every man, on the foundation of his own sufferings and joys, builds for all. (*Resistance* 272)

Note

[1] I am very grateful to Allan Antliff of the University of Victoria who suggested and encouraged my pursuing this relationship between Camus's work and the aspirations of a non-violent anarchism. I also want to thank James Gifford of Fairleigh Dickinson University in Vancouver for pointing me to the link between Camus and Alex Comfort. Finally, the input of the participants in the "Modernism's Anarchisms" seminar at the 2009 *Modernist Studies Association* meeting in Montreal, Quebec, Canada—organized by Allan and James—was invaluable in conceptualizing this work.

Works Cited

Antliff, Allan. *Anarchist Modernism: Art, Politics, and the First American Avant-Garde.* Chicago: U of Chicago P, 2001.

Bronner, Stephen Eric. *Camus.* Chicago: U of Chicago P, 1999.

Camus, Albert. *Notebooks 1942-1951.* New York: Paragon House, 1991.

Camus, Albert. *Notebooks 1951-1959.* Chicago: Ivan R. Dee, 2008.

Camus, Albert. *Between Hell and Reason: Essays from the Resistance Newspaper Combat, 1944-47.* Translated by Alexandre de Gramont. Hanover: Wesleyan UP, 1991.

Camus, Albert. *Caligula and Three Other Plays.* New York: Knopf, 1958.

Camus, Albert. *Resistance, Rebellion, and Death.* Trans. Justin O'Brien. New York: Knopf, 1960.

Camus, Albert. *The Stranger.* Translated by Matthew Ward. New York: Vintage, 1989.

Camus, Albert. *The Plague.* 1948. Translated by Stuart Gilbert. New York: Vintage, 1991.

Camus, Albert. *The Myth of Sisyphus and Other Essays.* Translated by Justin O'Brien. New York: Vintage, 1991.

Camus, Albert. *The Fall.* Translated by Justin O'Brien. New York: Vintage, 1991.

Camus, Albert. *The Rebel: An Essay on Man in Revolt.* Translated by Anthony Bower. New York: Vintage, 1991.

Carroll, David. *Albert Camus the Algerian: Colonialism, Terrorism, Justice.* New York: Columbia UP, 2007.

Comfort, Alex. *Writings against Power and Death*. London: Freedom Press, 1994.

Crowley, Martin. "Camus and Social Justice." In *The Cambridge Companion to Camus*. Edited by Edward J. Hughes. Cambridge: Cambridge UP, 2007. 93-117.

Guerin, Daniel. *No Gods No Masters: An Anthology of Anarchism*. Translated by Paul Sharkey. Oakland: AK Press, 2005.

Isaac, Jeffrey. *Arendt, Camus and Modern Rebellion*. New Haven: Yale UP, 1992.

Judt, Tony. *The Burden of Responsibility: Blum, Camus, Aron, and the French Twentieth Century*. Chicago: U of Chicago P, 1998.

Marin, Lou. "Camus and Gandhi: Essays in Political Philosophy in Hammarskjold's Times." *Critical Currents: Occasional Paper Series of the Dag Hammarskjold Foundation*. Uppsala: Dag Hammarskjold Foundation 3 (2008): 9-22.

Miller, Tyrus. *Late Modernism: Politics, Fiction, and the Arts Between the Wars*. Berkeley: U of California P, 1999.

Sprintzen, David. *Camus: A Critical Examination*. Philadelphia: Temple UP, 1988.

Wood, Paul. "Modernism and the Idea of the Avant-Garde." In *A Companion to Art Theory*. Edited by Paul Smith and Carolyn Wilde. New York: Blackwell Publishers, 2002. 215-228.

Woodcock, George. *Anarchism: A History of Libertarian Ideas and Movements*. 1962. New York: World Publishing, 1971.

CHAPTER EIGHT

Can One Be Harmed Posthumously?

Jack Lee

Can we be harmed by posthumous events? Can an event that occurs after a person's death count as a harm to her? It seems that plausible, well-considered arguments can be presented to support either affirmative or negative answers. However, as Ernest Partridge suggests, this question appears to be such that no answer can put us fully at ease. [1]

Certainly the Epicureans would advocate the negative answer. They would argue that it is impossible for one to be harmed posthumously. Epicureans would ask: Who exactly is the subject of the alleged *posthumous* harms? Is it the "*post*-mortem person" mouldering in his grave? But a decaying dead corpse can obviously not be harmed. It is only a mere thing. [2] This position seems reasonable.

On the other hand, we are also inclined to feel that one can really be harmed posthumously. Consider the following relevant case:

> After John's death, an enemy cleverly forges documents to "prove" very convincingly that John was a philanderer, an adulterer, and a plagiarist, and communicates this "information" to the general public that includes his widow, children, and former colleagues whose good opinion he coveted and cherished. [3]

In the face of this case, our intuition tells us that John has been badly harmed. I take it that John has been harmed by such libels. That is, I believe that a person can still be harmed by some events occurring after his death. However, in order to affirm the validity of my belief that a person can be harmed posthumously, it is necessary to offer further theoretical support. Let us now explore this theoretical reason.

Joel Feinberg correctly points out that some of a person's interests can survive the death of the person and are still capable of being blocked. [4] A person is able to have different kinds of interests. Some kinds of interests die with the person. They are the interests which can no longer be dashed or promoted by posthumous events. These include most of one's self-regarding (dependent) interests, those based, for example, on personal achievement and personal enjoyment. [5] However, there are still some other kinds of interests which can survive their owner's death, and can be defeated or promoted by events subsequent to that death. In other words, the fulfilment and thwarting of certain kinds of interests may still be possible after a person's death. [6] For example, a person's reputation or the well-being of her loved husband can still be harmed or promoted by events subsequent to her death.

Accordingly, it appears that the claim that the fulfilment and thwarting of certain kinds of interests may still be possible after a person's death is plausible. However, this claim is seriously challenged by Stephen E. Rosenbaum as follows:

(1) All interests are the interests of some person or another. In other words, only if there is a subject to be the subject of surviving interests, does it make sense to grant that there are surviving interests.

(2) Death is the cessation of one's existence, the first moment of a state of nonbeing. That is, after death, there is no longer a subject left.

(3) Therefore, there are no surviving interests. [7]

This is indeed a very powerful argument against the possibility of "posthumous harms." [8] We are obliged to admit both (1) and (2). We should accept that all interests are the interests of some person or another. And it is true, in a sense, that after death there is no longer a subject. In short, this argument *does* pose a problem for the claim that a person's interests can survive the death of the person and thus can be frustrated. Call this problem the *missing subject problem.*

I maintain that the solution of the missing subject problem is an essential task for writers who have strong theoretical incentives for defending the position that a person can be harmed posthumously. In other words, in order to justify the position that a person's interests can survive the death of the person and thus can be frustrated, our task, given (1) and (2), would be: *To find out where the subject of these surviving interests, and thus of the corresponding harms, is located.*

Clearly, these surviving interests are not themselves the true subjects of these surviving interests, and thus of the corresponding harms, because that suggests a bizarre ontological reification, as if each interest were a little person in its own right. [9] Likewise, it is also difficult to accept the view that some Absolute Mind is the true subject of these surviving interests, and thus of the corresponding harms. What we are concerned about here is certainly *not* whether some Absolute Mind can be harmed by some events occurring after someone P's death, but rather whether P can be harmed by some events subsequent to P's death.

Charles Tandy recently suggests a new principle of personhood to try to locate the true subject of these surviving interests, and thus of the corresponding harms. He writes:

> Persons can indeed be harmed without experiencing harm; moreover, persons can indeed be harmed after death. A dead person is a (dead) person. Considerations above suggest to me the following, a new principle of personhood: "Once a person, always a person." [10]

He continues to explicate:

> Dead persons exist now as real facts about the past. The past is a fixed (determined) unity that *will always exist* even if the universe dies...
> Catterson presents a plausible account of time where the past exists in the present. Indeed, Catterson argues that "...there could be no possible world where the dead do not exist." Accordingly, both dead persons and living persons have no choice but to be presently existing. A

presently existing dead person may be characterized as person-identity (fully complete and up-to-date) *fact-information* that: (1) lacks a mind that functions as a living person at the present moment (e.g., is an "experiential blank"); and, (2) lacks a body that functions as a living person at the present moment. [11]

Accordingly, Tandy concludes:

It is false that a dead person does not presently exist *at all*. Thus there is no *missing subject problem*. It is the (presently existing) dead person (fact-information, not cremated ashes) who is the subject of (present) posthumous harms *and benefits*. [12]

To Tandy's position above, the Epicureans would challenge: "Death is the cessation of one's existence, the first moment of a state of nonbeing. After death, there is *no longer* a subject left." I support this Epicurean view here. To explicate this view, let us first suppose that P was born at t_0 and died at t_1. In this case, I believe:

(1) There exists the living, breathing P between t_0 and t_1.
(2) From t_1 on, there is only P's remainders in whatever form (e.g., his corpse, his cremated ashes, or even his reputation).

However, Tandy believes: (i) there exists the living, breathing P between t_0 and t_1, and (ii) from t_1 on there is in addition to P's remainders (e.g., *his* corpse, *his* cremated ashes, or even *his* reputation) *a dead P*.

The notion of a person at death (and thereafter) is problematic. In other words, it is absurd for Tandy to claim that there still exists a dead person *at and after* death; "it is the presently existing dead person who is the subject of (present) posthumous harms." Tandy gives no metaphysical explanation to support this notion except saying:

...the fact that person P really existed in the past as a living person means that it really is empirically possible for person P to exist as a living person. (This is something that cannot be said of fictional- or fantasy-characters.)... [13]

In fact, "the presently existing dead person" is seen by Tandy as (fully complete and up-to-date) person-identity fact-information. Understood as "fact-information," *the dead person* does exist now in a sense. However, as Tandy contends, it lacks a mind that functions as a living person at the present moment. Besides, it lacks a body that functions as a living person at the present moment too. Given these two "lacks," how can *the dead person* be a subject of these surviving interests, and thus of the corresponding harms. Or, how can *the dead person* be a PERSON?

Tandy claims that P can be harmed by some events occurring after his death. Suppose that P is harmed by event S happening at t_2 (after his death). Surely, Tandy would assert that P is harmed at t_2 (after his death). It is true that P's interests (e.g., his reputation or the well-being of his children) is frustrated at t_2. However, it is *not* true that P is harmed at t_2. For at t_2, there is no more a subject (or person) to be harmed.

In short, there is no so called "the presently existing dead person" in terms of personhood. Therefore, Tandy's proposal is problematic.

Julian Lamont also suggests a proposal to try to locate the subject of posthumous harms. He writes:

> ...a person can have properties at times when she does not exist. This assumption seems plausible (e.g. 'Frege has become famous this century' does not seem particularly problematic)... [14]

He elucidates this point:

> To say that Frege became famous in the twentieth century

is not to postulate a mysterious entity, some form of currently existing Frege who has continued after death...To say that Cottlob Frege became famous after his death, is simply to say that the subject, a German logician who lived from 1848-1925, became famous after his death. [15]

Strictly speaking, it is not correct to claim, "a German logician who lived from 1848-1925, became famous after his death." This German logician, who lived from 1848-1925, is the *pre-mortem* Frege. And this *pre-mortem* Frege was completely annihilated at his death (in 1925). I suggest that the statement that "Frege has become famous this century" is merely a convenient way to express a more complex but precise statement, such as "the *pre-mortem* Frege's reputation is highly praised this century" or "the *pre-mortem* Frege's name is well recognized this century." Since the *pre-mortem* Frege has an interest in his posthumous good reputation, the *pre-mortem* Frege *is* benefited by *his* reputation's being highly praised now (or, put loosely, by his being famous now). However, it is not the case, as Lamont would maintain, that the pre-mortem Frege is *now* benefited by his reputation's being highly praised now. [16] Lamont writes: "To say that Cottlob Frege became famous after his death, is simply to say that the subject, a German logician who lived from 1848-1925, became famous after his death." But how can the subject (who lived from 1848-1925) become famous after his death, provided that the subject did *not* exist any more at and after his death? In short, there is no form of currently existing subject who has continued after death.

I suppose that Lamont has here conflated two very different notions—"an ontological subject" and "a linguistic subject." In this case, the ontological subject (i.e., the pre-mortem person) existed roughly from 1848-1925 (i.e., before *his* death); whereas it was annihilated at *his* death. However, for convenience, after his death, we still use a linguistic subject to refer to the annihilated ontological subject. But note that this linguistic subject cannot be harmed or benefited at all. That is to say, it is of no ontological meaning. Strictly speaking, to claim that "Frege has become famous this century" is to make a category mistake (commit a categorical confusion).

Accordingly, Lamont's proposal is compromised too. But then who is the real subject of posthumous harms?

To find out the true subject of these surviving interests, and thus of the corresponding harms, it is useful to examine Pitcher's illustration of two ways of describing a dead person.

Pitcher points out that there are two different ways a person might adopt if he sets out to describe a friend of his who is now dead: (1) He can describe the dead friend as a living person—a description of an *ante*-mortem person after his death,
(2) He can describe the dead friend as he is now, in death (mouldering, perhaps, in a grave)—a description of a *post*-mortem person after his death. [17]

Apparently, it would be utterly absurd to claim that the interests harmed by events that occur after the moment a person's non-existence commences are interests of the decaying body he left behind. Given this, it is reasonable for us to claim that the subject of the surviving interests, and thus of the corresponding harms must be the *living* person who no longer is with us—an *ante*-mortem person referred to after his death. Indeed, although both ante-mortem and post-mortem persons can be described after their death, only ante-mortem persons can be harmed after their death. And it is ante-mortem persons who are the subject of posthumous harms.

If the subject of posthumous harms is the ante-mortem person, then it is deduced that this subject is harmed before his death. Indeed, it is the *living* person (roughly between his birth and death) who is harmed by the posthumous harm-event. Although the posthumous harm-event happens after his death, he *is* harmed before death. Strictly speaking, the ante-mortem person is not harmed by the posthumous harm-event; he *is* harmed by the state of affairs that his (surviving) interests are going to be frustrated after he dies. In a sense, the posthumous harm-event makes this state of affairs true. Suppose Allen's loved son was killed today in an air crash just five minutes after Allen's death (say, at t_2). In this case, most of Allen's good friends would feel very sorry for Allen. Of course, this is a harm for Allen. If Allen were alive now, he would be very sad. But when *is* Allen harmed? Surely, the harm-event happens at t_2 (after his death). However, Allen was harmed before death. In fact, the living Allen was harmed by the state of affairs that his son was going to be killed soon after his death. [18]

At this point, someone might challenge my position by asking: "How could it be possible that Allen has already been seriously harmed while that harm-event has not happened yet?" Indeed, before t_2, we still could not claim that his son was going to be killed soon after his death. For this accident had then not yet happened. However, at and after t_2, we could claim that his son was going to be killed soon after his death. And the living Allen was harmed (before death) by the state of affairs that his son was going to be killed soon after his death.

Accordingly, it is concluded that the subject of the surviving interests, and thus of the corresponding harms is the ante-mortem person. That is to say, there is *no* "missing subject problem." Given this, it is maintained that one can be harmed by posthumous events. And in such case, one is harmed posthumously *before* death.

To sum up, the above argument for the position that a person can be harmed posthumously can be concisely shown as follows:

> (i) Some of a person's interests can survive the death of the person.
> (ii) The interests which survive the death of the person may be impaired by posthumous events.
> (iii) Harm is the impairment of interest.
> (iv) Therefore, a person can be harmed by posthumous events. That is, one can be harmed posthumously.

As for the questions:

> (1) Who exactly is the subject of posthumous harms?
> (2) When is one harmed posthumously?

My answers are:

(1) The Ante-mortem Person.
(2) Before Death. [19]

References

Callahan, Joan C., "On Harming the Dead," *Ethics* 97 (1987), pp. 341-352.

Epicurus, "Letter to Menoeceus," trans. C. Bailey, in *The Stoic and Epicurean Philosophers*, ed. Whitney Jennings Oates (New York: Modern Library, 1957).

Feinberg, Joel, "Harm to Others," in *The Metaphysics of Death*, ed. John Martin Fisher (Stanford: Stanford University Press, 1993).

Fischer, John Martin, ed. *The Metaphysics of Death* (Stanford: Stanford University Press, 1993).

Glannon, Walter, "Persons, Lives, and Posthumous Harms," *Journal of Social Philosophy* 32, No. 2 (2001), pp. 127-142.

Grey, William, "Epicurus and the Harm of Death," *Australasian Journal of Philosophy* 77, No. 3 (1999), pp. 358-364.

Grover, Dorothy, "Posthumous Harm," *Philosophical Quarterly* 39, No. 156 (1989), pp. 334-353.

Lamont, Julian, "More Solutions to the Puzzle of When Death Harms Its Victims?," in *Death and Anti-Death, Volume 1: One Hundred Years After N. F. Fedorow (1829-1930)*, ed. Charles Tandy (Polo Alto, California: Ria University Press, 2003).

Lamont, Julian, "A Solution to the Puzzle of When Death Harms Its Victims," *Australasian Journal of Philosophy* 76, No. 2 (1998), pp. 198-212.

Levenbook, Barbara Baum, "Harming Someone after His Death," *Ethics* 94 (1984), pp. 407-419.

Li, Jack, *Can Death Be a Harm to the Person Who Dies?* (London: Kluwer Academic Publishers, 2002).

Li, Jack, "Commentary on Lamont's When Death Harms Its Victims," *Australasian Journal of Philosophy* 77, No. 3 (1999), pp. 349-357.

Luper, Steven, "Mortal Harm," *Philosophical Quarterly* 57, No. 227 (2007), pp. 239-251.

Luper-Foy, Steven, "Annihilation," in *The Metaphysics of Death*, ed. John Martin Fisher (Stanford: Stanford University Press, 1993).

Marquis, Don, "Harming the Dead," *Ethics* 96 (1985), pp. 159-161.

McMahan, Jeff, "Death and the Value of Life," in *The Metaphysics of Death*, ed. John Martin Fisher (Stanford: Stanford University Press, 1993).

Partridge, Ernest, "Posthumous Interests and Posthumous Respect," *Ethics* 91, No. 2 (1981), pp. 243-264.

Pitcher, George, "The Misfortunes of the Dead," in *The Metaphysics of Death*, ed. John Martin Fisher (Stanford: Stanford University Press, 1993).

Rosenbaum, Stephen E., "The Harm of Killing: An Epicurean Perspective," in *Contemporary Essays on Greek Ideas: The Kilgore Festschrift*, eds., Robert M. Baird, William F. Cooper, Elmer H. Duncan, and Start E. Rosenbaum (Waco, Texas: Baylor University Press:1987).

Rosenbaum, Stephen E., "How to Be Dead and Not Care: A Defence of Epicurus," in *The Metaphysics of Death*, ed. John Martin Fisher (Stanford: Stanford University Press, 1993).

Waluchow, W. J., "Feinberg's Theory of 'Preposthumous' Harm," *Dialogue* 25 (1986), pp. 727-734.

Notes

1. See Ernest Partridge, "Posthumous Interests and Posthumous Respect," *Ethics* 91 (1981), pp. 243-244.

2. See Epicurus, "Letter to Menoeceus," trans. C Bailey, in *The Stoic and Epicurean Philosophers*, ed. Whitney Jennings Oates (New York: Modern Library, 1957), pp. 30-31; and Stephen E. Rosenbaum, "How to Be Dead and Not Care: A Defence of Epicurus," in *The Metaphysics of Death,* ed. John Martin Fischer (Stanford: Stanford University Press, 1993), pp. 120-122.

3. Cf. Joel Feinberg, "Harm to Others," in *The Metaphysics of Death,* ed. John Martin Fischer (Stanford: Stanford University Press, 1993), pp. 180-181.

4. Ibid., p. 176.

5. Ibid., p. 179.

6. Jack Li, *Can Death Be a Harm to the Person Who Dies?* (London: Kluwer Academic Publishers, 2002), p. 83.

7. See Stephen E. Rosenbaum, "The Harm of Killing: An Epicurean Perspective," in *Contemporary Essays on Greek ideas: The Kilgore Festschrift*, eds., Robert M. Baird, William F. Cooper, Elmer H. Duncan, and Start E. Rosenbaum (Waco, Texas: Baylor University Press:1987), pp. 214-219.

8. I take it that harm is the impairment of interest. For an extensive discussion of why harm is the impairment of interest, see Jack Li, *Can Death Be a Harm to the Person Who Dies?* (London: Kluwer Academic Publishers, 2002), pp. 67-74.

9. Joel Feinberg, "Harm to Others," in *The Metaphysics of Death,* ed. John Martin Fischer (Stanford: Stanford University Press, 1993), p. 176.

10. See Charles Tandy's book review of *Can Death Be a Harm to the Person Who Dies?*, in *National Central University Journal of Humanities* 31 (2008), p. 230.

11. Ibid., p. 231

12. Ibid.

13. Ibid., p. 232.

14. Julian Lamont, "More Solutions to the Puzzle of When Death Harms Its Victims?" in *Death and Anti-Death, Volume 1: One Hundred Years After N. F. Fedorow (1829-1903)*, ed. Charles Tandy (Polo Alto, California: Ria University Press, 2003), p. 383.

15. Ibid.

16. Ibid., p 384.

17. George Pitcher, "The Misfortunes of the Dead," in *The Metaphysics of Death,* ed. John Martin Fischer (Stanford: Stanford University Press, 1993), p. 161.
18. Note that it is not correct to claim at the time before t_2 that his son was going to be killed soon after his death. This is simply because that accident has not happened yet at that time (before t_2). However, after t_2, we can correctly claim that his son was going to be killed soon after his death. And this claim is about an actual fact which cannot be changed. Thus, it is after t_2 that we say: Before death the living Allen was harmed by his son's premature death occurring after Allen's death.

19. Part of this paper was originally published in my article "Search for the Possibility of Posthumous Harm," *Applied Ethics Review* 48 (2010), pp. 1-12.

CHAPTER NINE

The Gödelian Argument: Turn Over The Page

J. R. Lucas*

I want to start by quarrelling with Sir Roger Penrose. In 1990 the Journal of Behavioral and Brain Sciences published a large number of peer reviews of his book, The Emperor's New Mind. At the end he said in his response:

> All my adverse critics on this topic have jumped to conclusions and, in one way or another, have missed the point of what I am trying to say. None seem to have grasped the full import of the Gödelian argument. The fault is mine: I should have explained things more clearly.[1]

I have no quarrel with the first two sentences: but the third, though charitable and courteous, is quite untrue. Although there are criticisms which can be levelled against the Gödelian argument, most of the critics have not read either of my, or either of Penrose's, expositions carefully, and seek to refute arguments we never put forward, or else propose as a fatal objection one that had already been considered and countered in our expositions of the argument. Hence my title. The Gödelian Argument uses Gödel's theorem to show that minds cannot be explained in purely mechanist terms. It has been put forward, in different forms, by Gödel himself, by Penrose, and by me.

Gödel gave the Gibbs lecture on Boxing Day, 1951, twenty years after he had discovered his theorem, to an audience in the United States, but the lecture was not published or much known about,

* Reprinted by permission of J. R. Lucas
<http://users.ox.ac.uk/~jrlucas/Godel/turn.html>
Copyright © J. R. Lucas

until after his death. It appeared in the third volume of his Collected Works, which was published in 1995. I read a paper, "Minds, Machines and Gödel" to the Oxford Philosophical Society on October 30th, 1959, which was subsequently published in Philosophy, 36, 1961, pp.112-127, and reprinted in Kenneth M. Sayre and Frederick J. Crosson, eds., The Modeling of Mind, Notre Dame, 1963, pp.255-271; and in A. R. Anderson, Minds and Machines, Prentice-Hall, 1964, pp.43-59. In 1970 I published a fuller version in my The Freedom of the Will, in which I went in greater detail into objections to the Gödelian argument and how they should be answered. Roger Penrose had been thinking about the problem for many years before he published his The Emperor's New Mind in 1989, which attracted much attention. In 1994 he published Shadows of the Mind in which he countered some of the criticisms that had been levelled against the earlier version of the argument. There have been a large number of discussions, mostly critical, of both his and my versions of the argument.

Gödel argues for a disjunction: an Either/Or, with the strong suggestion that the second disjunct is untenable, and hence by Modus Tollendo Ponens that the first disjunct must be true.

> So the following disjunctive conclusion is inevitable: Either mathematics is incompletable in this sense, that its evident axioms can never be comprised in a finite rule, that is to say, the human mind (even within the realm of pure mathematics) infinitely surpasses the powers of any finite machine, or else there exist absolutely unsolvable diophantine problems of the type specified . . .**2**

It is clear that Gödel thought the second disjunct false, and that he was implicitly denying that any Turing machine could emulate the powers of the human mind.**3**

Roger Penrose uses not Gödel's theorem itself but one of its corollaries, Turing's theorem, which he applies to the whole world-wide activity of all mathematicians put together, and claims that their creative activity cannot be completely accounted for by any algorithm, any set of rigid rules that a Turing machine could be programmed to follow.

I used Gödel's theorem itself, and considered only individuals, reasonably numerate (able to follow and understand Gödel's theorem) but not professional mathematicians. I did not give a direct argument, but rather a schema, a schema of disproof, whereby any claim that a particular individual could be adequately represented by a Turing machine could be refuted. My version was, designedly, much less formal than the others, partly because I was addressing a not-very-numerate audience, but chiefly because I was not giving a direct disproof, but rather a schema which needed to be adapted to refute the particular case being propounded by the other side. I was trying to convey the spirit of disproof, not to dot every i and cross every t, which might have worked against one claim but would have failed against others. I also chose, in arguing against the thesis that the mind can be represented by a Turing machine, to use Gödel's theorem, not Turing's. This was in part because, once again, Gödel's theorem is easier to get the flavour of than Turing's theorem, but also because it involves the concept of truth, itself a peculiarly mental concept.

Other differences need to be noted. Gödel was a convinced dualist. He thought it obvious that minds were essentially different from, and irreducible to, matter; one reason, perhaps, why he did not make more of his argument was that he did not feel the need to refute materialism: why waste effort flogging a dead horse? Penrose is a materialist, but thinks that physics needs to be radically revised in order to accommodate mental phenomena. Quite apart from this, he reckons physics must be developed to account for the phenomenon of the collapse of the psi function in quantum mechanics. He hopes to produce a unified theory which will be both a non-algorithmic theory of quantum collapse and accommodate the phenomenon of mind. I acknowledge the importance of both these problems, but think they are separate. Instead of trying to expand physics in order to have a physical theory of mind, I distinguish sharply two different types of explanation, the regularity explanations we use to explain natural phenomena and the rational explanations we use to justify and explain the actions of rational agents. In that sense I am a dualist. I have difficulty with the full-blooded Cartesian dualism of different sorts of substance, which was, I think, Gödel's position, and am, as regards substance only a one-and-a-halfist at most. But, as regards explanation, I am at least a dualist, indeed, a many-times-more-than-dualist.

Many objections have been raised against the Gödelian argument. Many AI enthusiasts protest that they do not work with Turing machines, and have much more complicated and subtle connexionist systems. I do not dispute that, and specifically allowed in my original article that we might one day be able to create something of silicon with a mind of its own, just as we are able now to procreate carbon based bodies with minds of their own.4 To those who say I am therefore flogging a dead horse, I reply by citing Breuel, who explains

> the reason AI researchers, for practical purposes, adhere to the idea that brains are no more than computational devices is not philosophical stubbornness but the fact that no physical process is known to exist that can be used to build a device computationally more powerful than a Turing machine, and no concrete theories of psychological and cognitive phenomena have so far required any recourse to physical mechanisms that were more powerful than a Turing machine.5

We are dealing not only with practical attempts to build machines that can insert a collar stud or tie a bow tie, but with attempts to understand the workings of the human mind; and then to show that one very widely accepted schema of explanation is unavailable is well worth doing.

A much more serious objection is based on the assumption of consistency that I make. Hilary Putnam, when I first put the argument to him in a bar in Princeton, objected that in order to be in a position to know that the Gödelian formula was true, one needed to know that the system was consistent; he maintained that Gödel's second theorem showed that this was impossible.6 Many other critics have maintained the same:

> Thus the premise that the Gödel sentence is true (and unprovable) cannot be known unless it is known that arithmetic is consistent, and no Turing machine can know the latter. So who says that humans can know it either:7

Well, Gentzen for a start. He gave a convincing proof of the consistency of Peano Arithmetic (the simple arithmetic of the natural numbers), using transfinite induction. What Gödel's second theorem showed was that a system could not be proved to be consistent within itself: we cannot prove Peano Arithmetic consistent from the axioms and by means of just the rules of Peano Arithmetic, but we can give a consistency proof---a very convincing one---applying principles from outside Peano Arithmetic. Such proofs, of course, depend on the wider set of principles being consistent, and that assumption can be called into question, the more particularly since Russell's pointing out that Frege's set theory was inconsistent.**8**

These are proper objections, but not insuperable ones. Although by Gödel's second theorem we cannot prove the consistency of a formal system within that formal system, we can argue for the consistency of a formal system by means of wider considerations. And so when Putnam raised his objection, I countered that although a Turing machine could not, without inconsistency, prove its own consistency, we could affirm that any plausible representation of a mind must be consistent, since minds were selective and were unwilling to assert anything whatsoever, which an inconsistent machine will do. I discussed the matter both then and in my article and book,**9** but Putnam in his review of Shadows of the Mind in the New York Times Book Review (***, p.7) ignores the argument, and says simply:

> Mr Lucas's mistake was to confuse two very different statements that could be called "the statement that S is consistent." In particular Mr Lucas confused the colloquial statement that the methods of mathematicians use cannot lead to inconsistent results with the very complex mathematical statement that would arise if we were to apply Gödel's theorem to a hypothetical formalization of these methods.

But I did not confuse them. I argued in some detail, considering and countering various objections that might be put forward, that the two were connected (why else would the word "consistent" be applied in each case?), and that if a Turing machine was inconsistent it could not be a plausible representation of the mind.

Since many critics are unaware of the argument,[10] and are unlikely to look back at papers published some time ago, it is worth articulating the argument afresh. It is useful to borrow the terminology of First and Second Public Examinations at a university. The Mechanist claims to have a model of the mind. We ask him whether it is consistent: if he cannot vouch for its consistency, it fails at the first examination; it just does not qualify as a plausible representation, since it does not distinguish those propositions it should affirm from those that it should deny, but is prepared to affirm both undiscriminatingly. We take the Mechanist seriously only if he will warrant that his purported model of the mind is consistent. In that case it passes the First Public Examination, but comes down at the Second, because knowing that it is consistent, we know that its Gödelian formula is true, which it cannot itself produce as true. More succinctly, we can, if a Mechanist presents us with a system that he claims is a model of the mind, ask him simply whether or not it can prove its Gödelian formula (according to some system of Gödel numbering). If he says it can, we know that it is inconsistent, and would be equally able to prove that 2 and 2 make 5, or that 0=1, and we waste little time on examining it. If, however, he acknowledges that the system cannot prove its Gödelian formula, then we know it is consistent, since it cannot prove every well-formed formula, and knowing that it is consistent, know also that its Gödelian formula is true.

In this formulation we have, essentially, a dialogue between the Mechanist and the Mentalist, as we may call him, with the Mechanist claiming to be able to produce a mechanist model of the Mentalist's mind, and the Mentalist being able to refute each particular instance offered. Many critics have failed to note the dialectical character of the argument, and have rushed in to show that the Mentalist is not in all respects superior to all minds, but has his own limitations, and can often be beaten by a mind. But if only they had turned over the page, they would have seen that I acknowledged as much, and was not making a general claim of superiority, but only a particular one of some difference in each particular case. Other critics try to avoid the dangerous dialogue by having the mechanist not show his hand.**11** He does not tell us which precise model of Turing machine represents a particular mind, nor whether or not the purported mechanist model is consistent; and raises the question whether I, or any human mind, could really fathom the immense complexity of a representation of a human brain. But then why should I? It is for the Mechanist to make good his case. I cannot be just an abstract idea of a turing machine, a generalised Turing machine, I know not what. I must be a particular definite machine. Although Benacerraf may plead ignorance, it must in principle be knowable which machine it is that purports to represent me, and whether it is consistent. And then the argument proceeds.

But still, it may be objected, the Turing machine will be fiendishly complicated. Presented with an enormous printout of gobbledygook, how could I make out what it meant, or what its Gödelian formula was? But it does not have to be just me. As Michael Dummett pointed out when I read the original paper to the Oxford Philosophical Society, I could be helped, indeed helped by all the mathematicians in the world, who might be keen to see a mind, even mine, defeat mechanism. (In this point there is a similarity with Penrose's argument, which is concerned with the output of the entire community of working mathematicians.) Also, now, of course, I could be helped by computers. Not everything has to be in machine code: we can have programs to translate into higher-level, more transparent languages. It would, admittedly, still be difficult to identify all the transformation rules, all the inferences that the machine could make, all its initial assumptions, but not impossibly so, if the mind really were a machine, and really did proceed according to some algorithmic rules. And once we had done this, and chosen some suitable scheme of Gödel numbering, we could set about calculating what the Gödelian formula for that system under that scheme of Gödel numbering must be.

But, it is sometimes further objected, in order to be confident that I can always calculate the Gödelian formula, I must have an algorithm to do it by, in which case a machine could be programmed to do it too.**12** But this is not so. It is easiest seen if we pursue a criticism of I. J. Good, who pressed home an objection I had countered, that the Mechanist might incorporate a Gödelizing operator, which indeed he could, but at the cost of making the machine a different machine and hence with a different Gödel formula. But the move could be iterated, and though at each stage the machine would be different, we should be engaged in a game of catch-as-catch-can in the transfinite ordinals.**13** And there is no algorithm for naming the transfinite ordinals.**14** Every now and again we run out of existing names, and have to devise something new. I claim that this is something we can be confident of doing. My critics claim that I am being over-confident, and that only if there were an algorithm would I be justified in maintaining that I always could go on, and there is no such algorithm. But, say I, we do not have to have an algorithm; when we run out of standard names of transfinite ordinals, we invent new ones, not according to some rule but by the exercise of ingenuity. And, more generally, once we have got the hang of the Gödelian argument, we can adapt it as necessary to the needs of the particular case, and can go on producing appropriate Gödelian formulae, improvising as necessary when the going is not entirely straightforward.

At this stage the importance of originality and creativity begins to emerge. What I have offered is not one knock-down argument but a schema of refutation, which can be seen to work in some cases, but needs adaptation to the particular case. When we consider not a particular case, but all possible cases, I cannot offer a single, all-sweeping argument, but only an approach, relying on the mind's ability to improvise as needed in new circumstance. It is a difficult schema of argumentation, and critics often try to reconstrue it as if it were a single all-encompassing argument, and then find their reconstruction faulty. In spite of my specific disclaimer,**15** they suppose that I am trying to prove that the mind is better than all computers taken together. And then it is easy to point out that anything I do can be simulated by a computer suitably programmed.**16** All I can do is to repeat my original argument, that I do not claim to be better than all computers, but can show for each particular one that I am different from it. It is the Mechanist's claim that is being evaluated, and can be outwitted in each particular case and shown by the mind it claims to represent, to be an inadequate representation of that very mind.**17**

The argument bifurcates. The Mechanists refuse to acknowledge any originality of the mind. And I cannot make them. If the only thing that will budge the Mechanists is a rule-governed inference which cannot be resisted on pain of inconsistency, then they cannot be made to see the general applicability of Gödelian arguments. All that can be done is to refute each and every particular claim they put forward. But their obduracy is unappealing. Once we get the hang of the Gödelian argument, we see that it will be applicable in all cases, though its mode of application will have to be altered to fit each case individually. There is a way of arguing that commends itself to those possessed of minds, who get the hang of the Gödelian argument, and twig that they can apply it, suitably adapted, in each and every case that crops up. Mechanists may refuse to see the general case, and, acknowledging only knock-down arguments, will have to be knocked down each time they put forward a detailed case: minds can generalise, and will realise that defeat for the Mechanists is always inevitable.

The originality required of the mind is not very great, and it may still be objected that the Gödelian argument has not done much to vindicate real creativity in the face of sceptical doubt. So far as the argument thus far adduced, it is a fair complaint. All that has been achieved so far is to "defeat the defeaters"; a line of argument, supposedly supported by the success of science, which would lead to a reductionist view of the mind, and the elimination of all originality, has been defeated. We no longer have to think of the mind as an essentially dull automaton, but may still observe that many minds are nonetheless somewhat dull, not to say boring. The Gödelian argument may refute mechanism, but leaves the woodenness of ordinary, uninspired human life untouched.**18** But, although Gödel cannot make us scintillate, he does show that scintillation is conceptually possible. He shows us that to be reasonable is not necessarily to be rule-governed, and that actions not governed by rules are not necessarily random. Many thinkers have thought otherwise, because they supposed that rational actions and decisions were so because they were in conformity with some explicit or implicit rule. Even great creative artists have a style peculiarly their own, and it is natural to think that their style is characterized by some finite description, with some parameters left undetermined, and that random choices of these parameters is what produces different masterpiece of the artist's oeuvre. But this need not be so: the disjunction between the random variation and the finished specification is not exhaustive. For in the case of First-order Peano Arithmetic there are Gödelian formulae (many, in fact infinitely many, one for each system of coding) which are not assigned truth-values by the rules of the system, and which could therefore be assigned either TRUE or FALSE, each such assignment yielding a logically possible, consistent system. These systems are random variants, all satisfying the core description of Peano Arithmetic. But among them there is one, the one that assigns TRUE to all the Gödelian formulae which is reasonable, characterizing standard arithmetic, although not more in accordance with the specification of Peano Arithmetic than any of the others. So there is some sort of reasonableness, picking out this one instantiation of the specification in preference to all the others which is reasonable and right, though not any more in accordance with the antecedently formulated rules than any other instantiation.

The critic may still complain that doing arithmetic in non-non-standard models is still a pretty boring activity; but that is not the point. Gödel's theorem shows that there is conceptual room for creativity, by allowing that to be reasonable is not necessarily to be in accordance with a rule. We can see how it works out with the style of great creative artists. Titian, or Bach, or Shakespeare, develop a style which is peculiarly their own, and which we can learn to recognise. But it is not static. Up to a point they can produce works that are variations on a theme, but beyond that point we begin to criticize, and say that they are painting, composing, or writing, according to a formula; it is the mark of second-rate artists to be content to go on doing just that, but the genius is not content to rest on his laurels, but seeks to go further, and innovate, breaking out of the mould that his previous style was coming to constitute for him. Instead of there just being a formula to which all his work conformed, he produces work which differs in some significant respect from what he had been doing, and this difference is not just a random one, but one which ex post facto we recognise as essentially right, even though it was not required by the previous specification of his style.**19** Thus, though the

Gödelian formula is not a very interesting formula to enunciate, the Gödelian argument argues strongly for creativity, first in ruling out any reductionist account of the mind that would show us to be, au fond, necessarily unoriginal automata, and secondly by proving that the conceptual space exists in which it intelligible to speak of someone's being creative, without having to hold that he must be either acting at random or else in accordance with an antecedently specifiable rule.

Notes

1. Journal of Behavioral and Brain Sciences, 13:4, 1990, p.693.

2. Kurt Gödel: Collected Works, III, ed. Feferman, Oxford, 1995, p.310.

3. It is necessary to stress this point, as many critics try to distance Gödel's thought from that of Lucas and Penrose. Although there are significant differences between the three approaches, their conclusions are substantially similar. See H.Wang, From Mathematics to Philosophy, London, 1974, pp.324-326; and Reflections on Kurt Gödel, MIT Press, Cambridge, Mass., USA, 1987, p.48, pp.117-118.

4. Philosophy, 36, 1961, p.126/The Modeling of Mind, Kenneth M.Sayre and Frederick J.Crosson, eds., Notre Dame Press, 1963, pp.269-270/Minds and Machines, ed Alan Ross Anderson, Prentice-Hall, 1954, pp.58-59.

5. Thomas M.Breuer, Journal of Behavioral and Brain Sciences, 13:4, 1990, p.657; he continues:

> Penrose's argument may be cautious first steps towards changing both of these facts, but I feel they are still much too tentative and informal to require serious reconsideration of the marriage of AI and the Turing model of computation.

6. See Hilary Putnam "Minds and Machines", in Sidney Hook, ed., Dimensions of Mind. A Symposium, New York, 1960; reprinted in A. R. Anderson, Minds and Machines, Prentice-Hall, 1964, pp. 72-97.

7. Chris Mortensen, Journal of Behavioral and Brain Sciences, 13:4, 1990, p.678.

8. George Boolos, Journal of Behavioral and Brain Sciences, 13:4, 1990, p.655:

> I suggest that we do not know that we are not in the same situation vis-a-vis ZF that Frege was in with respect to naive set theory (or, more accurately, the system of his Basic Laws of Arithmetic) before receiving, in June 1902, the famous letter from Russell, showing the derivability in his system of Russell's paradox.

Martin Davis, Journal of Behavioral and Brain Sciences, 13:4, 1990, p.660:

> . . . convincing oneself that the given axioms are indeed consistent, since otherwise we will have no reason to believe that the Gödel sentence is true. But here things are quite murky: Great logicians (Frege, Curry, Church, Quine, Rosser) have managed to propose quite serious systems of logic which later have turned out to be inconsistent.

9. pp.120-124/263-268/52-56.

10. In addition to those noted in [note] 8 above, it is worth citing:

> [1.] S. Guccione, Journal of Behavioral and Brain Sciences, 16:3, 1993, p.612: *The most conclusive and immediate objection against this argument is Putnam's (1960) observation that (following Gödel's theorem) the human mind can only demonstrate the implication: "If S is consistent, then Gs is true.";*

> [2.] Alexis Manaster-Ramer, Walter J. Savitch and Wlodek Zadrozny: *But all this - and more - depends on granting Penrose's argument, and this we should not do. The error is a small but lethal flaw in his presentation, and application, of Gödel's theorem. For Gödel does not say that a certain proposition P is true but unprovable in a formal system F, but merely that P is true but unprovable if*

F is consistent. Penrose notes that if F is inconsistent then P is provable but false, but then makes the inexplicable mistake of assuming that, "Our formal system should not be so badly constructed that it actually allows false propositions to be proved!" (pp. 107-108). Without this, only the conditional can be proved (and this can be done algorithmically!).

[3.] G. Boolos, in Kurt Gödel: Collected Works, III, ed. Feferman, Oxford, 1995, pp.296, 295: *The classic reply to these views <of Ernest Nagel and James R. Newman (1958), J.R. Lucas (1961), and Roger Penrose (1989)> was given by Hilary Putnam (1960): Merely to find from a given machine M a statement S for which it can be proved that M, if consistent, cannot prove S is not to prove S--- even if M is consistent. It is fair to say that the arguments of these writers have as yet obtained little credence.*

[4.] David J. Chalmers, Review of Shadows of the Mind, in PSYCHE: an interdisciplinary journal of research on consciousness, 2(9), June, 1995. Filename: psyche-95-2-09-shadows-7-chalmers, (an elaboration of his review in Scientific American, June 1995, pp. 117-18).

11. Paul Benacerraf, "God, the Devil and Gödel", The Monist, 1967, pp. 9-32.

12. Judson C. Webb, Mechanism, Mentalism and Metamathematics; An Essay on Finitism, Dordrecht, 1980, p.230.

13. I. J. Good, "Human and Machine Logic," British Journal for the Philosophy of Science, 18, 1967, pp. 144-147; and "Gödel's Theorem is a Red Herring", British Journal for the Philosophy of Science, 19, 1968, pp. 357-8.

14. This point is conceded by Douglas R. Hofstadter, Gödel, Escher, Bach, New York, 1979, p.475, and R. W. Kentridge, Journal of Behavioral and Brain Sciences, 13:4, 1990, p.671, who seem to think it is a criticism, not a support, of my argument.

15. pp.117-118/262-262/49-50.

16. See R. W. Kentridge, Journal of Behavioral and Brain Sciences, 13:4, 1990, p.671:
What it <this demonstration> actually shows is that we can do better than one particular algorithm H. It is easy to see how to construct a new algorithm H', a modification of H in which H'(k;k)=O, which does just as well as we do. Therefore, the example does not show that we think nonalgorithmically; all it shows is that we can prove something that one particular algorithm cannot. Adina Roskies, Journal of Behavioral and Brain Sciences, 13:4, 1990, p.682: Briefly, Penrose conflates the ability to solve an instance of a noncomputable problem with the ability to solve all instances of that problem. Only the latter would entail solving it nonalgorithmically. (my emphasis)

17. Thus J. Higginbotham, Journal of Behavioral and Brain Sciences, 13:4, 1990, p.668:
Lucas's original argument to this effect is notoriously suggestive but vague, and critical literature on it has often taken the form of making the argument more precise, and then showing that in the precise form envisaged it fails to prove the case. (Penrose notes some of this literature, but does not discuss it directly in the text.)

18. See Crispin Wright, Realism, Meaning and Truth, 2nd ed., Blackwell, Oxford, 1993, p.351.

19. For further elucidation of the difference between there being a specification to which all cases belong and there being some case that differs from an antecedent specification, but is right to do so, see J. R. Lucas, "The Lesbian Rule", Philosophy, 1955, pp.195-213.

CHAPTER TEN

The Function Of Assisted Suicide
In The System Of Human Rights

Ludwig A. Minelli*

---

* The Editor thanks Ludwig A. Minelli for reprint permission. By Ludwig A. Minelli, DIGNITAS Forch-Zurich, at the Symposium of Netherland's Association for a Voluntary End of Life (NVVE) 35th anniversary, March 28, 2008, at Amsterdam.

Dear Chairman,
Ladies and Gentlemen,

Thank you for inviting me to Amsterdam: I am honoured to be allowed this opportunity to congratulate and address the NVVE on its 35th anniversary. Thirty five years of struggling to implement what I call the «last Human Right» in your country have been quite successful and you lead the BENELUX-nations. I think you have also worked hard in the struggle towards enlightenment as expressed by IMMANUEL KANT, the sense of man's emergence from his self-imposed immaturity. But, in this respect, the struggle is not over yet; it has only just begun.

In all European states with the exception of the last two dictatorial systems – Belarus and the Vatican – the European Convention on Human Rights (ECHR) is in force, and the European Court of Human Rights in Strasburg is the powerful watchdog of this Convention.

Two of the most important Human Rights contained in this Convention are the Right to Life in article 2 and the Right to respect one's Private Life in article 8. Article 2 says that «Everyone's right to life shall be protected by law. No one shall be deprived of his life intentionally...» Article 8 says that «Everyone has the right to respect for his private and family life...»

As a consequence of article 2, contracting states have a duty to protect life by law. The Convention's intention is to protect all citizens from being killed by any other person.

On the other hand, when one asks whether a human being has the right to decide to end his or her own life, Article 8 of the Convention was interpreted by the European Court of Human Rights in its decision in the case of DIANE PRETTY of April 29th 2002 as follows:

«Although no previous case has established as such any right to self-determination as being contained in Article 8 of the Convention, the Court considers that the notion of personal autonomy is an important principle underlying the interpretation of its guarantees.»

And the Court added:

«The very essence of the Convention is respect for human dignity and human freedom. Without in any way negating the principle of sanctity of life protected under the Convention, the Court considers that it is under Article 8 that notions of the quality of life take on significance. In an era of growing medical sophistication combined with longer life expectancies, many people are concerned that they should not be forced to linger on in old age or in states of advanced physical or mental decrepitude which conflict with strongly held ideas of self and personal identity.» About four and a half years later, on November 3rd 2006, the Swiss Federal Court rendered a decision containing the following statement, here translated from German: «The right of self-determination in the sense of article 8 § 1 ECHR includes the right to decide on the way and the point in time of ending one's own life; providing the affected person is able to form his/her will freely and act thereafter.» This was the very first time that the supreme court of any of the contracting states to the Convention positively acknowledged this fundamental Human Right: the Right to Suicide, the Right of self-determination even when that meant choosing to end one's own life.»

The Federal Court's decision, therefore, acknowledged what most eminent lawyers have been telling us for many years. As early as 1992, LUZIUS WILDHABER said in his comment on Article 8 that the Right to Life does not constitute a duty to live if it entails an unacceptable cost to the individual. «With his decision on the basis of his right of personality at the moment of his death, the sufferer is free to decide by himself, when and how he would make use of his Right to Life.» By the way, LUZIUS WILDHABER was the president of the European Court of Human Rights from 1998 to 2007.

If the Right to Suicide is a Human Right, and no doubt whatsoever can be raised after this outstanding Swiss decision, we must accept that, in order to make use of this right, there must be no legal requirements other than that the person has the mental capacity needed to decide to end his or her own life. Conditions which insisted that somebody must be terminally or severely ill would interfere with the essence of that Human Right. Human Rights are, inherently, unconditional. They cannot be made subject to conditions – this is fundamental to their being meaningful for the humans who bear them.

One objection to this view might be that the decision of the Swiss Supreme Court did not say that there is also a right to have access to help with suicide, nor that the state would have to provide such help. Of course, the Swiss Federal Court has said that there is no obligation on the state to help people to commit suicide. And the European Court of Human Rights, in its DIANE PRETTY decision, outlined that:

> «It is primarily for States to assess the risk and the likely incidence of abuse if the general prohibition on assisted suicides were relaxed or if exceptions were to be created. Clear risks of abuse do exist, notwithstanding arguments as to the possibility of safeguards and protective procedures.»

But neither the European Court of Human Rights nor the Swiss Federal Court has paid any attention to a very important problem: it is not at all easy to end one's own life. We all know that there is, in fact, a very high risk of failure when somebody attempts suicide; the risk of danger to oneself and others, the risk of surviving an attempt but suffering irreversible harm. We know of cases where people trying to commit suicide on a railway track have just lost their legs. We know of cases where people trying to commit suicide with a gun have survived but lost their mental capacity. We know of many cases of people who tried to commit suicide but failed, living in nursing homes for decades afterwards. What we do not know is the ratio between attempted and successful suicides.

The Swiss Federal Government, called the «Federal Council», told the lower chamber of the Swiss Parliament, the «National Council», on January 9th 2002 that Swiss statistics record about 1,350 cases of suicide every year. The Government added that the number of suicide attempts is not known: the suggestion is that there are between 10 and 50 times more attempts than successful suicides. The National Institute for Mental Health in Washington said that the number of attempts is up to 50 times higher than the number of successes.

Does the fact that there is a risk of about 9 to 1 or even of about 49 to 1 of failure with every suicide attempt have some impact on the consideration of this Human Right? I think it does.

If the European Convention on Human Rights really does guarantee the Right to self-determination and, thereby, the Right to Suicide, this right cannot exist, in reality, while there is such a high risk of failure. This is especially true given that the risk is not only that of failing to die but the far greater risk of ending up in a much more desperate condition than before.

Long ago the European Court of Human Rights ruled «that the Convention is intended to guarantee not rights that are theoretical or illusory but rights that are practical and effective». You may read this elementary consideration in paragraph 33 of the Judgement of the Court in the case of ARTICO vs. Italy of May 13, 1980.

Why didn't the European Court of Human Rights, in its DIANE PRETTY judgement, deal with this elementary problem? Their decision denied her husband the legal right to help her to commit suicide – she was unable to do so alone due to the impairments of Motor Neurone Disease which had paralysed her almost entirely.

It sounds rather unbelievable: neither DIANE PRETTY's lawyers nor the agents of the British Government discussed the problem of the risks of suicide attempts. Looking only at her case, it was clear to them that Mrs Pretty had no opportunity at all to attempt suicide on her own, being almost completely paralysed, so the question of risks could not be dealt with.

But this question is the crucial one, and it should have been discussed in relation to the argument put forward by the British Government that the Suicide Act of 1961 aims to protect the so-called «vulnerable». Instead of addressing this issue, the dangers of other pretended risks were emphasised. The European Court followed this view by saying:

«The more serious the harm involved the more heavily will weigh in the balance considerations of public health and safety against the countervailing principle of personal autonomy. The law in issue in this case, section 2 of the 1961 Act, was designed to safeguard life by protecting the weak and vulnerable and especially those who are not in a condition to take informed decisions against acts intended to end life or to assist in ending life. Doubtless the condition of terminally ill individuals will vary. But many will be vulnerable and it is the vulnerability of the class which provides the rationale for the law in question. It is primarily for States to assess the risk and the likely incidence of abuse if the general prohibition on assisted suicides were relaxed or if exceptions were to be created. Clear risks of abuse do exist, notwithstanding arguments as to the possibility of safeguards and protective procedures.»

We should acknowledge that, in the field of suicide and suicide attempts, our attitude is still governed by strong taboos and a broad lack of knowledge. We are quite happy to leave those problems to so-called specialists, namely psychiatrists. The effect is that suicidal problems are generally not discussed in society. Because of this neglect, the problem remains hidden.

We need to change this passive attitude, which reflects badly on the situation of people who are suicidal, in order to find an active solution.

The sad fact is that, at the moment, a person who has suicidal thoughts runs the risk of being put in a psychiatric institution if he or she talks to anyone else about wanting to die. This will result in them losing their freedom, so most people who have suicidal thoughts will not talk to other people about their feelings. Put simply, suicidal people are normally left alone with their suicidal thoughts. Under such conditions, people have very little chance of finding the answer to a desperate situation. If they had the opportunity to discuss their problems with others without the danger of being put in a psychiatric institution, there would be greater opportunity to find a solution to their problems. This would have an immediate impact on their suicidal thoughts, which would become less dominant. Society should accept the principal of suicide, and we need to change our attitude towards it.

If someone wants to end his life, society's reaction should not be immediately to decide that that person lacks full mental capacity and should be confined to a psychiatric institution to prevent their suicide: the only result of confinement is that the person has his sense of responsibility for himself taken away. Instead, we should offer the suicidal person the opportunity to discuss his problems with others to see whether a solution can be found. Having suicidal ideas should not, primarily, be seen as a mental disturbance but rather as a crisis in a person's life. Because suicidal problems usually arise out of interactions with other people, solutions should also be sought through interactions with other people. It is our duty, therefore, to offer help instead of isolation.

The aim of this help seems to be very simple: working with the suicidal person, we need to find ways to solve their problems and to look for a solution pointing towards life. If there is no acceptable solution to their problems, then we must be willing to offer assisted suicide with professional help, thus guaranteeing that the very real risks accompanying a failed attempt at suicide can safely be ruled out.

DIGNITAS' experience, following this working model of dealing with a person's life crisis, is a positive one. By accepting the idea of suicide in principal and by being prepared to offer professional help with suicide, DIGNITAS is recognised as a credible and trustworthy source of help for people in suicidal situations. Members who are suicidal are taken seriously and they do not relinquish responsibility for themselves to anyone else. As a result, most of them are able to discuss their problems and look for proposals directing them towards life rather than only looking for assisted suicide. The cooperation between DIGNITAS and the physicians who are ready to help its members has produced an interesting statistic. It shows that receiving the so-called «provisional green light» – meaning that a Swiss physician would be ready to write the prescription for the lethal drug – has an immediate effect on the majority of members.

About 70 per cent of members ask for a green light but, having been given it, most people never contact DIGNITAS again. For them, simply knowing that there is an emergency exit available, should they need it, is so comforting that they have a good chance of living until their life's «natural end».

We should be willing to consider this whole problem of suicide in our society, the suicide situation in general and the special situation of people with severe physical or mental problems who are looking for suicide free from risk and pain. If we do not change the law, we will perhaps prevent several dozen people having an assisted suicide. At the same time, we must be ready to accept that hundreds of thousands of people will attempt a lonely suicide with all the risks that entails for themselves and for others. We must also be prepared to accept that thousands of people will commit suicide without having had an opportunity, as a free person and free from pressure, to seek help with a potentially solvable problem.

Thus, the conclusion is very simple if we want to protect the maximum possible number of people from a premature death by suicide or from the terrible risks of failed suicide attempts. We must be prepared to offer professionally-supervised assisted suicide to those people whose problems cannot be solved, even after intense discussion with non-judgemental people free from any paternalism, whether medical, religious or governmental.

Considering all this, we may arrive at the conclusion that accepting the idea of freedom and autonomy of the individual in connection with the question of self-determination in its fullest sense is the very best way to protect the maximum number of people from premature death. Therefore, we should trust the idea of freedom, Human Rights and self-determination in the general public interest.

I told you, at the beginning, that the struggle is not over yet, that it has only just begun. Why is this true, not only in the Netherlands, but also in Belgium and, soon, in Luxemburg?

As long as residents of those countries have to travel to Switzerland for assisted suicide because the law of their own country does not allow them to ask for it at home, neither their freedom of choice nor their right to suicide can be said to correspond with the guarantees of the European Convention on Human Rights. The Convention gives a free choice and its goal is, indeed, the autonomous individual. As a result, contracting states have only limited freedom to establish safeguards protecting the lives of people with full mental capacity who want to end to their lives.

We should also be aware that a large part of the opposition to this last freedom of human beings is not based on the idea of protecting weak and vulnerable people, but simply by considerations of power: power of the state, power of the church, power of doctors.

Wherever and whenever a struggle to achieve Human Rights is necessary, it will inevitably mean a fight against those three very influential circles. If people want to defend or enlarge their range of rights and freedoms, the state, the church and the doctors always have to surrender some of their former power. There are two ways to win the fight for this «last Human Right»: we can either try to get majorities in parliaments for more or less good or bad laws to permit assisted suicide, or we can take cases to court and ask the legal authorities to make decisions. Based on my opinion that the Right to Suicide is a guaranteed Human Right, I am convinced that the route through the courts will be quicker than through parliaments.

In fighting for this goal you are not only working towards another Human Right, but you will also be making an important contribution towards enlightenment. To return to KANT's view, this will be a major contribution to man's emergence from his self-imposed immaturity.

And that is certainly a goal worth fighting for!

CHAPTER ELEVEN

Death, Resurrection, And Immortality: Some Mathematical Preliminaries

R. Michael Perry*

1. Introduction

A mathematical formalism is developed that, it is argued, has relevance to the nature of personhood, and addresses certain fundamental problems including death, resurrection and immortality. An argument from principle is given that resurrection, even after total obliteration or "information death," is logically possible and not in violation of generally accepted physics. The argument is not based on the necessity of recovering a "hidden past" but, as a last resort, on the eventual likelihood of a duplicate of a lost person recurring fortuitously, supposing time and space are overall unlimited. This sort of argument for resurrection is an old one but here is newly strengthened by the formalism that is developed. A hoped-for resurrection, in this sense, could still be delayed to an all-but inconceivably remote future, and be preceded by very many near-misses who, considered as persons, are nonhistorical fantasies. However prospects are considerably improved by assuming a multiverse cosmology, where more than one version of a resurrection are authentic and many authentic versions are produced concurrently. If resurrection is possible then it might happen more than once, sequentially, after different fatal incidents in the life of the person. In this way, the immortality of any past or present individual becomes a physical possibility.

* Note: This article was in the main written about 1995, but never actually completed or published, and then lay largely forgotten, until I recently rediscovered it in an old computer folder. I have now revised and abridged it in an effort to produce what I hope can serve as a kind of progress report. The list of references has not been updated. Further work is in progress. RMP

A person is viewed, essentially, as an aggregate of mental events or experiences and is modeled accordingly as a type of "process." An exact characterization of a person as distinct from other processes is not attempted. However properties of a certain class of process, arguably inclusive of persons, are investigated, and results are offered as a possible resolution of problems connected with personal existence.

Formally a process is defined as a mapping from certain subsets or "hyperregions" of a partially ordered set (poset) representing "spacetime," into a set of "histories." (The allowable hyperregions form a topology of spacetime.) Intuitively, the value (history) assigned to the process, for a given hyperregion in spacetime, is a summary of the behavior of the process, as it might be viewed by an imagined observer able to see and record all that is happening. The process thus associates a body of "information" with each hyperregion. The set of histories itself is ordered (is a poset) with certain special properties, a greater history indicating more information.

This modeling is chosen to facilitate expression of causal connections: A greater history can be regarded, in an appropriate context, as a causal derivative of a lesser history. This in turn is important for certain notions such as recall, which in a person involves a causal connection between past and present perceived events. The poset model of histories fits several, rather divergent ideas, one being to regard history as a set of possible event sequences, with a larger set of event sequences indicating greater uncertainty in the records (so a smaller set would be "greater," as ordered in the poset). The notion of process as here developed, it should be noted, generalizes certain other ideas of processing that have been proposed as a model of human mental activity, including but not limited to digital computation.

Although a given history summarizes behavior over a particular hyperregion of spacetime, this behavior could itself involve recall of events outside the hyperregion and thus could furnish information about a more distant past. The cases considered here assume a "linearity": the larger the hyperregion over which a process is evaluated, the greater the associated history, indicating more recoverable information.

An "observer" is a kind of process that has complete recall of its past (events on time-bounded hyperregions of spacetime are

236

later recapitulated on other time-bounded hyperregions). It thus could serve as an idealized person, if one interprets its "recall" as involving present "awareness" of past "states of consciousness." Such an interpretation seems justified under some conditions. Another property of the observer is that recall of past events is always possible at a later date; the observer thus is "immortal," and in particular is a non-terminating process. Such a process will be composed of terminating subunits called "episodes," which in turn (usually) correspond to phenomena observable over finite volumes of spacetime. Another kind of process, the "preobserver," may lack perfect recall but is nonterminating or "enduring" and contains an observer as a subprocess, thus building up a "database" that is immune from later erosion or alteration. A preobserver might be taken as a model of a person in the process of becoming immortal, who is not required to retain every past experience but does accumulate a permanent fund of experiences, whether quickly or slowly. Properties of the observer thus are important in exploring the possibility of persons as entities subject to resurrection and immortalization.

The domain of a process is the set of hyperregions of spacetime for which the process is defined. By convention, these hyperregions consist of subsets of an enclosing hyperregion or "subdomain" which is also in the domain. A subprocess is obtained by restricting the subdomain. Any process can be so restricted; on the other hand "consistent" processes can be "added" to obtain a single process whose subdomain is the union of the original subdomains. The ability to recapitulate can serve as a basis for identification of a later episode or fragment of an observer or preobserver with an earlier fragment. In this way a process can be started which serves as a continuer of an earlier, terminated process. An observer or preobserver can thus be constructed using an original, terminating process as foundation. By adding a later process to the terminating foundation we obtain a third process, having the foundation as a subprocess, whose existence in turn extends beyond the termination point. In this way it is possible to model mathematically the resurrection and immortalization of persons.

A resurrected person will recall events that occurred before death and thus, in an internalized sense, experience a causal connection with these events. The resurrection need not involve a causal connection in the usual, externalized sense, however; the connection can be purely an artifact of the construction. The viewpoint advocated here is that a resurrection is always authentic, provided authentic memories are recalled by the resurrectee, even if there is no way outsiders can verify authenticity. An authentic resurrection could thus, in principle, happen purely fortuitously, though a more straightforward reanimation might be preferred, when possible.

A word should now be said about the arrangement of material. There is a short historical note (section 2), then the mathematical theory, the bulk of the paper, is presented in detail (sections 3-20), followed by a conclusion (section 21). The theory, in effect, argues for the "memory criterion" of personal identity, this being necessary to justify the claim that resurrection is always possible. Since the memory criterion has been criticized by philosophers, some argument is needed to answer these criticisms; this is not addressed here but should be the subject of a later work. More generally there is much that is not addressed in this preliminary work that should be in a study of this nature; hopefully a useful beginning has been made.

2. Historical Note

Death, resurrection and immortality have exercised the minds of philosophers and theologians, among others, since time immemorial. However it is only in comparatively recent times that death has been examined seriously as a problem that might be solved scientifically. The first to do this seems to have been the Russian moral philosopher Nikolai F. Fedorov (1829-1903), who, following Newtonian-Laplacian physical theory, argued that persons of the past could be resurrected, in principle, by minutely tracking and then repositioning atoms.1 This now seems unlikely, due to quantum uncertainty, though a few diehard physicists still hold out hope for recovery of the "hidden past" that would allow implementation of Fedorov's idea, which would eliminate hyperonticity in favor of enonticity.

More recent developments have explored some other possibilities. A small movement began in the 1960s known as cryonics, to freeze the recently deceased in anticipation of future technology that would enable reanimation. Again this form of resurrection would ideally be entirely enontic based on the premise that sufficient information was stored in the frozen remains to fully reconstruct the original person. Some among the cryonics community (which includes the present writer) have also tried to address more philosophical issues. David Krieger, for example, developed a "you-ness" function to assess the extent to which a given individual might be said to survive in a later version of the self.[2] Robert Ettinger, Max More and Lee Corbin have also considered issues relating to identity and survival.[3] Ettinger, in particular, noted some years ago the frequent speculations that resurrection would require only recreation of a copy or the "pattern" of the original, and could happen by accident.[4] (Ettinger himself did not accept this point of view, which however is supported by the formalism developed here.)

Outside of cryonics there also is interesting theoretical work relating to the possibility of immortal life in the universe, notable proponents being the physicists Frank Tipler, John Barrow and Freeman Dyson.[5,6] Philosopher Taras Zakydalsky has taken up the cause of Fedorov, systematizing the earlier work and exploring criteria by which the validity of a purported resurrection might be judged.1 Robert Nozick presented a notion of "continuer" that has been adapted and developed here.[7] Other efforts to better elucidate the nature of personhood have sometimes approached the viewpoints offered here. A paper by Daniel Kolak and Raymond Martin, for example, argues against the necessity for causality (in the usual, enontic sense) between the later and earlier stages of a person, supporting the possibility of resurrection after information death.[8]

The above brief survey is by no means comprehensive, but will give some idea of efforts being made in the developing field of immortalism—which treats of solving the problem of death scientifically. The existing literature does not appear, however, to contain work closely related in detail to what is reported here; to this we now turn.

3. Mathematical Theory: Basics

We start with the apparatus of Zermelo-Frankel set theory, with axiom of choice (ZFC, implying Zorn's lemma). This in turn offers a standard treatment of the notions of relation (set of ordered pairs) and function (relation in which no two ordered pairs have the same first element), and general topology.[9,10] For a function f the set of first elements, which is the set of values over which the function is defined, is the *domain*, here denoted by Dom(f). Similarly the set of second elements, values assumed by f as it varies over its domain, is the *range*, Rng(f). Given $S \subseteq$ Dom(f), $f[S]$ denotes the set of values $f(x)$ where x ranges over S, that is, $f[S] = \{f(x) \mid x \in S\}$. Similarly, for $T \subseteq$ Rng(f), $f^{-1}[T] = \{x \in$ Dom(f) $\mid f(x) \in T\}$.

We write "$f{:}X{\to}Y$" to mean "$X =$ Dom(f), $Y \supseteq$ Rng(f)," while if it always happens that $f(x) = y_x$ we write "$f{:}x{\mapsto}y_x$," or simply, "$x{\mapsto}y_x$." In case we have two functions, f_1, f_2, with Rng(f_2)$\subseteq$Dom(f_1), the *composition* $f_1 {\circ} f_2$, is defined, a function with domain equal to that of f_2, range included in that of f_1, such that $f_1 {\circ} f_2 {:}\ x \mapsto f_1(f_2(x))$. Composition is *associative*, that is, $f_1 {\circ} (f_2 {\circ} f_3) = (f_1 {\circ} f_2) {\circ} f_3$. Each set X, moreover, has a unique identity function $e_X : x \mapsto x$, with domain and range equal to X. Given any function $f{:}X{\to}Y$, $f {\circ} e_X = e_Y {\circ} f = f$.

We will also have occasion to consider partial functions which may be defined for some points of an indicated set but not necessarily all. We write $f{:}X{\rightharpoonup}Y$ to mean $X \supseteq$ Dom(f), $Y \supseteq$ Rng(f).

Some further notational conventions will be useful, particularly for more complicated mathematical objects. In general, such an object will consist of an "underlying set" with some specified structure. An object will be given in the usual form of a list of items among which different relations hold. By convention an underlying set of object **O**, written |**O**|, will be the first element of the list.

One of the principal objects is the partially ordered set or *poset*, which is needed to model both spacetime or a "staging space," and a data structure for histories or "history space," wherein events will unfold and processes will be expressed. A poset will be defined as an ordered pair $\mathbf{K} = \langle K, \geq \rangle$ where K is the actual set and $\geq$ ("greater than") is the partial order. (Thus $K = |\mathbf{K}|$.) A partial order has the usual properties of reflexivity ($x \geq x$), antisymmetry (if $x \geq y$ and $y \geq x$ (or $x \leq y$) then $x = y$), and transitivity (if $x \geq y$ and $y \geq z$ then $x \geq z$). $x > y$ ($y < x$) iff $x \geq y$ and $x \neq y$. Formally, the partial order is a relation or set of ordered pairs of elements chosen from K, a subset of the set of all ordered pairs, K^2. Thus we have $\langle x, y \rangle \in (\geq) \subseteq K^2$ just in case $x \geq y$; in particular by reflexivity we must have $\langle x,x\rangle \in \geq$ for all $x \in K$, so that K can be recovered from any partial order on K. In effect then, K is the "underlying set" of the partial order and is denoted by $|\geq|$.

Given any subset $K_0 \subseteq K$, there is a natural or *canonical* restriction, $\geq_0$, of the partial order $\geq$ to the elements of K_0, that is, $\geq_0 = \geq \cap K_0^2$. It is easy to see that $\geq_0$ satisfies the properties of a partial order (thus $\geq_0 = \geq \cap |\geq_0|^2$) so that, if x, $y \in |\geq_0|$, then $x \geq_0 y$ iff $x \geq y$. We see also that for each K_0 there is exactly one canonical restriction $\geq_0$, which will be referred to as *the* canonical restriction of $\geq$ to K_0. The partial orders of any set K *themselves* form a poset under the ordering of canonical restriction, that is, where $\geq_0$ is "less than or equal to" $\geq$. Many different partial orders are possible; some will be subscripted, as here, or otherwise marked to distinguish them; where no confusion seems likely the distinguishing marks will be omitted.

Another needed concept is that of *topological space*: an object $\langle S, \mathfrak{I}\rangle$, where S is an arbitrary set, sometimes called the *point set*, and $\mathfrak{I}$, the *topology* on S, is a family of subsets of S, considered to be all the "open" subsets. $\mathfrak{I}$ must be closed under arbitrary unions and finite intersections, and must contain both S and the empty set $\varnothing$. Thus, as with the partial order, the original set S can be recovered from the topology $\mathfrak{I}$; in this case, $S = \cup \mathfrak{I}$. A set $T \in \mathfrak{I}$ is an (open) *neighborhood* of a point $y \in S$ in case $y \in T$. Similarly, T is a neighborhood of an arbitrary subset $U \subseteq S$ iff $U \subseteq T$. Thus S itself is always a neighborhood of U as well as any point $x \in S$ and, on the other hand, T is a neighborhood of x iff it is a neighborhood of $\{x\}$.

The family of subsets of T in $\mathfrak{I}$, $\mathfrak{I}(T)$, forms a topology of T, that is to say, with T as the join. This is the *relative* or *subspace topology* of T under $\mathfrak{I}$. On the other hand, a *subtopology* $\mathfrak{I}_0 \subseteq \mathfrak{I}$ is any subset which itself is closed under arbitrary unions and finite intersections, and also contains the empty set. In particular, the relative topology of T is a subtopology, for which it happens that $\mathfrak{I}_0 = \mathfrak{I}(T) = \mathfrak{I}(\cup \mathfrak{I}_0)$. This property will not hold in general, but instead only $\mathfrak{I}(T) \subseteq \mathfrak{I}(\cup \mathfrak{I}_0)$, so that the relative topology is the largest possible (or most "refined") subtopology with the given point set. One interesting property that does hold in general is that an intersection of any nonempty family of subtopologies is a subtopology, as is easily seen, and in fact is a greatest lower bound under the inclusion order.

Returning to topology $\mathfrak{I}$, given arbitrary $T \subseteq S$ (T not necessarily open) the set $S-T$ consisting of points in S that are *not* in T is called the *complement* (in S) of T. A subset $T \subseteq S$ is *closed*, under $\mathfrak{I}$, in case its complement is open. The union of all open subsets of T is the *interior* of T, here denoted by $T^{\downarrow}$, and is the largest possible open subset of T. The complement of the interior of the complement of T is the *closure* of T, $T^{\uparrow}$, and is the smallest closed set that includes T. The closure of T can also be shown to equal the set of all $x \in S$ such that every neighborhood of x contains a point in T. The family of closed subsets of S itself is closed under arbitrary intersections and finite unions. The interior of the closure of T, or *closure-interior*, $T^{\uparrow\downarrow}$, is the largest open subset included in the closure of T, and includes, but is not necessarily equal to, T itself. Easy properties that will be useful in what follows are that (1) an open set is nonempty iff its closure-interior is nonempty, and that, in this case, (2) any open subset of the closure-interior has a nonempty (open) intersection with the original set.

4. Staging Spaces and Subspaces

Combining some of the above notions, a *staging space* will be a topological poset: an object $S = \langle \mathfrak{I}, S, \geq \rangle$ with $\langle S, \mathfrak{I} \rangle$ a topology on an arbitrary underlying set $S = \cup \mathfrak{I} = \cup |S|$, referred to as *spacetime*. $\mathfrak{I}$ is also referred to informally as the topology of S, where hopefully no confusion is likely. (The topology $\mathfrak{I}$ is placed first in the definition list of staging space S rather than second as in the preceding definition of a topology to emphasize the importance that will be attached to open sets rather than individual points in spacetime—see below.) Associated to $\mathfrak{I}$ is a partial order $\geq$, the *time order*, which is extendable to a partial order on S and satisfies "sufficiency conditions" as explained below.

Intuitively, a staging space models a structure like the physical universe in which observable events occur. Points in spacetime, x and $y \in S$, are locations of "events," with (when the time order is extended to points as promised) $x > y$ ($x \geq y$ and *not* $y \geq x$) whenever x occurs *later* than y. It may happen, of course, that $x \neq y$ but that *neither* is $x \geq y$ nor $y \geq x$; that is, x and y are not comparable. This situation occurs in relativistic spacetime, when x and y have spacelike as opposed to timelike separation. A simple example of a staging space, that will be useful in illustrations later, is just the positive integers with the usual order and the discrete topology under which every set of points is open.

The *sufficiency conditions* for the time order $\geq$, which will allow its extension from topology $\mathcal{T}$ to the underlying point set S, are as follows, where U, V are elements (open sets) in $\mathcal{T}$, and $\mathcal{F}$, $\mathcal{G}$ are subsets of $\mathcal{T}$.

(1) $U > V$ only if U is nonempty and U, V are disjoint.

(2) If some $U \in \mathcal{F}$ is nonempty then: $\cup \mathcal{F} > \cup \mathcal{G}$ iff $U > V$ for all nonempty $U \in \mathcal{F}$ and all $V \in \mathcal{G}$.

(3) $S \geq \varnothing$.

In particular, (2) has the consequence that if $U > V$ and $W \subseteq U$ is nonempty, then $W > T$ for arbitrary $T \subseteq V$. Thus we have $U \geq \varnothing$ for all U. A rudimentary time order that satisfies these conditions and is available in any topology is obtained if we require that $U > V$ iff $U \neq \varnothing$, $V = \varnothing$. On the other hand, if spacetime is the positive integers with the usual order, we can define the time order by $U > V$ iff $U \neq \varnothing$, V is finite, and either $V = \varnothing$ or the minimum element of U is greater than the maximum element of V.

Next we extend the time order to points x, $y \in S$ as follows: $x > y$ iff there exist U_{xy}, $V_{xy} \in \mathcal{T}$ such that $x \in U_{xy}$, $y \in V_{xy}$, and $U_{xy} > V_{xy}$. Then by (1) $x \neq y$ (and in fact are separable in the topology) and we can verify the necessary property of transitivity as follows. If, in addition, $y > z$ for $z \in S$ then $U_{yz} > V_{yz}$ for some U_{yz}, $V_{yz} \subseteq S$ such that $y \in U_{yz}$, $z \in V_{yz}$. By (2) we must then have $U_{xy} > W = V_{xy} \cap U_{yz}$. On the other hand, since $y \in W \subseteq U_{yz}$, W is nonempty so that, again by (2), $W > V_{yz}$. It then follows that $U_{xy} > V_{yz}$, which establishes $x > z$ as desired.

It is also easy to show that, for any U, $V \in \mathcal{T}$, $U > V$ iff U is nonempty and $x > y$ for all $x \in U$, $y \in V$. In particular, if $x > y$ for some $x \in U$, $y \in V$, we can find U_{xy}, $V_{xy} \in \mathcal{T}$ such that $x \in U_{xy}$, $y \in V_{xy}$, and $U_{xy} > V_{xy}$. By (2) we then have that $U \cap U_{xy}$, $V \cap V_{xy}$ are nonempty subsets of U, V respectively such that $U \cap U_{xy} > V \cap V_{xy}$. Considering unions over all x, y of such sets and again applying (2) we obtain $U > V$.

We can further extend $\geq$ to arbitrary subsets A, $B \subseteq S$ by the requirement that $A > B$ iff A is nonempty and $x > y$ for all $x \in A$, $y \in B$. Transitivity will then hold, and A, B will be disjoint as is required for open sets.

By convention also, $x >$ ($<$) $A \subseteq S$ iff $\{x\} >$ ($<$) A; that is to say, $x > A$ (and $A < x$) iff $x > y$ for all $y \in A$, $x < A$ (and $A > x$) iff A is *nonempty* such that $x < y$ for all $y \in A$. This completes the extension of the time order to arbitrary points and subsets of spacetime. We assume, then, that in any staging space the partial order satisfies the sufficiency conditions.

A subset of spacetime will be called a *hyperregion*. Unless otherwise noted, such sets will assumed to be open. (This, of course, does not apply when the closure of a set is explicitly indicated.)

Given staging space $S = \langle \mathfrak{I}, S, \geq, \rangle$, a *(staging) subspace* of S is $S_0 = \langle \mathfrak{I}_0, S_0, \geq_0 \rangle$, such that $\mathfrak{I}_0 \subseteq \mathfrak{I}$ is the subspace topology on $S_0 = \cup \mathfrak{I}_0$ ($\mathfrak{I}_0 = \mathfrak{I}(S_0)$), and $\geq_0$ is the canonical restriction of the time order to $\mathfrak{I}_0$ ($\geq_0 = \geq \cap \mathfrak{I}_0^2$). Thus $\mathfrak{I}_0$ must be closed under unions and finite intersections and must contain the empty set. It is also straightforward that $\geq_0$ satisfies the sufficiency conditions on $\mathfrak{I}_0$. Each hyperregion S_0, then, has its own unique, associated subspace.

We can order the subspaces of S by inclusion on the associated hyperregions: for subspaces S_0, S_1, $S_1 \geq S_0$ iff $\cup |S_1| \supseteq \cup |S_0|$, in which case $|S_1| \supseteq |S_0|$, so the subspaces of S form a poset under this order, which will be called the *subspace order*.

5. History Spaces and Subspaces

By a *history space* is meant a poset $\mathbf{H} = \langle H, \geq \rangle$ in which (1) H is nonempty, (2) every nonempty subset $H_0 \subseteq H$ has a greatest lower bound in H, written $\inf_H H_0$ or more usually just $\inf H_0$, and (3) every chain or totally ordered subset of H has an upper bound in H. A subset H_0 that has an upper bound will be called *consistent*; two elements g, h are consistent iff they have a common upper bound, that is if $\{g, h\}$ is consistent. If no upper bound exists the two histories are *inconsistent*, and we write $g \perp h$.

One consistent subset of histories is just the empty set; also singleton sets are consistent. Both the empty set and singleton subsets have *least upper bounds*. (For the empty set the least upper bound is just inf H.) More generally, by (2), every consistent subset H_0 can be easily shown to have a least upper bound (it is the greatest lower bound of the nonempty set consisting of all the upper bounds of H_0); it will be written sup H_0 or occasionally and more strictly, $\sup_{\mathbf{H}} H_0$.

A simple example of a history space is just any topology $\mathfrak{T}$, with the elements (sets) ordered by inclusion. $\mathfrak{T}$ is nonempty since it must contain the empty set. Every nonempty subset $\mathfrak{F} \subseteq \mathfrak{T}$ has a greatest lower bound, the (possibly empty) union of all open subsets of the intersection $\cap \mathfrak{F}$, that is, the intersection-interior of $\mathfrak{F}$. Finally, every chain $\mathfrak{F}$ (and every subset whatever, in fact) has an upper bound, in particular $\cup \mathfrak{F}$, which happens to be the least upper bound.

Points in a history space are intended to model "histories" (though, as in the case of the arbitrary topology above, the modeling may not be particularly intuitive). If g and h are histories then $g \geq h$ means, intuitively, that (1) g is consistent with h, and (2) g contains at least as much information as h. In general it may happen that *neither* $g \geq h$ nor $h \geq g$, so that g and h are not comparable. This can be because the histories are contradictory or not consistent, or the two histories may be consistent but each has information lacking in the other.

By hypothesis any chain of histories must have an upper bound, so, by Zorn's lemma, there must be maximal histories. Every history will be upper-bounded by some maximal history, and a family of histories will be consistent iff every history in the family has the same maximal upper bound. For maximal histories g, h, either $g = h$ or $g \perp h$.

History space $\mathbf{H}'$ is a (*history*) *subspace* of $\mathbf{H}$ if (1) $|\mathbf{H}'| \subseteq |\mathbf{H}|$, (2) for all g, $h \in |\mathbf{H}'|$, $g \geq_{\mathbf{H}'} h$ iff $g \geq_{\mathbf{H}} h$, (3) for all nonempty $H_0 \subseteq |\mathbf{H}'|$, $\inf_{\mathbf{H}'} H_0 = \inf_{\mathbf{H}} H_0$, and (4) if nonempty $H_0 \subseteq |\mathbf{H}'|$ is consistent in $\mathbf{H}'$, so that it is also consistent in $\mathbf{H}$, then $\sup_{\mathbf{H}'} H_0 = \sup_{\mathbf{H}} H_0$. In particular if H_0 is a nonempty chain in $\mathbf{H}'$ (and thus also in $\mathbf{H}$), H_0 must be consistent in $\mathbf{H}'$ by the requirements of a history space, consequently, must have the same least upper bound in $\mathbf{H}'$ as in $\mathbf{H}$. (Only for the degenerate case of empty H_0 can the two least upper bounds differ.) On the other hand, though, it is perfectly possible that H_0 is consistent in $\mathbf{H}$ but is not consistent in the smaller space

H′. In this case **H′** is a "more discriminating" space able to separate histories that can be conjoined in **H**, a property that will carry over to processes to be defined. This is of particular significance in delineating the "personlike" character of certain processes, which are separate entities though parts of a larger whole. It will be noted that the partial order of a history subspace is the canonical restriction of the original partial order, to the corresponding subset of the original histories, which seems a natural requirement. Again, as for staging spaces, history spaces form a poset under the subspace order, so we write **H′≤H**.

In fact, it is an interesting, straightforward property that the family of staging subspaces of a given space, forms a history space, under the subspace order. In particular, the greatest lower bound of any nonempty set of subspaces is determined by the (nonempty) intersection of the associated topologies, in this case, the subspace topologies for the associated hyperregions. (Each topology must contain the empty set.) The least upper bound of any nonempty set of subspaces is similarly determined by the nonempty intersection of all the containing topologies, including that of the original space. For history spaces the subspaces need not form a history space because a family of subspaces need not contain a lower bound. (This can be seen by considering singleton subspaces.) Every chain of history spaces has a least upper bound, however, whose underlying set is the union of the underlying sets of the member subspaces.) On the other hand, if we choose any point (history), then the family of subspaces whose underlying sets contain the point does form a history space. In this case, for a nonempty family of subspaces, the greatest lower bound space will have as underlying set the intersection of the underlying sets of the subspaces. Put another way, it means that any nonempty family H_0 of histories is contained in a unique, minimal subspace whose underlying set is the intersection of the underlying sets of all subspaces containing H_0.

Both staging and history spaces, then, exhibit certain properties of subspaces which, it turns out, will also carry over to process spaces to be defined. It will be useful, then, to consider these properties briefly in a more general context. An object **O** becomes a "space" when certain requirements are satisfied, and *subspaces* are also defined, which then form a poset (subspaces are less than or equal), with **O**, the *superspace*, being the unique, maximal element. Moreover, under the subspace ordering, $\mathbf{O'} \geq \mathbf{O}$ $\Rightarrow |\mathbf{O'}| \supseteq |\mathbf{O}|$.

For the two cases considered, staging and history spaces, the subspaces are *simple-structured* in that, if $\mathbf{O'}$, $\mathbf{O''}$ are subspaces, then $\mathbf{O'} \geq \mathbf{O''}$ iff $|\mathbf{O'}| \supseteq |\mathbf{O''}|$, as is straightforward to establish. In particular, $|\mathbf{O'}| = |\mathbf{O''}|$ iff $\mathbf{O'} = \mathbf{O''}$. The associated subspace order is then said to be *simple-structuring*. To object $\mathbf{O}$ there corresponds a family of *favored subsets*, each of which is the underlying set of some subspace. In view of simple structuring, for each favored subset there is a unique, associated subspace. $|\mathbf{O}|$ itself, is, of course, always favored.

For staging spaces the associated topologies of subspaces are the favored subsets, so that the family of all favored subsets is closed under arbitrary intersections, though not unions. For history spaces it is also violated for unions, as simple examples will show. The *minimal subspace property* means that, for *any* nonempty subset T of the underlying set of the superspace, there is a unique, minimal enclosing subspace that includes T in *its* underlying set. This property holds in history spaces as we have seen. For staging spaces it also holds, since in this case the subset T is just any nonempty family of open sets, which can be extended to a unique, minimal topology consisting of the subspace topology on the union of the sets. The minimal subspace property will also hold in process spaces, as will be shown.

Some further properties of history space $\mathbf{H}$ will be found useful and are now summarized. A subset $H_0 \subseteq |\mathbf{H}|$ is *prevalent* in $\mathbf{H}$ in case every $h \in |\mathbf{H}|$ is the least upper bound of some nonempty subset of H_0. In particular then, $|\mathbf{H}|$ itself is prevalent, and every prevalent subset must be nonempty. For each $h \in |\mathbf{H}|$ we define the *downward closure* $dc(h;H_0)$ as $\{h_0 \in H_0 | h \geq h_0\}$. Then it is clear that $h = \sup dc(h;H_0)$, and that this latter set is the largest subset of H_0 with h as the least upper bound. Since the least upper bound is unique, no two h have the same downward closure. If we order the downward closures by inclusion (more inclusive sets are "larger") then the downward closures become a history space $\mathbf{H'}$ which is isomorphic to $\mathbf{H}$, that is, there exists a 1-1, onto mapping $\mathfrak{m}:|\mathbf{H}| \to |\mathbf{H'}|$ such that, for $g,h \in |\mathbf{H}|$, $g \geq_{\mathbf{H}} h$ iff $\mathfrak{m}(g) \geq_{\mathbf{H'}} \mathfrak{m}(h)$. (This is the standard definition of isomorphism for an arbitrary poset; note that a poset that is isomorphic to a history space is itself a history

247

space.) Inasmuch as isomorphism is an equivalence relation, we then see that any two history spaces with the same prevalent subset are isomorphic, that is, identical up to renaming the elements of the underlying set.

Given an arbitrary subset, $G\subseteq|\mathbf{H}|$, the inf-*closure* of G, G^c, is obtained by adding to G all the points in $|\mathbf{H}|$ which are the greatest lower bounds of nonempty subsets of G. It is straightforward to show that taking the inf-closure is idempotent: $(G^c)^c = G^c$. A little reflection will then show that, if nonempty G is *chain-closed*, that is, contains the least upper bound, in $\mathbf{H}$, of every nonempty chain in G, then G^c is always a favored subset, that is, forms a history subspace of $\mathbf{H}$, under the canonical restriction of the partial order. Though this condition is not also necessary, it will hold in the more important cases considered here, for example in process spaces to be defined.

6. Processes

Let $S=\langle\, \mathfrak{I},\, S,\, \geq_S\, \rangle$ and $\mathbf{H}=\langle H,\, \geq_{\mathbf{H}}\rangle$ be staging and history spaces, respectively. A mapping $p\colon \mathfrak{I}{\to}H$ is a *process* in case it happens that, for any $\mathfrak{F}\subseteq\mathfrak{I}$, the set $H_0 =\{p(V)|V\in\mathfrak{F}\}$ is consistent, and $p(\cup\mathfrak{F})=\sup H_0$. This latter property will be referred to as the *linearity* of a process (justification given shortly). In particular it means that, if V and V' are hyperregions in the domain of p, with $V\subseteq V'$, then $p(V')\geq p(V)$. $\mathfrak{I}$, then, is the domain of the process; the set $S=\cup\mathfrak{I}$ will be called the *subdomain*, written subd(p), and $p(S)$ will be called the *summary*, written smr(p). Thus we have that smr(p)$\geq p(V)$ for all $V\in\mathrm{Dom}(p)$. A simple example of a process is just p = the identity map, for $\mathbf{H}$ the topology $\mathfrak{I}$, ordered by inclusion, For this case smr(p) = subd(p) = S.

Intuitively, a process assigns a history to a hyperregion, which faithfully reflects the behavior of the process over that hyperregion. Indirectly, this could include events that happen outside the hyperregion. For example, if the process is a person, part of the behavior could consist of recalling earlier events in that person's life, or more distant historical events of which the person has knowledge. For a larger hyperregion consisting of the original plus other hyperregions the process naturally will exhibit more ample behavior and a correspondingly more ample history, including the original behavior/history as a subset. On the other hand, the allowed increase is conservative: the history of the whole is only the *least* upper bound of histories assigned to the separate hyperregions, or in effect the "sum" of these separate histories, hence the "linearity." The summary, in particular, describes the whole behavior of the process, and is the least upper bound of the value of the process over all the hyperregions of its domain.

Given process p, and a hyperregion $V \in \mathfrak{I} = \mathrm{Dom}(p)$, the mapping obtained by restricting p to the subspace topology $\mathfrak{I}(V)$ is likewise a process, whose subdomain is V and whose associated staging space is the corresponding subspace of S. This process will be denoted by $p_{\|V}$ and is called a *primitive masking* of p. This primitive masking, whose subdomain is a subset of the original process, will be called a *subprocess*. More generally, for any process q whose subdomain is some hyperregion of S, the primitive masking $q_{\|V}$ is the mapping with domain $\mathfrak{I}(V)$ given by $q_{\|V}: T \mapsto q(\mathrm{subd}(q) \cap T)$. It is easy to show that $q_{\|V}$ is a process with subdomain V, which is identical to q itself on $\mathfrak{I}(\mathrm{subd}(q) \cap V)$, but whose subdomain need not be a subset of q's. In particular, $q_{\|S}$ is a unique process whose subdomain is all of spacetime, or an *entire* process, but whose summary is identical to that of q; $q_{\|S}$ then will be called the *entire extension* of q.

A (general) *masking* of process q is obtained by taking a primitive masking of a primitive masking. For convenience $(q_{\|V})_{\|V'}$ will be denoted by $q_{\|V,V'}$. This process has V' as subdomain but $q(\mathrm{subd}(q) \cap V \cap V')$ as summary. It thus cannot, in general, be a primitive masking of q since, for example, its summary is not simply $q(\mathrm{subd}(q) \cap V')$ as would be required. It is straightforward to show, on the other hand, that no further generality is achieved by taking additional, successive primitive maskings: that is, $(q_{\|V,V'})_{V''} = q_{\|V \cap V', V''}$.

Another useful concept along these lines will be the *certified masking*, defined for process q as a masking $q_{\|V,V'}$ whose subdomain is a subset of that of q, that is, $V' \subseteq \mathrm{subd}(q)$. A subprocess, then, is an example of a certified masking, though the concept is more general than this. "Certified" too is used in a more general sense, which is detailed shortly.

A process whose subdomain is a hyperregion in S and which maps into $|\mathbf{H}|$ will be called an *associated process* of S *into* $\mathbf{H}$; the family consisting of all associated processes will be denoted by $AP(S,\mathbf{H})$. We have seen, then, how the associated processes are *closed under masking*. In what follows we normally assume all processes are associated to some specific staging and history space. The family of associated processes is an example of what will be called a "process space," which is considered in the next section.

A set P_0 of processes is *consistent* iff the set of corresponding histories, $\{p(T)|p \in P_0, T \in \text{Dom}(p)\}$, is consistent. In view of linearity, it suffices if the set of summaries of all the processes is consistent. In particular, $p \perp p'$, as processes, iff $\text{smr}(p) \perp \text{smr}(p')$, as histories. In any case, for nonempty, consistent P_0 we can define a mapping p^+ on the set $\mathfrak{I}(V)$ where $V = \cup\{\text{subd}(p)|p \in P_0\}$, by $p^+ : T \mapsto \sup\{p(T \cap \text{subd}(p))|p \in P_0\}$. It is then straightforward to show that p^+ is a process with subdomain V; p^+ will be called the *sum of* P_0 and be denoted by ΣP_0. The sum of several process $p_1, p_2, \ldots$ is also denoted by $p_1 + p_2 + \ldots$. $AP(S,\mathbf{H})$ then is closed under summation of consistent families of processes, or *closed under consistent summation*, in addition to being closed under masking.

Next, it will be convenient to define an ordering of processes as follows. For processes p, p', $p \geq p'$ iff (1) $\text{subd}(p) \supseteq \text{subd}(p')$, and (2) for all $V \subseteq \text{subd}(p')$, $p(V) \geq p'(V)$. We then say that p *certifies* p', or that p' is *certified* or *p-certified*. In particular, the certified masking introduced above is easily seen to be a certified process. It can be seen that, in keeping with the notation, a partial order is indeed defined by these conditions; it will be called the *certifying order*. Certification then is a generalization of the notion of process enclosure, a subprocess or enclosed process being a special case of a certified process. Just as any nonempty set of subprocesses is consistent, and their sum is a subprocess, a nonempty set of certified processes is consistent, and their sum is a certified process.

If $p \geq p'$ then $p + p' = p$; more generally, if p certifies every process of a family P_0, then $\Sigma(\{p\} \cup P_0) = p$. p and p' are consistent if comparable under the certifying order—though consistency could also hold under more general conditions. If P_0 is any nonempty, consistent family, it easily follows that ΣP_0 is the least upper bound of P_0, and thus is the unique, minimal certifier of every process in P_0. On this basis it can be seen that every chain of processes has a least upper bound so that, by Zorn's lemma, a maximal process exists in AP(S,H). On the other hand, any P_0 must have a lower bound; in fact the process $p_\varnothing$ given by subd($p_\varnothing$) $= \varnothing$, $p_\varnothing(\varnothing) = \inf |\mathbf{H}|$, is clearly a universal lower bound on all the processes. It is straightforward to show that the set of lower bounds of any nonempty P_0 is consistent and, moreover, its sum is also a lower bound, which must then be the greatest lower bound. So, AP(S,H) *is a history space* in its own right, under the certifying order.

It is straightforward also to show that the subdomain of the greatest lower bound of any nonempty family $P \subseteq$ AP(S,H), that is, subd(inf P), is just the intersection-interior of the subdomains of the processes in P. On the other hand, as the foregoing discussion shows, the union of these subdomains is the subdomain of sup $P = \Sigma P$.

A process q' with the same summary as q will be called a *variant* of q. In particular, given hyperregion $V \supseteq$ subd(q), $q_{\|V}$ is a variant, which in this case is $\geq q$. The entire extension of q, then, is such a variant. It can be seen that the variants of a process form an equivalence class. It is clear too, that any set of variants is consistent, and sums to a process that is also a variant. The sum of all variants of a process, then, is a *maximal variant* of that process. A maximal variant is an entire process, since by appropriate masking we can transform any process into an entire process with the same summary. Any two maximal variants must thus be identical.

In the case of AP(S,H) it is easy to see that each maximal variant is just a constant map. (In the more general process spaces to be considered, which are subsets of AP(S,H), maximal variants need not be constant.) Here are some additional properties that hold for AP(S,H). For every history there is a unique maximal variant with that history as summary, and the certifying order of the maximal variants corresponds exactly to the ordering of the associated summaries. The maximal variants, in fact, form a history space $\mathbf{H}'$ that is *isomorphic* to $\mathbf{H}$, under the mapping h: $p \mapsto$ smr(p). That is, h: $|\mathbf{H}'| \to |\mathbf{H}|$ is one-to-one and onto, with $p \geq p'$ (certifying order), iff smr(p)$\geq$smr(p'), under the history order.

The history order, then, naturally induces an order of the associated processes. These in turn, it was already noted, form a "process space," a topic which will now be addressed.

7. Process Spaces and Subspaces

For the definition of a process space we start with staging and history spaces S and $\mathbf{H}$, and a nonempty family of processes, $P \subseteq AP(S,\mathbf{H})$. By convention, subd($P$) is the union of subdomains, $\cup\{\text{subd}(p)|p \in P\}$; similarly, $\text{Dom}(P) = \cup\{\text{Dom}(p)|p \in P\}$, and $\text{Rng}(P) = \cup\{\text{Rng}(p)|p \in P\}$. The object $\boldsymbol{P} = \langle P, S, \mathbf{H}\rangle$ will be a *process space* for S, $\mathbf{H}$, in case P satisfies the following three properties: (1) Rng(P) is prevalent in $\mathbf{H}$, (2) P is closed under masking, and (3) P is a history subspace of $AP(S,\mathbf{H})$, under the certifying order of processes induced by $\mathbf{H}$. Dom(P), then, is the topology (underlying set) of the associated space S. Rng(P), on the other hand, defines the underlying set of $\mathbf{H}$ up to isomorphism. A process space is obtained for $P = AP(S,\mathbf{H})$ itself, for which P is as large as possible (and $\text{Rng}(P) = |\mathbf{H}|$), but other process spaces are possible too. It is also clear that P, as a history subspace, must be chain-closed, which means that the family of summaries of processes, as elements of the associated history space containing Rng(P), must also be chain-closed. Since P is closed under masking, these summaries must include all of Rng(P) so that the latter too is chain-closed. Following an earlier result, then, $\text{Rng}(P)^c$ is a favored set under the history order, as is required.

A straightforward example of a maximal process space $AP(S,\mathbf{H})$ is obtained by considering the history space $\mathbf{H}$ that orders the topology $\Im$ of S by inclusion. As we noted earlier, the identity map on $\Im$ becomes a process under this history order, and in fact is an entire process. The general form of a process, on the other hand, is any mapping $p:\Im(V) \to \Im$, for some hyperregion V of S, that satisfies $p(\cup\Im) = \cup\{p(V)|V \in \Im\}$; V is then the subdomain and $p(V)$ is the summary. The maximal variant of any process p is just the constant, entire process p' that assigns smr(p) to each hyperregion of S. A simple example of a non-maximal process space can be obtained from this space, by dropping all but the constant processes. These will form a proper subset of $AP(S,\mathbf{H})$ in case S is nontrivial ($|S| \neq \{\varnothing\}$). Note, on the other hand, that the ranges of the constant processes exhaust the set of histories, thus more than meeting the requirements of a prevalent subset.

In general, given process space $\mathcal{P} = \langle P, S, \mathbf{H} \rangle$, a second process space, $\mathcal{P}' = \langle P', S', \mathbf{H}' \rangle$ is a *subspace* of $\mathcal{P}$ in case (1) $P' \subseteq P$, (2) S' is a staging subspace of S, and (3) $\mathbf{H}'$ is a history subspace of $\mathbf{H}$. In keeping with the definition of process space, P' must form a history subspace of $AP(S', \mathbf{H}')$ under the certifying order, with $\text{Dom}(P') = |S'|$. It follows straightforwardly, from the fact that $\mathbf{H}'$ is a history subspace of $\mathbf{H}$, that (1) the certifying order under $\mathbf{H}'$ is the canonical restriction of the certifying order under $\mathbf{H}$, to the subset P', (2) the greatest lower bound of any nonempty family, $P_0 \subseteq P'$, under $\mathbf{H}'$, is also the greatest lower bound under $\mathbf{H}$, and similarly, (3) a P_0 that is consistent under $\mathbf{H}'$ is consistent under $\mathbf{H}$. From this we conclude that $AP(S', \mathbf{H}')$ forms a history subspace of $AP(S, \mathbf{H})$ under the certifying order, consequently, P' also forms a history subspace of $AP(S, \mathbf{H})$, by transitivity of the history subspace order. It is also straightforward to establish that process spaces form a poset, under the subspace order, so that, as before, we may write $\mathcal{P}' \leq \mathcal{P}$.

Next, we can show that process subspace order is simple-structuring, as follows. Let $\mathcal{P}'' = \langle P'', S'', \mathbf{H}'' \rangle$ be a second subspace of $\mathcal{P}$, with $P'' \subseteq P'$. It then follows that both the domain and range of P'' are subsets, respectively, of the domain and range of P'. Both S' and S'' are staging subspaces of one space S, thus, by simple-structuring, $S'' \leq S'$, and similarly, $\mathbf{H}'' \leq \mathbf{H}'$. This establishes that $\mathcal{P}'' \leq \mathcal{P}'$, as required.

As with staging and history spaces, then, there are certain favored families of processes, that are the underlying sets of subspaces, and which also can be formed into subspaces in only one way. As we have seen, a favored family P_f must form a history subspace under the certifying order. It is evident too that P_f must be *closed under local masking*; that is, given $p \in P_f$, we must have $p_{\|V, V'} \in P_f$, whenever $V, V' \in \text{Dom}(P_f)$. In fact these two conditions can be shown sufficient for a favored set as follows. For a family Q of processes which forms a history subspace and is closed under local masking, consider the object $\mathcal{Q} = \langle Q, S_Q, \mathbf{H}_Q \rangle$ where S_Q is the staging subspace generated from the set $\text{Dom}(Q)$, which is favored in S, and $\mathbf{H}_Q$ is similarly the history subspace generated from the favored set $\mathrm{Rng}(Q)^c$. Then clearly, relative to S_Q, Q is closed under (global) masking, and the other necessary properties are straightforward, so that Q is a process space, which then is a subspace of $\mathrm{AP}(S_Q, H_Q)$. The latter in turn, is a subspace of $\mathrm{AP}(S, H)$, so that by transitivity, Q is a subspace of $\mathrm{AP}(S, H)$ also, and consequently Q is a favored set. Since by hypothesis Q forms a history subspace of P, Q must also be a subset of P, so that finally, in view of simple-structuring, Q must be a subspace of P.

It can now be shown that process subspace order has the minimal subspace property. This follows easily by considering the intersection of favored families of processes. Such an intersection, if nonempty, must form a history subspace since history subspace order has the minimal subspace property. Since all of the favored families of processes are closed under local masking, the domain of the intersection must equal the intersection of the domains (topologies) of the intersecting families, and must also be closed under local masking. Consequently, this intersection is itself a favored family of processes. In view of this result, for any nonempty family P_0 of processes there is a unique, minimal, enclosing process space, generated by the favored set consisting of the intersection of all the favored sets that include P_0.

8. Some Simple Illustrations

Some further cases of process spaces and subspaces are considered here. A better illustration is given (though still rudimentary) of the notion of process space as it might be applied to the physical universe, and also for a multiverse cosmology with multiple universes.

We considered the example of a process space in which the associated history space is defined directly from the staging space, "histories" merely consisting of hyperregions ordered by inclusion. A family of histories, we noted, always has a greatest lower bound (the intersection-interior), and a least upper bound (the union). For a second illustration we now consider a "dual" space in which the histories are reverse ordered by inclusion, as in the case of the prevalent sets in section 5. H then shall consist of the *closed*, complement sets of the form $h = \cup \mathfrak{I} - T$, where T ranges over $\mathfrak{I}$, with $h \geq h'$ iff $h \subseteq h'$. Again, a family of histories has both a least upper bound, in this case the intersection, and a greatest lower bound, the closure of the union. There is a maximal history, the empty set, that upper-bounds every family of histories. There is the intuitively interesting property that a greater history is a *smaller* set, which might be interpreted as a set exhibiting less ambiguity or more information. Intuition, however, breaks down where the empty set must be considered as having "more" information than any other history.

This could be remedied if only *nonempty* closed sets are allowed as histories (as was specifically required for the prevalent sets). Assuming the original point set of spacetime is more than a singleton, with some set intersections empty, then, not every family of histories will be consistent, which is also in keeping with intuition. Instead only lower bounds are always defined. The smallest sets again contain the most information while larger sets have greater ambiguity, hence less information. This viewpoint is suggested, in particular, by the many-worlds formulation of quantum mechanics (a form of multiverse cosmology) in which "loss of information makes the past ambiguous." Assuming $|S|$ is nonempty, then, we always have the process space $AP(S,\mathbf{H})$, with $\mathbf{H}$ as defined, together with its subspaces. In particular, if the topology of S is at least T_1 so that singleton sets are closed, then the maximal histories are all singletons of the points in spacetime.

More generally, maximal histories will consist of minimal, nonempty closed sets M whose existence in turn is guaranteed by Zorn's lemma, inasmuch as $\mathbf{H}$ is a history space. Maximal processes will consist as usual of constant processes, in this case which evaluate to the sets M. This situation is not realistic in terms of the physical universe, where it is reasonable that a process, evaluated for a larger set in its domain, will return a *larger* amount of information.

As one possible remedy out of many, the allowable processes can be limited to mappings of the form $p_C : T \mapsto C \cup (\cup S - T)$, and the associated minimal subspace, where C ranges over the closed sets. Each p_C is then an entire process, and is consistent with a second such process, $p_{C'}$, iff $C \cap C' \neq \varnothing$. If C is a minimal closed set we obtain a maximal process. Two such processes, then, are either identical or inconsistent, just as with maximal histories.

9. Monogenic Enduring and Spaces

This section develops some ideas that will be useful later, particularly as regards the "immortalization theorem," the "preobserver," and the embedding of personlike processes in a surrounding universe. There is also consideration of some features that seem relevant to modeling physical reality and indefinite survival.

A process space $\langle P, S, H \rangle$ is *monogenic* iff there exists a process $p \in P$ that certifies every process in P; p then will be called a *universal* process. A universal process, if it exists, upper-bounds every process and thus is unique. In this case all processes, and consequently all histories, are consistent. It is also clear that the universal process, being maximal in the space, must be entire. In any process space, a monogenic subspace is obtained as the family of all processes certified by any given process.

One example of a monogenic space, that will be of use later, is obtained by considering the family P of all certified *maskings* of a process p, where p is chosen in turn from some process space, $\mathcal{P}_0 = \langle P_0, S_0, H_0 \rangle$. Every process in P must have the form of some masking $q = p_{\|V,V'}$ where $V' \subseteq \text{subd}(p)$ and we can assume, without loss of generality, that $V \subseteq V'$. It is then straightforward to show that a nonempty family of q's, $Q \subseteq P$, will have, as least upper bound *in* $\mathcal{P}_0$, $p_{\|W,W'}$, where W is the union of the V's and W' is similarly, the union of the V''s. Q will also have as greatest lower bound in $\mathcal{P}_0$, $p_{\|X,X'}$ where X, X' are, respectively, the interior intersection of the V's and the V''s. These upper and lower bounds, it should be emphasized, are independent of the embedding space $\mathcal{P}_0$ outside of p and its local maskings, so that P becomes a history subspace of P_0, under its certifying order. Moreover, by process linearity we can establish that $\text{Rng}(P)$ is chain-closed, as a subset of $|H_0|$, so its inf-closure becomes a history subspace, $H \leq H_0$, and a history space in its own right. P of course is closed under local masking so that its domain becomes a staging subspace, $S \leq S_0$. We can then conclude that $\mathcal{P} = \langle P, S, H \rangle$ is a process space in its own right, and a subspace of the original $\mathcal{P}_0$. In a context where the process p is specified, along with associated staging and history spaces S_0, H_0, such that $p \in \text{AP}(S_0, H_0)$, $\mathcal{P}$ will be referred to as the *masking space* of p.

A monogenic space can similarly be formed from the set of *all* maskings of any one process p, not necessarily entire, in an arbitrary process space, as simply the masking space of the entire extension of p. This extension will then serve as the universal process. It is more dependent, however, on the embedding process space, and in fact, less useful here.

From one point of view, a monogenic space makes a good model of reality: the observable universe could be regarded as one universal process, each other process (including each person) being an element in the masking space. From another perspective, however, it would be desirable to allow substantially different processes to share the same subdomain. A person, for example, will appear as very different processes depending on whether one is looking on the level of psychology (as may be more usual) or physiology, biology, or atomic interactions. Though all levels can be reconciled as part of one "process" it is useful to be able to consider them separately. More generally, it would be desirable to allow mutually inconsistent processes, families of which would give rise to inconsistent families of histories, to model such possibilities as a multiverse cosmology.

In addition to the universe as a whole, we wish to model certain features of specific processes, among which is a form of "endurance" or indefinite persistence or survival over "time." The time order, which after our preliminary treatment has not featured much in our development, now becomes important. Points in hyperregions also have greater significance, as opposed to open sets, though the latter still have prominence.

For staging space S with spacetime S, a hyperregion $V \subseteq S$ is *terminating* in case there exists $x \in S$ such that $x \geq y$ for all $y \in V$. The space S is terminating in case every hyperregion is terminating, which means that the whole of spacetime is terminating, so that there exists a maximal element. For this case the spacetime is also (trivially) *upward-directed*: for any two points x, y there exists z with $z \geq x$, $z \geq y$. In particular a staging space in which spacetime has only one point is terminating, as any staging space becomes if we add a "point at infinity": an x not already present with the time order extended so that $x > y$ for all points $y \neq x$, say with $\{x\}$ as an open set and the topology $|S|$ extended by union-closure. It is worth noting that, in general, being terminating is relative to an enclosing space. A hyperregion may itself fail to contain any maximal element and thus may not be terminating, relative to itself, that is to say, within its associated staging subspace, though in fact it may be terminating (along with the associated subspace) within the larger, enclosing space.

Staging space S is *enduring* iff (1) spacetime S forms a nonempty, upward-directed set under time order: that is, if for any two points x, y there exists z such that $z \geq x$, $z \geq y$; and (2) no point in spacetime is maximal. Thus an enduring space cannot be terminating. Applying the axiom of choice, we can well-order S and choose a subset consisting entirely of those elements which are greater, in the time order, than all their predecessors in the well-ordering. In this way we obtain a well-ordered subset of S, $\mathbf{T} = \{x_i \mid i \in I\}$ for index set I of ordinals, with the properties that (1) $x_j > x_i$ whenever $j > i$, and (2) for all $x \in S$, $x_i > x$ for some $i \in I$, and thus, for all $j > i$. $\mathbf{T}$ then is *cofinal* in S under the time order; $\mathbf{T}$ will be called a *timeline* for S. (We shall also have occasional use for an "unrefined" timeline that is not well-ordered but is a chain from which a well-ordered subset meeting the requirements of a timeline could be chosen.) A hyperregion in S is enduring iff it contains a timeline for S; thus spacetime itself is enduring. A staging subspace is enduring in S, similarly, if its spacetime contains a timeline for S. It is perfectly possible, of course, that a staging subspace S' could be enduring, so far as itself is concerned, that is, contain a timeline that is cofinal in its own spacetime, but not be enduring in the enclosing space S, because the timeline (and others in the subspace) are not cofinal in S.

An enduring hyperregion is "never-ending" and must include infinitely many points. A family of hyperregions is enduring (again, in the main, enclosing space) iff its union, a single hyperregion, is enduring, so that the family could be enduring though some or all of its constituents are not. In fact it is straightforward to show that every point is contained in some terminating hyperregion, so that the family consisting of all such hyperregions has the property that its union is enduring while none of the constituents are. Moreover, given any hyperregion U with $x \in U$, there is a terminating subset of U that contains x, obtainable by intersection, so that every hyperregion is the union of terminating hyperregions. (Terminating hyperregions form a topological base for the staging space.)

Going now to process spaces, a process is enduring iff its subdomain (a family of hyperregions) is enduring for the associated staging space. A family of processes is enduring, similarly, iff its domain is enduring, in which case its subdomain is also enduring. Finally, a process space, including a subspace, is enduring iff its domain, defined as the domain of its underlying set (of processes), is enduring for the original, enclosing space. It is, of course, possible for a process subspace to be enduring without any of its constituent processes being enduring, as in the case of staging spaces. A subspace could also be enduring, relative to itself, but not to the enclosing space.

Besides simple endurance, a personlike, immortalized process should exhibit past recall. This will be taken up later. In the next section we consider a specific example of a process space which, though quite crude by the standards of physics, illustrates an enduring process space; this will be important in developing results pertaining to immortalization.

10. A Space Based on Ternary Strings

A process space is defined in which spacetime is the positive integers and histories are strings over a 3-letter alphabet. Histories need not be consistent. The space is not monogenic, and many processes can share a subdomain. This will be particularly useful in what follows, and thus is considered in some detail.

The staging space S will consist of the positive integers, N^+, with the usual order and the discrete topology or powerset of $|S|$, under which every subset is open. A "history" will be an infinite ternary expansion or string η consisting of 0's, 1's and u's, the latter indicating "bit unspecified." Such a string is defined formally as a mapping from N^+ into the set $\{0,1,u\}$. If all the bits are specified (0 or 1, not u) the string is *binary*. Histories are consistent iff the specified bits agree. Histories are ordered so that those with additional specified bits rank higher: thus $\eta_1 \geq \eta_2$ means that, for each positive integer n, either $\eta_2(n)=u$ or $\eta_1(n)=\eta_2(n)$. It is, of course, entirely possible for two histories to be unordered; this can happen if the specified bits disagree or if neither history is obtainable from the other by specifying additional bits. On the other hand, it is easy to see that every chain has an upper bound, a history in which a given bit is specified iff it is specified for some element in the chain, and thus, for every greater element. Any nonempty family of histories will have a greatest lower bound, a history obtained by specifying just those bits that are specified in *every* history in the family, *and* on which all the histories agree. A *consistent*, nonempty family will also have a least upper bound, a history obtained by specifying only those bits that are specified in one or more histories in the family, and as they are specified. (In this case the specified bits must agree, just as in the case of the greatest lower bound, the difference being that for a bit to be specified in the lower bound, *all* the histories in the family must specify the given bit, not merely *at least one*, as in the upper bound.) The maximal histories are precisely the histories having every bit specified, that is, the binary strings. These, and the other histories with some bits unspecified, form history space **H**.

In what follows it will be convenient to represent an arbitrary, ternary history by a "quotient" of binary strings: $\eta=\alpha/\beta$ with the meaning that, for given n, $\eta(n)=\alpha(n)$ if $\beta(n)=1$, $\eta(n)=u$ if $\beta(n)=0$. Another useful operation between binary strings will be a "product": $\alpha*\beta$ will denote the binary string γ whose nth bit is 1 just in case the nth bit of *both* α and β is 1; otherwise $\gamma(n)=0$, or in other words, $\gamma(n)=\alpha(n)\beta(n)$. (Note that the product of binary strings here defined is also a binary string while the quotient need not be. The product is also commutative and associative, and can be iterated.) The product easily generalizes to an arbitrary number of binary strings: the product of strings α_i is γ where $\gamma(n)=1$ iff $\alpha_i(n)=1$ for all i, and 0 otherwise. A hyperregion of spacetime will be denoted by S_τ for binary string τ, with the interpretation that positive integer $n\in S_\tau$ just in case $\tau(n)=1$; thus an arbitrary hyperregion can be represented. Two special binary strings of interest will be **0** and **1** in which every bit is 0 and 1, respectively.

The most general form of process will be $p_{\alpha\beta\gamma}$ with subdomain S_γ, defined for binary strings α, β, γ and hyperregion $S_\tau\subseteq S_\gamma$ by $p_{\alpha\beta\gamma}(S_\tau)=\alpha/(\beta*\tau)$. Such a mapping satisfies linearity and thus qualifies as a process; $p_{\alpha\beta\gamma}(S_\tau)$ is in fact the least upper bound of histories of the form $\alpha/(\beta*\tau_i)$, where the strings τ_i correspond to the singleton subsets of S_τ, that is, are zero except at a single value $i\in S_\tau$. The $p_{\alpha\beta\gamma}$ can then be ordered by the certifying order induced by **H**. It is straightforward to show that the greatest lower bound of an arbitrary, nonempty family of processes, $P_\bullet=\{p_{\alpha_i\beta_i\gamma_i}|i\in I\}$, exists and equals $p_{\alpha\beta\gamma}$, where α is the greatest lower bound of the α_i, and β and γ are the product of the β_i and the γ_i, respectively. If the family forms a chain, on the other hand, then both upper bounds, $\eta=\sup\{\alpha_i/\beta_i\}$ and $\gamma=\sup\{\gamma_i\}$ are defined. γ is a binary string, while η is a ternary string equal to α/β for appropriately chosen binary strings α,β, and we then have $\sup P_\bullet = p_{\alpha\beta\gamma}$. The processes thus form a history space, and will now be shown to form a process space.

Consider, first, a less general form of process, p_α, whose sub-domain is all of spacetime, and for which $p_\alpha(S_\tau)=\alpha/\tau$; this, we see, is just $p_{\alpha\beta\gamma}$ with $\beta=\gamma=1$. It is clear then, that for the more general case of β and γ, $p_{\alpha\beta\gamma}$ has the form of a masking of p_α: $p_{\alpha\beta\gamma}=(p_\alpha)_{\parallel S_\beta, S_\gamma}$. The processes $p_{\alpha\beta\gamma}$ thus are closed under masking, and, together with S and $\mathbf{H}$ as defined, form a process space, henceforth referred to as $\mathbf{P}_0$. This space and modifications of it will furnish useful illustrations in what follows. It is worth remarking that $\mathbf{P}_0$, though containing a richly varied assortment of processes (as the discussion below should make clear), is far from inclusive of all processes in $AP(S,\mathbf{H})$, for example, there are no constant processes, with the exception of the processes in which no bits are ever specified, that is, $p_{\alpha\beta\gamma}$ with $\beta=0$. Another point worth making is that the processes p_α are those among the $p_{\alpha\beta\gamma}$ which are entire, *and* whose ranges consist of entirely specified histories; thus they are precisely the maximal processes in $\mathbf{P}_0$ under the certifying order.

The following illustrations are limited to p_α for simplicity, but could easily be extended to the more general case. If S_β is empty ($\beta=0$) then $p_\alpha(S_\beta)$ is just the totally unspecified history consisting of a string of all u's; effectively there is no information. (Any two processes p_α and $p_{\alpha'}$ assign the same history to this S_β and cannot be distinguished.) If on the other hand S_β is all the positive integers ($\beta=1$) then p_α is just α and there is *complete* information. (Any two distinct processes assign a different, maximal history and *can* be distinguished.) Similarly, any two distinct processes with subdomain S_γ can be distinguished on the set S_γ. In general, the larger the subset of spacetime over which a given process operates, the greater, that is, more specified or less ambiguous, is the associated history.

An interesting alternative representation of histories, equivalent to the above, is to regard a history η as a *nonempty set* of binary strings consisting of *all* those strings that agree with η on its specified bits. For example if η has the first bit unspecified and all remaining bits $= 1$ then, in this representation, η would correspond to the two-element set $\{\alpha,\beta\}$ in which $\alpha=\mathbf{1}$ and β is the string with first bit 0 and all other bits 1. The set of all binary strings, on the other hand, would denote the history with all bits unspecified (effectively, no information), while singleton sets would specify every bit (complete information or no ambiguity). Two histories η_1, η_2 would be reverse-ordered by inclusion: $\eta_1 \geq \eta_2$ if and only if $\eta_1 \subseteq \eta_2$. Similarly, η_1, η_2 would be consistent if and only if $\eta_1 \cap \eta_2 \neq \varnothing$; for this case the least upper bound would be given by the intersection. On the other hand, the greatest lower bound would be defined regardless of consistency, and would include (but not necessarily equal) the union $\eta_1 \cup \eta_2$. More generally, the least upper bound of any family of histories, if the family is consistent, would be the intersection of the sets representing the individual histories, while the greatest lower bound, again, would include the union. This representation is of interest because, like the earlier one involving closed subsets in a topology it corresponds to a multiverse cosmology in which lack of information is equivalent to a multiplicity of possible (complete) histories.

A process essentially resides in one of two states (0 or 1) at each point in its subdomain. This may seem rudimentary, but in fact has considerable representational power, supposing the subdomain is infinite. For example, we could group the subdomain points into blocks of 2^{28} and interpret the blocks as high-definition gray-scale video images (1024×1024 arrays of 8-bit bytes or pixels allowing 256 intensities each, or one megabyte of information per image). These in turn could correspond to a robot's visual memory, or arguably, to the "states of consciousness" of a personlike entity possessing sentience. The transition from one state (image) to another might be governed by specifiable rules, as in digital computation, but this would not be required. Entirely unpredictable transitions could also be represented, or those showing an intractable logic, as might be expected from an "oracle" whose behavior is not Turing computable, or perhaps from a quantum device whose behavior is not *efficiently* computable by the more usual, classical devices. More representational power would be achievable through a larger block size. An entire, active human brain could thus be represented down to atomic resolution in each image, or that portion of the brain that participates directly in consciousness. By an appropriate choice, the time interval between consecutive images could also be made very short. Multiple data tracks could also be represented, a second video source, for example, a sound track, et cetera. With another interpretation, the original data could depict events in an n-dimensional light cone to high resolution. In this way, phenomena resembling those in the universe could be modeled with considerable fidelity, including the life and death of persons, and other possibilities to be considered.

For each interpretation it would be expected that only a small fraction of the many possible processes $p_{\alpha\beta\gamma}$ would be meaningful and interesting, for example, in the case of the 2^{28}-size images representing the robot visual memory. However, it is clear in this case that arbitrary video "movies" could be represented for suitable choices of α. A similar flexibility will hold for other interpretations.

11. Continuers and Episodes in Process Spaces

We return now to the general case of process space $\mathcal{P}=\langle P, S, H\rangle$. Process p is a *continuer* of a process p_0 in case (1) subd(p)>subd(p_0), and (2) smr(p)>smr(p_0). For (1) we are using the time order for the staging space S and for (2) the history space order for H. Intuitively, then, p occurs later than p_0, and also *recapitulates* all the activity of p_0, *with additional activity*, or in other words, *elaborates* the activity of p_0. The two processes are, of course, consistent, and their sum is defined. We shall write $p \triangleright p_0$ ("p continues p_0"). From the order relations on the subdomains and the summaries of the two processes, it can be seen that a partial order on processes, the "continuer order," has been defined, which is quite distinct from the certifying order previously defined.

A process is terminating iff its subdomain is terminating. A terminating process will be called an *episode*. (An episode, then, is not enduring.) One example of an episode is a *degenerate* process, in which the subdomain is empty. Of the non-degenerate cases, it is clear that a process having only one point in the subdomain is also always upper-bounded. Further, if spacetime is upward-directed then every finite set is also upper-bounded, which carries over to the corresponding processes: those with finite subdomains are episodes. In case a hyperregion has both upper and lower time bounds, we say it is *time-sandwiched*, and similarly for processes having the hyperregion as subdomain. Time-sandwiched sets and processes will be of some interest, although as usual there is more emphasis on upper than lower bounds.

To consider an example, for the space $\mathbf{P}_0$, a hyperregion V is terminating iff it is finite; V then is also time-sandwiched (every set being lower-bounded) and, if nonempty, has both a greatest lower and a least upper bound, and contains both these bounds. In fact $\mathbf{P}_0$ is enduring, since its subdomain has a timeline: any infinite set of points in spacetime, in this case, infinitely many positive integers, can serve as a timeline for $\mathbf{P}_0$, though this is not true in general for process spaces.

We turn now to the more general case of an enduring process space, $\mathbfcal{P}=\langle P, S, \mathbf{H}\rangle$. In view of the sufficiency conditions, given point x in spacetime, the *predecessor set* consisting of all $y<x$ is open, thus a hyperregion, as is also the *successor set* consisting of all $y>x$. Let $x_0, x_1, \ldots$ be the points of a timeline for S. The associated predecessor sets $V_0, V_1, \ldots$ form a *cofinal chain* for the terminating hyperregions, ordered by inclusion, in the sense that every such hyperregion is included in (is less than or equal to) some V_i, and for V_j with $j>i$. This will be called a *precursor chain*. Since every terminating hyperregion must be included in some element of the precursor chain, and since the terminating hyperregions form a topological base, the union of all sets in the chain must be all of spacetime.

One consequence is that, if Q is any family of processes such that every upper-bounded masking of Q (every $Q_{\|V,V'} = \{p_{\|V,V'} \mid p \in Q\}$ where V' is upper-bounded) is consistent, then Q itself is consistent. To show this, consider a precursor chain. By linearity we can limit consideration to the case that $V = S = $ all of spacetime, and V' is in the chain. Let $q_{V'} = \Sigma\, Q_{\|S,V'}$. Then the $q_{V'}$ form a chain under the certifying order ($q_{V'} \geq q_W$ if $V' \supseteq W$) and thus have an upper bound, q_S. On the other hand, for fixed $p \in Q$, the processes $p_{\|S,V'}$ also form a chain, with $p_{\|S,S} = p_{\|S}$ as the *least* upper bound, inasmuch as the union of the V' is S. q_S, then, must upper-bound $p_{\|S}$ which in turn must upper-bound the original p. So q_S must upper-bound every $p \in Q$, that is, Q itself, which establishes the desired consistency; in fact it is straightforward to show that $\Sigma\, Q = (q_S)_{\|\mathrm{Dom}(Q)}$.

The fact that upper-bounded sets form a topological base means that a process, essentially, is determined by its behavior on terminating sets. The consistency result additionally means that a process exists when it "ought to," that is, when all terminating maskings exist.

As suggested in the Introduction, the episode is intended to allow the representation of processes that observably come to an end, thus corresponding to the more usual phenomena of observation. On the other hand, we shall consider ways of extending an episode through a "continuer," so that, in the case of a mortal human being, arguably representable as an episode (in this case, time-bounded from below too), the logical possibility of resurrection is given a mathematical foundation. More generally, it is possible to view an episode as a subprocess of an "observer," which in turn is a kind of idealized, immortalized person. Properties will be developed for an "observer space," including the existence of continuers of episodes. In such a space, every episode is a subprocess of some observer. Some arguments are offered suggesting that the physical universe may be an observer space, providing one pathway through which mortal persons, as upper-bounded processes, can be both resurrected and immortalized. Later another possibility will be considered, in which resurrection and immortalization of persons is accomplished through processes which themselves do not operate in any one process space only but whose actions nevertheless combine synergistically to give rise to an immortal individual.

12. The Observer

In a process space an *observer* is a process p with the property that given any terminating masking $p_{\|V,V'}$, there exists a second terminating masking $p_{\|W,W'}$ which is a continuer of $p_{\|V,V'}$. It is clear, then, that $\mathrm{subd}(p) \cap W \cap W'$ is a nonempty, upper-bounded set that is greater than V', and that the primitive masking $p_{\|W \cap W'} = p_{\|W \cap W',W}$

$_{\cap W'}$ is also a continuer of $p_{\|V,V'}$, and also of the primitive masking $p_{\|V \cap V'}$. We then say that p *elaborates* $V \cap V'$ on $W \cap W'$, and that p is *elaborative*. One simple consequence is that every upper-bounded, primitive masking $p_{\|V}$ has an upper-bounded, primitive masking continuer $p_{\|W}$. It is significant, however, that this property is not *equivalent* to the defining property of an observer. Equivalence requires, additionally, that *whatever* upper-bounded hyperregion X I may choose, I can choose $W > X$, or equivalently, whatever point x in spacetime I pick, I can choose $W > \{x\}$. The subdomain of an upper-bounded continuer is also lower-bounded, by the ordering requirements.

It is evident, since p is elaborative, that the subdomain of p is not upper-bounded for otherwise there would exist $W \cap W' >$ subd(p), on which p would elaborate subd(p), clearly an impossibility. p, then, is a enduring process, which infinitely often recalls every earlier experience it has had, with additional activity each time, and thus is "immortal."

As envisioned here, then, an observer is a special sort of process that "survives" through information that recurs in the sets of histories associated with the process as time progresses. An observer thus must be part of a process space for which spacetime is "timelike" or not upper-bounded or terminating. The definition of observer is intended to capture two properties that seem particularly critical to a notion of robust or perfect survival: (1) recall of past information, and (2) addition of new information. Both of these properties are expressed in the notion of being elaborative.

Let us now consider a simple example of an observer, for a masking space $\mathcal{P} = \langle P, S, H \rangle$ defined as follows. As with P_0, S is the positive integers with the usual ordering and the discrete topology. H, however, consists of the positive integers with the usual ordering, together with an additional value, ∞, regarded as being greater than any integer. H thus forms a simple history space in which any subset has both a greatest lower and a least upper bound. Next, P shall consist of a single process p together with all its maskings, where p is defined, for a set of integers T, by $p(T) =$ the least upper bound, in H, of T. $p(T)$ then is the maximum value in T if T is finite (or 0 if $T = \varnothing$), and otherwise $p(T) = \infty$. In particular we see that, given a family of sets T_i of integers, $p(\cup T_i)$ is the least upper bound of $\cup T_i$, which in turn is just the least upper bound of all the $p(T_i)$; p thus qualifies as a process, and $\boldsymbol{\mathcal{P}}$, formed from the maskings of p, qualifies as a process space. $\boldsymbol{\mathcal{P}}$ in fact is an enduring space, having the same spacetime as $\mathbf{P}_0$. Finally, it is easy to show that p is an observer. For any terminating T and integer n we can choose m both greater than n and greater than any element of T. $\{m\}$, then, is a terminating hyperregion greater than both T and $\{n\}$, and furthermore, $p_{\|\{m\}} \rhd p_{\|T}$. As noted above, this suffices to show that p is an observer. In fact, since every episode is a subprocess of p, $\boldsymbol{\mathcal{P}}$ itself is an observer space.

In effect, in this simple case process p, over *each instant* of spacetime, is able to recall its entire past and add additional information! Though satisfying the formal properties of an observer, this ability clearly is counterintuitive and unnatural. A more realistic example of an observer would thus be instructive.

In fact it can be asked whether $\mathbf{P}_0$ contains any observers. The answer is *no*; none of the processes in this space are elaborative, hence none can be an observer. To show this, first consider some process $p_{\alpha\beta\gamma}$, supposed to be an observer, which in turn must have an infinite subdomain. Choose $i \in \mathrm{subd}(p_{\alpha\beta\gamma})$. Then, since singletons in $\mathbf{P}_0$ are terminating, open sets, there must be a hyperregion $V > \{i\}$ such that $(p_{\alpha\beta\gamma})_{\|V}$ elaborates $(p_{\alpha\beta\gamma})_{\|\{i\}}$. But from the definition of $p_{\alpha\beta\gamma}$ it is clear that for *no* hyperregion V in the subdomain that excludes i can it be true that $p_{\alpha\beta\gamma}(V) \geq p_{\alpha\beta\gamma}(\{i\})$, let alone $p_{\alpha\beta\gamma}(V) > p_{\alpha\beta\gamma}(\{i\})$, so that no such elaboration is possible.

In the next section a process space is defined in which observers exist. Spacetime (including the associated topology) is the same as in $\mathbf{P}_0$, histories are a subset of those in $\mathbf{P}_0$, and processes are defined differently. As a consequence, a process is able to "remember" its past arbitrarily far into the future, and elaborate its recollections as required. Unlike the observer just defined, full recollection does not happen in an instant but is more realistically spread over a hyperregion whose size is proportional, approximately, to the amount of the past that is recalled. Some formal groundwork is also laid that will be useful in later illustrations.

13. A Space with Observers

As before, spacetime is the positive integers with the usual ordering and the discrete topology; the staging space S, then, will remain the same. Histories are ternary strings, indicating specified and unspecified bits as before; however, in this case, the set of allowable histories is restricted. Let m^Δ denote the greatest odd divisor of positive integer m. An allowable history h must have the "r" property (for "redundancy") that, *if* for some m, $h(m)$ is specified, then $h(2^k m^\Delta)$ is specified for $k \geq 0$ and is always equal to $h(m)$. $h(2^k m^\Delta)$, then, simply repeats the information stored at $h(m)$, and this is done infinitely often, though sparingly. Such an h will be called an r-*string*.

It is easy to show that the inf of any nonempty family of r-strings, treated as histories, is an r-string, as is the sup of any consistent family, including a chain; the r-strings, then, form a history space $\mathbf{H'}$ which happens to be a subspace of the history space $\mathbf{H}$ of $\mathbf{P_0}$.

In addition to the r-strings, another class, the "pre-r" strings will be important. h is pre-r iff, whenever $h(2^k m^\Delta)$, $h(2^{k'} m^\Delta)$ are specified for integers k, $k' \geq 0$, $m > 0$, then $h(2^k m^\Delta) = h(2^{k'} m^\Delta)$. The bits in question need not be specified, however. A pre-r string, then, is not always an r-string, but is *minimally* extendible to an r-string, in just one way, by setting unspecified bits $h(2^{k'} m^\Delta)$ to $h(2^k m^\Delta)$ when the latter is specified for *some* k. The simplest examples of pre-r strings are strings in which only odd-numbered bits are specified. If η is a pre-r string, η^+ will denote its minimal extension to an r-string.

We can now define the mappings that will serve as processes in the new space. As before, these mappings will be of the form $p_{\alpha\beta\gamma}$, where each of α, β, γ are binary strings (ternary strings in which *every* bit is specified). In addition, we shall require, unlike the case of $\mathbf{P_0}$, that α is an r-string; β and γ are not similarly restricted, and can be arbitrary binary strings just as before. With this restriction, it can be seen that, *in* $\mathbf{P_0}$, the value of $p_{\alpha\beta\gamma}$ is always a pre-r string, and thus is amenable to extension to an r-string. For the new space, we apply the minimal extension automatically: $p_{\alpha\beta\gamma}(S_\tau) = (\alpha/(\beta * \tau))^+$.

Using the properties that the inf and sup of a family of r-strings, where defined, are r-strings, it is straightforward to show that (1) each $p_{\alpha\beta\gamma}$ is a process that maps into $|\mathbf{H'}|$, (2) the family P'

of processes forms a history space, under the certifying order induced by $\mathbf{H}'$, and (3) $\mathrm{Rng}(P')=|\mathbf{H}'|$. As before, a maximal process is obtained for $p_\alpha=p_{\alpha\beta\gamma}$ for $\beta=\gamma=\mathbf{1}$. Again we can show straightforwardly that, for more general β, γ, $p_{\alpha\beta\gamma}$ has the form of a masking: $p_{\alpha\beta\gamma}=(p_\alpha)_{\|S_\beta,S_\gamma}$. Once again, the $p_{\alpha\beta\gamma}$ are closed under masking. From these properties, then, we conclude that the object $\mathbf{P}_1= \quad \langle P', S, \mathbf{H}'\rangle$ is a process space. (Because of the additional complication, in evaluating $p_{\alpha\beta\gamma}$, of extending the result to an r-string, $\mathbf{P}_1$ is not simply a subspace of $\mathbf{P}_0$, however.) Though clearly related to $\mathbf{P}_0$, $\mathbf{P}_1$ differs in important respects, one of which is that it has observers.

In fact, each p_α is an observer, as can be seen by considering a hyperregion S_τ: $p_\alpha(S_\tau)=(\alpha/\tau)^+$. Let mS_τ denote the set obtained by multiplying each element (positive integer) in S_τ by m. We then must have $p_\alpha(2^kS_\tau)=p_\alpha(S_\tau)$ for any $k\geq0$. If S_τ is upper-bounded, hence finite, we can multiply each element by a large power of 2 and introduce a large odd number to obtain finite $S_{\tau'}>S_\tau$ for which $p_\alpha(S_{\tau'})>p_\alpha(S_\tau)$. The other properties establishing p_α as an observer are straightforward. This, it should be noted, is true regardless of the choice of α. More generally, $p_{\alpha\beta\gamma}$ will be an observer, regardless of α, for many choices of β and γ, including the case that each has all but finitely many bits $=1$. (And in fact, whether $p_{\alpha\beta\gamma}$ is an observer is entirely dependent on β and γ, not α.)

It was noted in connection with $\mathbf{P}_0$ that each process could be regarded as furnishing a sequence of images to provide a "movie" of past events. To extend this notion to $\mathbf{P}_1$ we could group the *odd-numbered* bits into blocks, say of 2^{28} each, and proceed analogously. Each video "frame" then would be interspersed with even-numbered bits providing information about past frames. In general any past frame, or finite sequence of frames, would recur and thus be "recalled" infinitely often. In effect, scanning even-numbered bits would be recalling the past, while scanning odd-numbered bits would be observing the present. In this rudimentary way then, we see how a process could carry out forms of observation and reflection. Other, more realistic possibilities would follow either by redefining the processes or by considering special classes of processes. (A process for instance, in the course of "observing the present" might also carry out deductions, initiate actions, et cetera.)

14. Observer Spaces

Like P_0, P_1 is an enduring space, since this property depends only on the staging space, which is the same in both cases. In addition, though, it has the property that *every episode has an enclosing observer*, for which the given episode is a subprocess, that is, a primitive masking over a subset of the original subdomain. This may be seen as follows. Every upper-bounded process must have the form of $p_{\alpha\beta\gamma}$ where γ corresponds to a finite set (only finitely many bits $=1$), and is identical to any process $p_{\alpha\beta'\gamma}$ for which $S_{\beta'}\cap S_\gamma = S_\beta\cap S_\gamma$. In fact, since S_γ is finite, a string β' can be chosen with all but finitely many bits $=1$. $p_{\alpha\beta'\gamma} = p_{\alpha\beta\gamma}$ will then be a subprocess of $p_{\alpha\beta'\gamma'}$ where $\gamma'=1$. As noted above, such a process is an observer. Clearly it is also an enclosing process of the original $p_{\alpha\beta\gamma}$.

An enduring space in which every episode has an enclosing observer will be called an *observer space*. In particular, if p is an observer, then the masking space of p is an observer space, or in other words, any process space with an observer has an enduring observer subspace, which in fact is a masking space.

The principal result of this section, which will be referred to as the *Immortalization Theorem*, is as follows. An enduring space is an observer space if and only if every episode has a continuer episode.

Proof. In an observer space, every episode has an enclosing observer, which can furnish infinitely many continuer episodes. If, on the other hand, every episode in an enduring space has a continuer episode, it can be shown to have an enclosing observer, as follows. Let p be an episode, and let V_0, V_1, ... be a precursor chain. From Zorn's lemma there exists a maximal chain of episodes p_i, with $p_0=p$, and $p_{i+1} \triangleright (p_i)_{\|V_{j(i)}}$, where the values $j(i)$ are chosen sufficiently large so that $\mathrm{subd}(p_i)\subseteq V_{j(i)}$, with j a strictly increasing function of i. Let p_+ be the sum of the p_i. From the construction and linearity it is evident that, for a terminating set $W\subseteq V_{j(i)}$, $\mathrm{smr}((p_+)_{\|W})\leq\mathrm{smr}(p_i)<\mathrm{smr}(p_{i+1})$, while also $W<\mathrm{subd}(p_{i+1})$. Now, consider a terminating masking $(p_+)_{\|V,V'}$. By linearity and the definition of a masking, $h=\mathrm{smr}((p_+)_{\|V,V'})=\mathrm{smr}((p_+)_{\|V\cap V'})$. On the other hand, for i sufficiently large, $V\cap V'\subseteq V'\subseteq V_{j(i)}$, so that $h\leq\mathrm{smr}(p_i)$, and we then have $p_{i+1}=(p_+)_{\|\mathrm{subd}(p_{i+1})}\triangleright(p_+)_{\|V,V'}$. Every terminating masking of p_+ thus has a terminating masking continuer, so that p_+ is an observer.

This result, the Immortalization Theorem, deals mainly with extending a terminating process (episode) to a enduring process (observer). The extension succeeds, mainly, because it is possible to design a process "fragment" (episode) that "fits onto" and extends the previous, terminated fragment, which is also, of course, an episode. "Immortalization" amounts to constructing a sufficiently lengthy succession of these extending fragments. Each successive extension is, in a sense, a resurrection. It requires, first of all, embodying a process with an adequate recollection of a "past life" corresponding to a process that previously existed, and enabling it, in a reasonable sense, to add to that recollected past in the manner of an individual storing new memories and updating that past experience. Such a resurrection may happen more than once; each time the original episode plus the previous continuers must all, in a sense, be restored before further elaboration can take place.

15. The Universe as an Observer Space

The formalism of a process presented here invites comparison with natural phenomena, and leads to the question of whether the observable universe might be an observer space, in a reasonable sense. On the other hand, if the universe *is* an observer space then the claim that resurrection is possible has a logical foundation, according to the results just derived. This is a problem of very large scope and full treatment is certainly not possible in this study, though its importance demands that it be considered, at least at a starting level. The main question clearly involves three sub-issues: (1) whether the universe is to be considered a process space, under a reasonable notion of "process," (2) supposing the answer is "yes," whether the universe is an enduring space, and (3) again, if so, whether it is, additionally, an observer space. Much about the universe, of course, remains unknown, particularly as it might apply to the possibilities of eternal survival. The following, however, will offer some suggestions how reality as we know it might qualify as an observer space and thus allow for unlimited personal existence and other possibilities that have been delineated mathematically.

In general, happenings in the universe are regarded as yielding "histories"—finite chunks of information (finite bit strings for example) which are assigned to bounded volumes of spacetime. By adding histories covering different hyperregions in spacetime we map the process over a larger hyperregion. The whole then is regarded as the sum of its parts—this is the principle of linearity. A larger mass of records corresponds to a greater history. The "least upper bound" for consistent records is obtained simply by lumping all the records together. If records are inconsistent, however (in which case they should be regarded as *possible* history only) then, with proper assessment, a maximal, consistent subset should be extractable; this is the "greatest lower bound." Certainly this approach involves some simplifications; uncertainties in records, for example, may not always be dealt with so straightforwardly. Such possibilities might still be accommodated, however, under the general paradigm of history as an order-complete poset. Extrapolating to a limiting case, a consistent, infinite collection of records could be combined into a meaningful history of infinite size; every chain thus would have an upper bound.

This approach seems intuitive and is also supported by the examples of P_0 and P_1, particularly if the capacity for more naturalistic modeling of real phenomena (for example the n-dimensional light cone) is taken into account. We can agree to call a subprocess, as defined here, a process in its own right, and more generally extend the notion of process to what has been called a masking, if mainly for technical reasons.

Spacetime reasonably forms a poset; certain events clearly precede others. According to relativity, an event precedes another if a light signal could be sent from the spatial location of the first event to arrive at the location of the second, not later than the second event occurs. (Actually, there is a distinction between causal precedence and chronological precedence. One event is earlier chronologically if a light signal would arrive *before* the second event happens, a distinction that is not critically important if one is only dealing, as here, with open subsets of spacetime ordered temporally as in process spaces described earlier.) A point in spacetime (event location) is specified as a 4-vector $\langle x,y,z,t \rangle$

where x, y, and z are spatial coordinates and t is time. (A particular coordinate system thus must be chosen; the precedence of events, however, will be independent of this.) A second point $\langle x',y',z',t' \rangle$ is later, chronologically, in case $t'>t$ *and* $(x-x')^2+(y-y')^2+(z-z')^2-c^2(t-t')^2 < 0$, where c is the speed of light. (If only equality holds and still $t'>t$ then the first event causally but not chronologically precedes the second, again not an important distinction here.) Some points, then, are not comparable even though their times are.

Spacetime also has a topology, which with the other properties just detailed, would complete the requirements for a process space. We can define an open sphere of radius r about the "center" $\langle x,y,z,t \rangle$ as the set of points $\langle x',y',z',t' \rangle$ for which $(x-x')^2+(y-y')^2+(z-z')^2+c^2(t-t')^2 < r^2$. Open spheres are terminating and can serve as a topological base (or the base can consist of just the open spheres of radius $\leq r$ for some fixed r, or of nonzero maximum radius that varies with the location of the center). Actually the open spheres are also lower-bounded (thus are time-sandwiched), and terminating hyperregions in general seem important. There also exists a timeline, assuming time is not upper-bounded. Any set of points $\langle x,y,z,t \rangle$ where x, y, and z are fixed and t is not upper-bounded, for example, would serve as an "unrefined" timeline from which a well-ordered subset could be chosen. In fact, by restricting t to positive-integer values we would obtain a countable timeline that is well-ordered or "refined." (The "zero point" of time, where $t=0$, could, of course, be arbitrary with respect to actual history, as with other details such as the units of time and distance.) Our universe then, could arguably be an enduring space, and it remains to consider whether it might also be an observer space.

A basic plausibility argument is possible, I think, if we assume that terminating processes can be simulated in a digital computer. This in turn seems reasonable: in a relativistic model of spacetime, every terminating hyperregion has a finite volume geometrically, and arguably supports only a finite number of events. With a little additional effort a time bounded process would have a terminating continuer: a consistent, finite history could always be added to in a meaningful way. (As a technicality, a *finite* set of events could be simulated computationally even if they were part of a process that is not Turing computable.) In the case of a deceased individual, for instance, it should be possible to create a duplicate able to scan his/her memories, and resume conscious life as a continuer of the earlier person. It might be argued that an infinite, upper- but not lower-bounded episode could not be recapitulated or elaborated with a terminating process as here envisioned; however such episodes do not seem part of observable reality and thus can be ignored. So it appears that, in the universe as we know it—possibly an enduring space—every episode could have an episode continuer, making the universe an observer space. This is far from guaranteed, of course—the universe could eventually self-destruct or otherwise become inhospitable to life, however cleverly an advanced civilization of the future might strive to meet the threat to its existence. But at least this sort of outcome is not guaranteed by what is presently known.

The creation of a duplicate person would involve, at worst, a finite number of lucky guesses, on the part of say, a programmer of the future working with a very large computer. (This does not overlook the fact that a *very* large number of trials would be necessary to recreate any specific person!) A similar resurrection could, if necessary, happen over and over, each time involving only a finite number of steps and a finite effort (and appropriate additional information covering the most recent states of consciousness). For immortalization or creation of an observer this sort of resurrection might have to happen infinitely often (though maybe not). In a general enduring space the order of infinity might itself be very large. In the observable universe, however, it need only be countable, since there exists a countable timeline.

The above can be taken as evidence supporting the position that immortalization of a personlike process would require at most a countable infinity of resurrections, each in turn requiring simulation of finitely many events for recapitulation, a countable number of operations altogether. All then, should be achievable if time and space are unlimited, or possibly in more restricted spatial domains in which time alone is infinite, as with P_1.

One issue that arises in comparing an observer space as defined here with the observable universe is that the former has properties *guaranteed*, while the latter appears to be *permissive* only. Thus, while it may be argued that, in physical reality, a process corresponding to an "episode" *may* have an "episode continuer" this is not foregone. On the other hand though, an argument might be advanced that *sooner or later*, in an unbounded universe, any particular *finite* process will happen, at least in some equivalent or isomorphic form, and even be repeated infinitely often for arbitrarily remote future times. This will guarantee arbitrary episode continuers, and as a consequence, enclosing observers of every episode. The thrust of this argument would be that immortality is inevitable, though one should still have a choice as to which particular characteristics would be more *likely* to occur in one's continuers.

Another, more interesting consideration is that the notion of histories lends itself to a multiverse formalism, as was remarked earlier in connection with the poset of histories used in P_0 and P_1. This would follow in a more general space if the elements of the history space H themselves take the form of nonempty subsets of some set H. As before, the subsets would be reverse-ordered by inclusion. A family of such subsets would be consistent if its intersection were nonempty, and the intersection would then serve as the least upper bound. Under such a scenario, the possibilities for resurrection are greatly expanded: essentially, what is permitted is guaranteed, so that again, immortality is inevitable, and all the more variedly and forcefully so, than in the single world scenario above. Nevertheless, one still has control over which scenarios are more likely, and this can be used as the basis of choices about matters affecting one's future; more on this later.

16. The "Soul"

The certifying order of processes, which is one of the fundamentals in the notion of process space we have considered, is used here to develop a concept of the "soul" of a process. It will be seen that, for those cases in which a "process" here defined could reasonably be regarded as a personlike entity, this soul bears a remarkable resemblance to traditional notions of the concept of soul. It is that portion of the entity that can be said to persist throughout time and will never suffer permanent loss. More formally, the soul is a process that is obtained as the sum of processes certified by the original process, and thus itself is a process that is certified by the original process. It is always defined, and is "self-summarizing" in a sense implying having total recall of its past at the end of time. In appropriate cases the soul will be an observer even though the certifying process is not. In this way the notion of an immortal, conscious component of an otherwise perishable, personlike entity can be modeled mathematically.

A hyperregion in staging space S is *stable* if it contains a point x together with all $y > x$; x will then be called a *free point* of the hyperregion. The family of stable, (open) hyperregions in $\mathfrak{I}$ will be denoted by $\text{Sh}(\mathfrak{I})$. (In case S is a directed set, as in an enduring space, two stable hyperregions must have a free point in common, and it is then straightforward to show that $\text{Sh}(\mathfrak{I}) \cup \{\varnothing\}$ is a topology.) $|S|$ is a stable hyperregion in $\mathfrak{I}$; thus, except in the trivial case of an empty spacetime, a stable hyperregion always exists and $\text{Sh}(\mathfrak{I})$ is nonempty. (It is worth noting here, that in an enduring staging space a stable hyperregion is also always enduring, though the converse need not hold, that is, an enduring hyperregion may contain "holes" or consist of disjoint subsets, so that stability is violated.)

For a process p the *final history* $\text{fh}(p)$ is defined as $\inf\{\text{smr}(p_{\|T}) | T \in \text{Sh}(\mathfrak{I})\}$. Intuitively, $\text{fh}(p)$ is the "surviving history" delineated by p. In general $\text{smr}(p)$ will be greater than $\text{fh}(p)$; the difference of the two histories will comprise the lost information.

In particular, then, in an observer space (where spacetime is unbounded), we see that, if p is an episode, $\text{fh}(p) = p(\varnothing) \leq p(X)$ for all $X \in \text{Dom}(p)$. On the other hand, if p is an observer, consider any free point x in a stable hyperregion T. Any precursor chain must have an element (hyperregion) V_i that contains x. For this case we must have $V' > V_i$ such that $p_{\|V'} \vartriangleright p_{\|V_i}$. Since it then must follow that $V' > \{x\}$ we must have $V' \subseteq T$, and consequently, $\text{smr}(p_{\|T})$ must upper-bound $\text{smr}(p_{\|V_i})$, for all V_i in the chain. Since the union of all the V_i is $|S|$, this shows that $\text{smr}(p_{\|T}) = \text{smr}(p)$. Since this holds for *every* stable T, we have that $\text{fh}(p) = \text{smr}(p)$. So, for an episode or terminating process, all information (other than the "name" $p(\varnothing)$) is eventually lost, while for an observer there is no lost information and the summary and final history are identical. In general a process whose final history and summary are equal will be called *self-summarizing*.

Besides an observer, a simple example of a self-summarizing process is just $p_{\|\varnothing}$ for any process p. In other words, *every process certifies a self-summarizing process*. Secondly, it is not difficult to show that *an arbitrary sum of consistent, self-summarizing processes is itself a self-summarizing process.* Since the family of all processes certified by p is consistent, we may sum the self-summarizing, certified processes to obtain a unique, maximal, self-summarizing, certified process, which will be called the *soul* of p. Intuitively, the soul captures the information about p that is to endure for all time. It is interesting that every process whatever has a soul in this sense, and also that an observer in an observer space is equal to its soul.

17. The Preobserver

A *preobserver* is a process p that certifies an observer. The observer in turn will be a process in an enduring space whose subdomain must contain a timeline, which in turn must be in subd(p). The intersections of hyperregions in the embedding space in turn with subd(p) must form a topological base for $\Im(\text{subd}(p))$. The upshot then is that the family of processes certified by p itself forms an enduring space. It is also natural to restrict attention to this space; in particular, then, a hyperregion is normally assumed to be a subset of the subdomain of p.

Although every observer is a preobserver, a preobserver need not be an observer, and in fact can be "forgetful." For this reason the preobserver seems a more realistic modeling of a personlike process than an observer. (Examples of preobservers that are not observers are easily specified in $\mathbf{P}_1$; such illustrations, and others relating to material covered here, are considered in the next section.) More than one observer, on the other hand, could be certified by p. With a reasonable assumption relating to histories, any sum of consistent observers is an observer (though this property does not hold in general). Preobserver p will then have a unique, maximal, certified observer consisting of the sum of all its certified observers. With another reasonable assumption relating to spacetime (though again, it does not hold in general), this maximal observer will also equal the soul.

A history h is *realized* in a process p' iff it is upper-bounded by the summary of p'. If h is realized then so is any *predecessor* history, $h' \leq h$. (Here h is regarded as a predecessor of itself.) If h is realized in an episode that is a subprocess of p' then it is *episodic* in p', along with its predecessors. (This applies to arbitrary

277

processes, not necessarily in an observer space.) By a *cover* of h is meant a set of histories H_0 such that $h \leq \sup H_0$. A subset of H_0 that is also a cover of h is a *subcover*. The assumption about the histories that are realized in preobserver p will be that *every cover of an episodic history has a finite subcover*. By analogy with the usual topological definition of compactness based on order by inclusion, this property will be called *order compactness*. (The histories in fact, play the role of open sets, while sup is the "union" and the episodic histories, though "open" like other histories, are also "closed and bounded" or "compact" and thus have a finite subcover.) The cover H_0 is *exact* in case $h = \sup H_0$; for this case every subcover must also be exact.

Intuitively, episodic histories are constrained in how much information they can contain. A straightforward example of such constraint would be the case of a process that has a beginning in time, in which each episode, occupying a hyperregion with both an upper and a lower time bound, assigns only a finite amount of information in its summary. A cover should be a family of histories that collectively contain all the information contained in the given history. The latter in turn will consist of a concatenation of some finite number of indivisible units (bits, say), each of which is assigned to some definite location or position. A given history in the cover either has the information contained in a given bit position or does not. If collectively all bit positions are covered, it means that for each position some specific history in the cover contains the information. Only finitely many such histories, then, are needed for a complete cover, so that all the information is contained in some finite subset of the original cover.

Given preobserver p, a p-certified process p' is *bounded* iff it has an episodic final history. If p' is bounded, then so is any process certified by p'. Moreover, if a process is bounded, then so is any masking whose subdomain is included in that of p. *An observer, on the other hand, cannot be bounded in this sense.* To show this, let p_0 be a p-certified observer, so that in particular, $\mathrm{fh}(p_0) = \mathrm{smr}(p_0) \geq \mathrm{smr}((p_0)_{\|T})$. We can then define an infinite sequence of terminating hyperregions, T_0, T_1, ..., such that the associated histories, $h_i = \mathrm{smr}((p_0)_{\|T_i})$, are totally ordered, with $h_i > h_j$ whenever $i > j \geq 0$. The h_i, then, are infinite in number and there is no maximal h_i. For the hyperregion $T=\cup\{T_i\}$ let $h=\text{smr}((p_0)_{\|T})=\sup H$ where $H=\{h_i|i\in I\}$. If p_0 is bounded then clearly h is episodic, so that by order compactness the exact cover H must have a finite, exact subcover. Since, on the other hand, the h_i are totally ordered, there must be a maximal h_i, or in other words h itself must be among the h_i, contradicting the property that no h_i is maximal. From this result we see that *no process that certifies an observer is bounded*, so that in particular, the preobserver itself is not bounded.

We shall now show that any sum of one or more p-certified observers is an observer, so that in particular, the sum of all the p-certified observers is an observer. Let $p_+=\Sigma P_0$ where P_0 is a nonempty family of p-certified observers, let V be a terminating hyperregion, and let x be a point of spacetime. To show that p_+ is an observer we must show (cf. section 8) that there exists terminating $W>\{x\}$ such that $(p_+)_{\|W} \rhd (p_+)_{\|V}$.

To show this in turn, we first note that the episodic history, $\text{smr}((p_+)_{\|V})$, has an exact cover, $\{\text{smr}((p_0)_{\|V})|p_0\in P_0\}$, which must thus have a finite, exact subcover. This means there exists a sum $p_\&$ of a finite subset of observers, $P_f\subseteq P_0$, such that $\text{smr}((p_+)_{\|V})=\text{smr}((p_\&)_{\|V})$. For each $p_f\in P_f$ in turn there exists terminating $T_f>\{x\}$ such that $(p_f)_{\|T_f} \rhd (p_f)_{\|V}$. Let W_0 be the union of the (finitely many) T_f. Then, since spacetime is an upward-directed set in an enduring space, W_0 is both terminating and $>\{x\}$, such that $(p_f)_{\|W_0} \rhd (p_f)_{\|V}$ for all $p_f\in P_f$. This shows that $\text{smr}((p_\&)_{\|W_0})\geq\text{smr}((p_+)_{\|V})$. If inequality holds, then we set $W=W_0$. Since, by construction, p_+ must certify $p_\&$, we then have $(p_+)_{\|W} \rhd (p_+)_{\|V}$ as desired. On the other hand, if only equality holds, we can construct an infinite sequence of terminating hyperregions W_1, W_2, ... all $>\{x\}$, such that, for any $p_f\in P_0$, $h_i=\text{smr}((p_f)_{\|W_i}) > h_j=\text{smr}((p_f)_{\|W_j})$ whenever $i>j\geq 0$. The histories h_i then, are totally ordered with no maximal element. Consequently, for the hyperregion $T=\cup\{W_i\}$, $(p_f)_{\|T}$ is unbounded. For this case, we again must have $\text{smr}((p_+)_{\|W_i})\geq\text{smr}((p_\&)_{\|W_i})\geq\text{smr}((p_+)_{\|V})$. If equality always holds, then both $(p_+)_{\|T}$ and $(p_\&)_{\|T}$ are bounded. This, however, is contradicted by the fact that both certify the unbounded process $(p_f)_{\|T}$. For some

W_i, then, we must have $\mathrm{smr}((p_+)_{\|W_i})>\mathrm{smr}((p_+)_{\|V})$, so that, setting $W=W_i$, $(p_+)_{\|W} \rhd (p_+)_{\|V}$, showing that p_+ is an observer as desired.

In particular the sum of all certified observers is a unique, maximal, certified observer. Intuitively, we could think of the maximal certified observer as an immortal being that preobserver p is giving rise to, over the course of infinite time. This rationalizes certain pruning operations or other alterations p might make in the course of existence. p itself, in view of these changes, may not strictly qualify as an observer, yet could still experience a meaningful immortalization.

We turn now to the second major objective of this section, to establish reasonable conditions under which the maximal certified observer will also equal the soul. One additional assumption will be made, that *every point in spacetime is a lower bound of some stable hyperregion*. This property, which applies to the subdomain of the preobserver (and by reasonable extension, to all of spacetime), will be called *stable time-bounding*. Clearly it is reasonable in a modeling of the physical universe: the set of points greater than (later than) a given point in spacetime (the forward light-cone) is complementary to the set of those that are less than or earlier and contains stable hyperregions.

We now show that, given preobserver p in an enduring space with the assumptions of order compactness and stable time-bounding, the soul of the preobserver is equal to the maximal, certified observer.

Let p_s denote the soul and let p_f be an observer certified by p. Since p_f is self-summarizing, it must be certified by p_s ($p_f \leq p_s$). So, if we can show that p_s is an observer, then it must be the maximal observer. Again, let V be terminating and let x be a point in spacetime. In an enduring space we can choose y so that $\{y\}>\{x\}$. By stable time-bounding y is a lower bound for some stable T so that T is also greater than $\{x\}$. Let h be the episodic history $\mathrm{smr}((p_s)_{\|V})$, and let $h_s=\mathrm{smr}(p_s)$. Again we seek $W>\{x\}$, in this case so that $(p_s)_{\|W} \rhd (p_s)_{\|V}$. As in the previous proof, the present argument proceeds by showing that, if no such W existed, a bounded process would certify an unbounded process.

Let V_0, V_1, ... be a precursor chain. T is then the union of all the sets of the form $V_i \cap T$, each of which is terminating. Since p_s is self-summarizing, $\mathrm{smr}((p_s)_{\|T})=h_s \geq h$. On the other hand, $\mathrm{smr}((p_s)_{\|T})$

is the least upper bound of all histories of the form $\mathrm{smr}((p_\mathrm{s})_{\|V_i \cap T})$, which in turn thus form a cover of the episodic history h. By order compactness, there must be a finite subcover, which shows that, for i sufficiently large, $\mathrm{smr}((p_\mathrm{s})_{\|V_i \cap T}) \geq h$. If inequality holds we can set $W = V_i \cap T$ and are done. If just equality holds for all i sufficiently large, then $\mathrm{smr}((p_\mathrm{s})_{\|T}) = \mathrm{smr}(p_\mathrm{s}) = h \geq \mathrm{smr}(p_\mathrm{f})$, so that p_f, an observer, is bounded. That, however, is impossible; thus some such terminating W exists, with $W > \{x\}$ and $(p_\mathrm{s})_{\|W} \rhd (p_\mathrm{s})_{\|V}$. So p_s, the soul, is also the maximal certified observer.

18. Examples of Preobservers in Enduring spaces.

Some simple examples will illustrate the properties of a preobserver, both in the "favorable" case where both order compactness and stable time-bounding hold, and in some interesting "unfavorable" cases where one or the other of these assumptions is violated.

In the space $\mathbf{P}_1$, the positive integers comprise all of spacetime. All subsets of spacetime are open, so that for any x, the set of $y \geq x$ is a stable, open hyperregion with x as lower bound. An episodic history h of any process has only a finite number, n, of specified *odd-numbered* bits. (The number of even-numbered, specified bits may be infinite, however, though these simply echo or repeat the corresponding odd-numbered bits.) For any family of histories that covers h, at most n histories are needed to cover the n odd bits, together with the even-numbered bits that echo them. So the two assumptions hold. Because of the convention that an even-numbered point m of spacetime recapitulates the information in $m/2$, any process that has all but a finite subset of spacetime as subdomain, is an observer. Similarly, an observer results if we omit from the subdomain any "elemental" set consisting of an odd-numbered bit position together with all power-of-2 multiples of the position, but leave all other elemental sets intact. We obtain an observer even if infinitely many elemental sets are omitted, provided infinitely many remain.

A preobserver that is not an observer can be obtained by severely depleting rather than omitting some elemental sets, leaving a finite, nonzero residue of elements in each, as long as infinitely many elemental sets remain intact. By restricting the process to the subdomain consisting of the union of the intact elemental sets we obtain a subprocess which is both the maximal, certified observer and the soul of the original process, as is straightforward to show. It is also possible to deplete any of the intact sets, so long as infinitely many elements remain, and still obtain a preobserver—a depleted set, if still infinitely populated, will function as a fully intact set.

Space P_1, then, illustrates the "favorable" case in which the maximal observer exists for a preobserver and equals the soul. There are some interesting and instructive "unfavorable" cases in which one or the other of the assumptions is violated; representatives will now be considered.

For the first such case, in which order compactness will be violated, we define a monogenic process space, $P_a=\langle P_a, S_a, H_a\rangle$, as follows. S_a is all the integers (including negative integers) with the usual ordering and the discrete topology. H_a is the nonnegative integers 0, 1, ... with the usual ordering, together with a point, ∞, which is greater than all the integers. H_a then is a history space, in which every subset is consistent. The processes will consist of the maskings of a process, p_0, whose subdomain is all the integers and which is defined, for hyperregion V, by $p_0(V)=\infty$ if V is infinite, 0 if V is empty, and (max V) − (min V) ≥ 0 if V is finite and nonempty.

It is straightforward, then, to show that P_a is an enduring space (though the history space with its ordering is not a particularly natural example of "histories") and that a masking $(p_0)_{\|V,V'}$ is an observer just in case $V\cap V'$ is infinite but lower-bounded. The sum of all such maskings clearly is p_0. p_0, however, is *not* an observer as can be seen by considering a hyperregion V that is infinite but terminating. For this episodic case we have $p_0(V)=\infty$, a value that cannot be elaborated for any $V'>V$, as would be required for an observer. Thus p_0 "knows too much already" or is "omniscient" meaning it cannot learn new things with the passage of time, as an observer must do. In this case it is order compactness that breaks down: ∞ is an episodic history. Any infinite set of nonnegative integers covers this history, but it has no finite subcover. With the discrete topology stable time-bounding holds, however, as with P_1.

For the second "unfavorable" example we define another masking, enduring space, $P_b=\langle P_b, S_b, H_b\rangle$, that violates stable

282

time-bounding but preserves order compactness. The sum of observers will be an observer in this space, but will not equal the soul of the universal process.

As with P_1, the staging space S_b will have the positive integers as underlying set, with the usual ordering. The associated topology, $\mathfrak{T}_b$, will differ however, and consist of (1) all sets of *even* positive integers, together with (2) all *finite* sets of the form $\{1,2,...n\}$, and (3) all unions of (1) and (2). Besides the cases of $\varnothing$ and N^+, then, an open set will consist of consecutive integers 1,2, ...n plus an unrestricted set of larger, even integers. It is easy to see that the only *stable* set is $\cup\mathfrak{T}_b=N^+$, so that stable time-bounding does not hold.

Next we introduce a notational convention: $n^\bullet$, for positive integer n, has the value n if n is odd, and otherwise equals twice the greatest odd divisor, $2n^\Delta$. As with n^Δ, we then have $n^\bullet{\le}n$. Where the range of n^Δ, on the other hand, consists of the odd integers, that of $n^\bullet$ is all the positive integers *except* the multiples of 4, or in other words, the odd integers plus twice the odd integers.

A set of integers will be called *even* iff each of its elements is even, and similarly *odd* iff each of its elements is odd. (The empty set then is both even and odd, an anomaly that does not otherwise occur.)

For a ternary string η, the set of specified bit positions will be called the *active set* and be denoted by $\mathfrak{a}(\eta)$; η is *acceptable* iff $\mathfrak{a}(\eta)\in\mathfrak{T}_b$. An acceptable string η is an r-string iff, for each *even n*, $\eta(n)$ is specified and equal to $\eta(n^\bullet)$; the odd bits are not similarly constrained. Acceptable η is a pre-r string iff, whenever, for even values of m, n, it happens that $m^\bullet=n^\bullet$ and both $\eta(m)$ and $\eta(n)$ are specified, then $\eta(m)=\eta(n)$. As in the definition of P_1, then, it is not difficult to show that the r-strings form a history space, and moreover, that a pre-r η can be minimally extended to an r-string η^+ in just one way. It is important to note that this extension redundantly duplicates the even-numbered bits, copying them as necessary to other even-numbered positions, *but does not copy or duplicate the odd-numbered bits*. Put another way, the even subset of the active set is enlarged, while the odd subset is unchanged. In particular, if η is a pre-r string with an even active set, then η^+ likewise will have an even active set.

A binary string τ is *q-acceptable* iff the quotient (ternary string) $1/\tau$ is acceptable. τ is q-acceptable, then, iff $S_\tau \in \mathfrak{I}_b$. In particular, then, **0** and **1** are q-acceptable, and, if β and γ are q-acceptable, then so is $\beta*\gamma$. If η is any pre-r string (including the case of an r-string) and β is q-acceptable, then η/β is a pre-r string.

Given binary strings α, β, γ, with α assumed to be an r-string, and β and γ q-acceptable, mapping $p_{\alpha\beta\gamma}$ is defined by $p_{\alpha\beta\gamma}(S_\tau) = (\alpha/(\beta*\tau))^+$. Here we also must assume that τ is q-acceptable, to limit the hyperregions S_τ to allowable values. As before, p_α denotes the (maximal) case that $\beta=\gamma=\mathbf{1}$. Again as before, the necessary properties hold so the $p_{\alpha\beta\gamma}$ will form a process space for the given staging and history space. In particular, the $p_{\alpha\beta\gamma}$ once again have the form of a masking: $p_{\alpha\beta\gamma}=(p_\alpha)_{\|S_\beta,S_\gamma}$, and every masking of p_α can be obtained by the appropriate choices of β and γ. An episodic history, in this case, has only a finite number of specified bits in positions n or $2n$, from which it is straightforward to show that order-compactness holds, much as with $\mathbf{P}_1$. Another property of significance, which follows from the general case of maskings, is that no change is made in $p_{\alpha\beta\gamma}$ if we replace β by $\beta*\gamma$, inasmuch as $S_{\beta*\gamma}=S_\beta\cap S_\gamma$.

It will be convenient to define the active set of a process, $\mathfrak{a}(p)$ where $p=p_{\alpha\beta\gamma}$, as the active set of the ternary string, $\mathrm{smr}(p)$. It is easy to see, then, that this is identical to the active set of $(1/(\beta*\gamma))^+$ and must include $S_\beta\cap S_\gamma$. (It is, of course, also independent of α.) If, however, $S_\beta\cap S_\gamma$ is even, then so is the active set of the process, and p itself is then referred to as an even process. p is even, in particular, if the active set of $1/\beta$ is even, regardless of γ; that is, the subdomain of an even process is not necessarily even. However, even in the more general case where p need not be even, as long as $S_\tau\in\mathrm{Dom}(p)$ is even then so is $p(S_\tau)$. (Corresponding properties hold for odd active sets too, but are not as important here.)

We now focus on a particular, maximal process p_α and the associated mappings $p_{\alpha\beta\gamma}$; this then is the set P_b and it forms the masking space of p_α, which in turn is to be the process space $\mathcal{P}_b$. For the cases that $\beta(n)=0$ whenever n is odd, or in other words, the even processes, we clearly obtain an exact analogue of a masking space in $\mathbf{P}_1$. The only difference in the new case is that all histories in the range of the processes have the odd-numbered bits unspecified, that is, for elements—sets of integers—in the domain of $p_{\alpha\beta\gamma}$, the odd-numbered integers are simply ignored. By considering only the even processes, then, observers can be formed from the $p_{\alpha\beta\gamma}$ as in $\mathbf{P}_1$. The maximal observer obtainable this way, whose active set is all the even integers, will be obtained for the case that $\beta(n)^*\gamma(n)=1$ iff n is even.

Suppose, on the other hand, that a process $p=p_{\alpha\beta\gamma}$ contains an odd number in its active set, which means that $\beta(n)=\gamma(n)=1$ for some odd n. Then p cannot be an observer, as can be seen by considering the requirements in this case. In the first place, by the q-acceptability of β and γ we can assume that $n=1$, which means that $\{1\}\in\mathrm{Dom}(p)$ and moreover, $1\in\mathfrak{a}(p(\{1\}))$. On the other hand, if p is an observer, there must exist stable $W\in\mathrm{Dom}(p)$ such that both $W>\{1\}$ and $p(W)>p(\{1\})$. The latter will hold iff $\mathfrak{a}(p(\{1\}))\subseteq\mathfrak{a}(p(W))$. However, from the construction of topology $\mathfrak{I}_b$, it is clear that $W>\{1\}$ only if W is even so that also $\mathfrak{a}(p(W))$ must be even, and $\mathfrak{a}(p(\{1\}))$ cannot be a subset. This shows that p cannot be an observer. The class of observers $p_{\alpha\beta\gamma}$, then, is exhausted by the cases above consisting of even observers, which in turn sum to an even, maximal observer as we have seen.

Finally, to show that this maximal observer is not the soul, consider p_α itself, which is the maximal process in the space under consideration. Since the active set of p_α, N^+ itself, is not even, p_α cannot be an observer. As noted earlier, on the other hand, N^+, which is also the subdomain of p_α, is the *only* stable hyperregion, thus p_α is clearly self-summarizing and thus equals its own soul. Again, this strange situation, in which the certified observers sum to a maximal observer which, however, is not the soul, occurs through a violation of stable time-bounding.

19. Parent and Image Spaces

Earlier the notion of subspace was introduced, applied to both staging and history spaces, and then extended to process spaces. The subspace proved useful, for example, in the case of the preobserver which gave rise to a monogenic subspace. The subspace furnishes one way of generating, from a given space (object) $\mathbf{O}$, a space of similar type $\mathbf{O}'$; for this case we have

$|\mathbf{O}'|\subseteq|\mathbf{O}|$. Here we shall consider another way of generating a similar type space $\mathbf{O}'$, without, however, requiring that $\mathbf{O}'$ be a subspace or that its underlying set be a subset of the underlying set of $\mathbf{O}$. Instead $\mathbf{O}'$ will be the "image space" of a partial (not always defined), onto map, $\hat{\imath}:|\mathbf{O}|\rightharpoonup|\mathbf{O}'|$ which will associate additional structure so that $\mathbf{O}'$ will be uniquely specified from its underlying set. The mapping $\hat{\imath}$ will be called an *imager* from the *parent* space $\mathbf{O}$ to the *image* space $\mathbf{O}'$.

The main motivation for considering image spaces is to broaden the conditions under which a personlike process or observer could be said to survive and endure. An observer in general will then be represented as a sum of constituent processes, each possibly terminating, and each of which is the image of some "source" process and thus "sustained" by it. The source processes themselves could also be all terminating or otherwise unable to sustain the entire observer, yet the observer will in fact be sustained by the entire collection of the source processes. No special connections between the different source processes are assumed, beyond the property that the constituents they image must all "fit" together appropriately, to sum up to the desired process, in this case, the observer.

Philosophically, we are committed to the position that the different constituents or "pieces" of the observer, though possibly sustained in very different ways, as for example, in different universes, still add up to one, unified entity. A given constituent could also be multiply sustained, by again, very different processes. This stance may pose a philosophical challenge for some but at least seems unproblematic from a mathematical standpoint, provided of course that appropriate formal requirements are met.

One property that will hold for the spaces considered is that all are posets. The imagers, as defined, will all be order-preserving: for $\hat{\imath}:|\mathbf{O}|\rightharpoonup|\mathbf{O}'|$, if $y\geq x$, and if both $\hat{\imath}(x)$, $\hat{\imath}(y)$ are defined, then $\hat{\imath}(y)\geq\hat{\imath}(x)$. For staging spaces, however, there are two partial orders to consider: inclusion and time-ordering. The imager in this case must be specified to preserve both; for history spaces there is only one partial order and the problem is more straightforward. An important feature of imagers on process spaces will be to naturally assign imagers on the associated staging and history spaces. Some additional complications arise which can be straightforwardly handled.

We proceed now to consider the individual cases, starting with staging and history spaces.

20. Imagers on Staging, History, and Process Spaces

For staging spaces $S=\langle \mathfrak{I}, S, \geq \rangle$, $S'=\langle \mathfrak{I}', S', \geq' \rangle$, an imager $\mathfrak{s}: \mathfrak{I} \to \mathfrak{I}'$ will have the form of a partial, 1-1, onto map or bijection between the two topologies. $\mathrm{Dom}(\mathfrak{s})$ will be a subtopology $\mathfrak{I}_\mathfrak{s}$ of $\mathfrak{I}$ over the same spacetime (point set) S, with $\mathrm{Rng}(\mathfrak{s})=\mathfrak{I}'$. Thus we actually have $\mathfrak{s}: \mathfrak{I}_\mathfrak{s} \to \mathfrak{I}'$, with $\mathfrak{s}^{-1}: \mathfrak{I}' \to \mathfrak{I}_\mathfrak{s}$, on an equal footing, though, again, $\mathfrak{I}_\mathfrak{s}$ is only a subtopology of the original topology $\mathfrak{I}$. We further require that $\mathfrak{s}$ be topology-preserving, so that, for a family of hyperregions $\mathfrak{F} \subseteq \mathfrak{I}_\mathfrak{s}$, and U, $V \in \mathfrak{I}_\mathfrak{s}$, $\mathfrak{s}(\cup \mathfrak{F})=\cup \mathfrak{s}[\mathfrak{F}]$, and $\mathfrak{s}(U \cap V)$ $=\mathfrak{s}(U) \cap \mathfrak{s}(V)$. In particular for the case that $\mathfrak{F}=\varnothing$ we have $\cup \mathfrak{F}=\varnothing$ also, from which it follows that $\mathfrak{s}(\varnothing)=\varnothing$ as desired. Similarly, if $\mathfrak{F}$ $= \mathfrak{I}_\mathfrak{s}$ so that $\cup \mathfrak{F} = S$, then, since $\mathfrak{s}$ is onto, $\mathfrak{s}[\mathfrak{F}] = \cup \mathfrak{I}'=S'$, again as desired. It is also straightforward to show that, if $U \supseteq V$ then $\mathfrak{s}(U) \supseteq \mathfrak{s}(V)$, and it also follows easily that $\mathfrak{s}^{-1}$ is a topology-preserving map from $\mathfrak{I}' \to \mathfrak{I}_\mathfrak{s}$. Finally, we shall require that $\mathfrak{s}$ be time-order preserving: if $U \geq V$ then $\mathfrak{s}(U) \geq' \mathfrak{s}(V)$. Then, since $\mathfrak{s}$ is a bijection, it follows that it strongly preserves both set inclusion and the time order: $U \supseteq V$ iff $\mathfrak{s}(U) \supseteq \mathfrak{s}(V)$, $U \geq V$ iff $\mathfrak{s}(U) \geq' \mathfrak{s}(V)$.

Intuitively, an imager transforms a staging space into a "dumbed down" version in which the topology is coarsened but, with this restriction, the order relations (time, inclusion) are maintained. It is straightforward to show that imagers can be composed, that composition is associative, and that identity imagers exist for all staging spaces (so imagers form a category). The particular "dumbing down" provisions that are here chosen for the imager rest on the idea of a topological space homomorphism, introduced by a continuous map on the point set, but there is an additional complication involving the time order.

For history spaces the situation is much simpler. Given $\mathbf{H} = \langle H, \geq \rangle$, $\mathbf{H}'=\langle H', \geq' \rangle$, an imager will be an onto (total, not partial, not necessarily 1-1) map $\mathfrak{h}:H \to H'$ that is order-preserving: for

$g, h \in H$, if $g \geq h$ then $\mathfrak{h}(g) \geq' \mathfrak{h}(h)$. Here it is straightforward to show that, given that H forms a history space under its associated partial order, the fact that H' is an image of an order-preserving map establishes that it too forms a history space under its associated order. (Another property that follows easily and is important for process space imagers considered below, is that an order-preserving map preserves prevalence of a set in a history space.) So, just as with staging spaces, the image space $\mathbf{H'}$ is "dumbed down" from the imaging space $\mathbf{H}$, though in a rather different way.

With imagers defined for both staging and history spaces, it is straightforward to proceed to process spaces. Given spaces $\boldsymbol{P} = \langle P, \boldsymbol{S}, \mathbf{H} \rangle$, $\boldsymbol{P'} = \langle P', \boldsymbol{S'}, \mathbf{H'} \rangle$, with $\mathfrak{s}$, $\mathfrak{h}$ imagers respectively from $\boldsymbol{S}$ to $\boldsymbol{S'}$ and from $\boldsymbol{H}$ to $\boldsymbol{H'}$, an imager will be an onto (again total, not necessarily 1-1) map $\mathfrak{p} : P \rightarrow P'$ which transforms a process $p \in P$ into a process $p' \in P'$ as follows. First, $\text{Dom}(p') = \mathfrak{s}[\text{Dom}(p) \cap \text{Dom}(\mathfrak{s})]$. So $\text{Dom}(p')$ is the subset of the topology $|\boldsymbol{S'}|$ consisting of those hyperregions U for which $\mathfrak{s}^{-1}(U) \in \text{Dom}(p)$. $\text{Dom}(p)$ in turn will have the form of the relative topology, in $\boldsymbol{S}$, on the associated hyperregion $\cup \text{Dom}(p) = \text{subd}(p)$. It may happen that $\text{Dom}(p) \cap \text{Dom}(\mathfrak{s}) = \varnothing$ so that $\text{Dom}(p') = \{\varnothing\}$, but this is a perfectly acceptable domain for a process (if not particularly interesting). Process p' is then defined, for hyperregion $U \in \text{Dom}(p')$, by $p'(U) = \mathfrak{h}(\mathfrak{s}^{-1}(U))$. That p' is a process as defined earlier, and that $\boldsymbol{P'}$ is a process space given that $\boldsymbol{P}$ is such a space, are easy consequences of the order-preserving maps used in the construction, particularly the history space imager, $\mathfrak{h}$. As before, process space imagers can be composed and form a category.

The notion of process space imager appears to provide options for modeling the "survival" of personlike processes under adverse conditions, as speculated earlier. As one possibility, a family of process spaces $\boldsymbol{P}_0$, $\boldsymbol{P}_1$, ..., none of which is enduring and which taken together do not form any larger "whole" that is, could individually image "sections" of an observer space. By the right choices of associated history space imagers $\mathfrak{h}_0$, $\mathfrak{h}_1$, ..., and other details, the resulting spaces $\boldsymbol{P}_0'$, $\boldsymbol{P}_1'$, ..., could achieve the desired fit and sustain an observer space. The original process spaces in turn might correspond to natural phenomena, even including entire universes, whose "real" presence, though possibly transient, is unproblematic. In this way, then, a rationale could be found for the persistence of a person, as a resurrected being, after arbitrary life-ending events, supposing, as seems plausible, that necessary constructions can eventually be made, in a suitable setting, after each disruption. Such possibilities at least call for further investigation.

21. Summary and Conclusions

A mathematical approach has been followed which has led to rationalizations of certain philosophical ideas. Under the formalisms that have been developed we have seen how personlike processes can be described, and how the death, resurrection and immortalization of such processes can be modeled. There were some further suggestions of possibilities for resurrecting and sustaining individuals under very adverse circumstances such as the destruction of universes in which they are "embedded." A process space provides a way for processes to be combined into larger processes, with some hints (only briefly noted) of how too much combining of processes can be avoided. In this manner, then, a personlike entity can be part of larger processes yet retain its individuality.

The theory presented here, needless to say, needs considerable development to accommodate physics and more accurately address conditions in the universe as we know it. A few of the many topics that seem in need of clarification or elaboration are as follows.

One is uncertainty in historical records. Errors, falsifications, and ambiguities occur. If a document *says* a thing happened, the proper interpretation may attach a high probability that it did, in fact, happen, but never absolute certainty.

Another issue concerns immortality. Arguably, the *quantity* of information inherent in a process should approach infinity with time. This property is suggested, in the formalism so far developed, in the fact that an observer must repeatedly add to the body of information it retains about its past. However the total "amount" it must add is not given because the concept of "amount" is never formulated for histories. A starting point for this could be had, it would seem, by providing a topology for histories, that is, for a history space, just as it is now supplied for spacetime.

Concerning spacetime, it would be highly desirable to relate the notion of "process" herein developed to happenings in the universe as we know it. In particular, making and saving or storing a record in the real world corresponds at least vaguely to a mechanism that is assumed to exist for many of the theoretical processes described in this study, but it is unclear what the precise connections must be.

These problems and many others would seem to offer an interesting field of study, with ample rewards for diligent efforts.

References

1. T. Zakydalsky, *N. F. Fyodorov's philosophy of physical resurrection.* Ph. D. thesis, Bryn Mawr College (University Microfilms International, 1976).

2. D. Krieger, "Isn't that you behind those Foster-Grants?" *Cryonics* **13(4)** 8, **13(6)** 12 (1992).

3. M. More, "Uploading, cryonics and the 'rapture,'" *Cryonics* **12(12)** 9 (1991).

4. R. Ettinger, *Man into superman* Ch. 11 p. 247 (St. Martins, 1972; p. 225 Avon ed., 1974).

5. F. Dyson, "Time without end: physics and biology in an open universe," *Review of Modern Physics* **51** 447 (1979).

6. J. Barrow and F. Tippler, *The anthropic cosmological principle.* (Oxford, 1988).

7. R. Nozick, *Philosophical explanations.* (Harvard, 1981).

8. D. Kolak and R. Martin, "Personal identity and causality: becoming unglued." *American Philosophical Quarterly* **24(4)** 339 (1987).

9. P. Halmos, *Naive set theory* (Springer-Verlag 1974).

10. J. Kelley, *General Topology* (Springer-Verlag 1975).

CHAPTER TWELVE

The Chinese Room Argument

John Searle*

0. Introduction

The **Chinese Room Argument** aims to refute a certain conception of the role of computation in human cognition. In order to understand the argument, it is necessary to see the distinction between *Strong* and *Weak* versions of Artificial Intelligence. According to Strong Artificial Intelligence, any system that implements the right computer program with the right inputs and outputs thereby has cognition in exactly the same literal sense that human beings have understanding, thought, memory, etc. The implemented computer program is sufficient for, because constitutive of, human cognition. *Weak* or *Cautious* Artificial Intelligence claims only that the computer is a useful tool in studying human cognition, as it is a useful tool in studying many scientific domains. Computer programs which simulate cognition will help us to understand cognition in the same way that computer programs which simulate biological processes or economic processes will help us understand those processes. The contrast is that according to Strong AI, the correct simulation really is a mind. According to Weak AI, the correct simulation is a model of the mind.

1. Statement of the argument

Strong AI is answered by a simple thought experiment. If computation were sufficient for cognition, then any agent lacking a cognitive capacity could acquire that capacity simply by implementing the appropriate computer program for manifesting that capacity. Imagine a native speaker of English, me for example, who understands no Chinese. Imagine that I am locked in a room with boxes of Chinese symbols (the database) together with a book of instructions in English for manipulating the symbols (the program). Imagine that people outside the room send in small batches of Chinese symbols (questions) and these form the input. I know that I am receiving sets of symbols which to me are meaningless. Imagine that I follow the program which instructs me how to manipulate the symbols. Imagine that the programmers who design the program are so good at writing the program, and I get so good at manipulating the Chinese symbols, that I am able to give correct answers to the questions (the output). The program makes it possible for me, in the room, to pass the Turing Test for understanding Chinese, but all the same I do not understand a single word of Chinese. The point of the argument is that if I do not understand Chinese on the basis of implementing the appropriate program for understanding Chinese, then neither does any other digital computer solely on that basis because the computer, qua computer, has nothing that I do not have.

The argument proceeds by a thought experiment, but the thought experiment is underlain by a deductive proof. And the thought experiment illustrates a crucial premise in the proof. The proof contains three premises and a conclusion.

Premise 1: Implemented programs are syntactical processes.

The implemented programs are defined purely formally or syntactically. This, by the way, is the power of the digital computer. The computer operates purely by manipulating formal symbols, usually thought of as 0s and 1s, but they could be Chinese symbols or anything else, provided they are precisely specified formally. To put this in slightly technical terminology, the notion *same implemented program* specifies an equivalence class defined purely in terms of syntactical manipulation and is completely independent of the physics of the realization of the implementation. Any hardware will do provided that it is stable enough and rich enough to carry out the steps in the program. This is the basis of the concept of **multiple realizability** whereby the same program can be realized in an indefinite range of different computer hardwares: electronic computers or people locked in Chinese Rooms or any number of other hardwares.

The claim that implemented programs are syntactical processes, is not like the claim that men are mortal. The essence of the program is identified by its syntactical features. They are not just incidental features. It is like saying triangles are three sided plane figures. There is nothing to the program qua program but its syntactical properties. Triangles may be pink or blue, but that has nothing to do with triangularity; analogously, programs may be in electronic circuits or Chinese rooms, but that has nothing to do with the nature of the program.

Premise 2: Minds have semantic contents.

In order to think or understand a language, you have to have more than just the syntax. You have to understand the meanings or the thought contents that are associated with the symbols. And the problem with the man in the room is that he has the syntax but he does not understand the appropriate semantic content because he does not understand Chinese.

Premise 3: Syntax by itself is neither sufficient for nor constitutive of semantics.

The Chinese Room thought experiment illustrates this truth. The purely syntactical operations of the computer program are not by themselves sufficient either to constitute, nor to guarantee the presence of, semantic content, of the sort that is associated with human understanding. The purpose of the Chinese Room thought experiment was to dramatically illustrate this point. It is obvious in the thought experiment that the man has all the syntax necessary to answer questions in Chinese, but he still does not understand a word of Chinese.

> **Conclusion**: Therefore, the implemented programs are not by themselves constitutive of, nor sufficient for, minds. In short, Strong Artificial Intelligence is false.

The Chinese Room Argument is incidentally also a refutation of the Turing Test and other forms of logical behaviorism. I, in the Chinese Room, behave exactly as if I understood Chinese, but I do not. One can see this point by contrasting the Chinese case with the case of a man answering questions in English. Suppose I, in the same room, am also given questions in English and I pass out answers to the questions in English just as I pass out answers to the questions in Chinese. From the point of view of the outside observer, my behavior in answering the questions in Chinese is just as good as my behavior in answering questions in English. I pass the Turing Test for both. But from my point of view, there is a huge difference. What exactly is the difference? The difference can be stated in common sense terms. In the case of English, I understand both the questions and the answers. In the case of Chinese, I understand neither. In Chinese I am just a computer. This shows that the Turing Test, or any other purely behavioral test, is insufficient to distinguish genuine cognition from behavior which successfully imitates or simulates cognition.

The Chinese Room Argument thus rests on two simple but basic principles, each of which can be stated in four words.

> **First**: Syntax is not semantics.

Syntax by itself is not constitutive of semantics nor by itself sufficient to guarantee the presence of semantics.

> **Second**: Simulation is not duplication.

In order actually to create human cognition on a machine, one would not only have to simulate the behavior of the human agent, but one would have to be able to duplicate the underlying cognitive processes that account for that behavior. Because we know that all of our cognitive processes are caused by brain processes, it follows trivially that any system which was able to cause cognitive processes would have to have relevant causal powers at least equal to the threshold causal powers of the human brain. It might use some other medium besides neurons, but it would have to be able to duplicate and not just simulate the causal powers of the brain.

2. The systems reply

There have been a rather large number of discussions and objections to the Chinese Room, but none have shaken its fundamental insight as described above. Perhaps the most common attack is what I baptized as the *Systems Reply*. The claim of the Systems Reply is that though the man in the room does not understand Chinese, he is not the whole of the system, he is simply a cog in the system, like a single neuron in a human brain (this example of a single neuron was used by Herbert Simon in an attack he made on the Chinese Room Argument in a public lecture at the University of California, Berkeley). He is the central processing unit of a computational system, but Strong AI does not claim that the CPU by itself would be able to understand. It is the whole system that understands. The Systems Reply can be answered as follows. Suppose one asks, Why is it that the man does not understand, even though he is running the program that Strong AI grants is sufficient for understanding Chinese? The answer is that the man has no way to get from the syntax to the semantics. But in exactly the same way, the whole system, the whole room in which the man is located, has no way to pass from the syntax of the implemented program to the actual semantics (or intentional content or meaning) of the Chinese symbols. The man has no way to understand the meanings of the Chinese symbols from the operations of the system, but neither does the whole system. In the original presentation of the Chinese Room Argument, I illustrated this by imagining that I get rid of the room and work outdoors by memorizing the database, the program, etc., and doing all the computations in my head. The principle that the syntax is not sufficient for the semantics applies both to the man and to the whole system.

3. Three misinterpretations

The Chinese Room Argument is sometimes misinterpreted. Three of the most common misunderstandings are the following. First, it is sometimes said that the argument is supposed to show that **computers can't think**. That is not the point of the argument at all. If a computer is defined as anything that can carry out computations, then every normal human being is a computer, and consequently, a rather large number of computers can think, namely every normal human. The point is not that computers cannot think. The point is rather that computation as standardly defined in terms of the manipulation of formal symbols is not by itself constitutive of, nor sufficient for, thinking.

A second misunderstanding is that the Chinese Room Argument is supposed to show that **machines cannot think**. Once again, this is a misunderstanding. The brain is a machine. If a machine is defined as a physical system capable of performing certain functions, then there is no question that the brain is a machine. And since brains can think, it follows immediately that some machines can think.

A third misunderstanding is that the Chinese Room Argument is supposed to show that **it is impossible to build a thinking machine**. But this is not claimed by the Chinese Room Argument. On the contrary, we know that thinking is caused by neurobiological processes in the brain, and since the brain is a machine, there is no obstacle in principle to building a machine capable of thinking. Furthermore, it may be possible to build a thinking machine out of substances unlike human neurons. At any rate, we have no theoretical argument against that possibility. What the Chinese Room Argument shows is that this project cannot succeed solely by building a machine that implements a certain sort of computer program. One can no more create consciousness and thought by running a computer simulation of consciousness and thought, than one can build a flying machine simply by building a computer that can simulate flight. Computer simulations of thought are no more actually thinking than computer simulations of flight are actually flying or computer simulations of rainstorms are actually raining. The brain is above all a causal mechanism and anything that thinks must be able to duplicate and not merely simulate the causal powers of the causal mechanism. The mere manipulation of formal symbols is not sufficient for this.

4. A brief history of the argument

The Chinese Room Argument had an unusual beginning and an even more unusual history. In the late 1970s, Cognitive Science was in its infancy and early efforts were often funded by the Sloan Foundation. Lecturers were invited to universities other than their own to lecture on foundational issues in cognitive science, and I went from Berkeley to give such lectures at Yale. In the terminology of the time we were called *Sloan Rangers*. I was invited to lecture at the Yale Artificial Intelligence Lab, and as I knew nothing about Artificial Intelligence, I brought a book by the leaders of the Yale group, in which they purported to explain story understanding. The idea was that they could program a computer that could answer questions about a story even though the answers to the questions were not made explicit in the story. Did they think the story understanding program was sufficient for genuine understanding? It seemed to me obvious that it was in no way sufficient for story understanding, because using the programs that they designed, I could easily imagine myself answering questions about stories in Chinese without understanding any Chinese. Their story understanding program manipulated symbols according to rules but it had no understanding. It had a syntax but not a semantics. These ideas came to me at 30,000 feet between cocktails and dinner on United Airlines on my flight East to lecture in New Haven. I knew nothing of Artificial Intelligence and because my argument seemed so obvious to me, I assumed that it was probably a familiar argument and that the Yale group must have an answer to it. But when I got to Yale I was amazed to discover that they were surprised by the argument. Everybody agreed that the argument was wrong but they did not seem to agree on exactly why it was wrong. And indeed most of the subsequent objections to the Chinese Room I heard, in early forms, in those days I spent lecturing at Yale. The article was subsequently published in *Behavioral and Brain Sciences* for 1980, and provoked twenty-seven simultaneously published responses, almost all of which were hostile to the argument and some were downright rude. I have since published the argument in other places, including my 1984 Reith Lectures book *Minds, Brains and Science*, *The Scientific American* and *The New York Review of Books*.

I never really had any doubts about the argument as it seems obvious that the syntax of the implemented program is not the same as the semantics of actual language understanding. And only someone with a commitment to an ideology that says that the brain must be a digital computer ("What else could it be?") could still be convinced of Strong AI in the face of this argument. As I was invited by various universities to lecture on this issue I discovered that the answers to it tended to fall into certain patterns, which I named respectively as the Systems Reply, the Robot Reply (if we put the computer inside a robot it would acquire understanding because of the robot's causal interaction with the world), the Wait 'Til Next Year Reply (better technology in the future will enable digital computers to understand), the Brain Simulator Reply (if we did a computer simulation of every neuron in the Chinese brain, then the computer would have to understand Chinese), etc. I had no trouble answering these and other objections. I assumed that the orthodox Strong AI people would fasten on to the Robot Reply because it seems to exemplify the behaviorism that was implicit in the whole project of assuming that the Turing Test was a conclusive proof of human cognition. But to my surprise the mainstream adopted the Systems Reply which is, I think, obviously inadequate for reasons I state in this essay. The Chinese Room Argument has had a remarkable history since its original publication. The original article was published in at least twenty-four collections and translated into seven languages. Subsequent statements of the relevant argument in Minds, Brains and Science were also reprinted in several collections and the whole book was translated into twelve languages. I have lost count of the publication, reprinting and translations of other statements. Two decades after the original publication of the article a book appeared edited by John Preston and Mark Bishop called *Views into the Chinese Room*. A web site ("Search results for 'chinese room'" <http://consc.net/mindpapers/search?searchStr=chinese+ro

om&filterMode=keywords>) currently cites many dozens of discussions (I assume mostly attacks) on the argument.

5. References

- Preston, John and Mark Bishop (eds.). Views into the Chinese Room: New Essays on Searle and Artificial Intelligence. Oxford/New York: Oxford University Press, 2002.

- Searle, John R. Minds, Brains, and Programs, The Behavioral and Brain Sciences, Vol. 3, 1980

- Searle, John R. Minds, Brains and Science. London: BBC Publications, 1984; Penguin Books, 1989. Cambridge, MA: Harvard University Press, 1984.

- Searle, John R. Is the Brain's Mind a Computer Program?, The Scientific American, January 1990.

- Searle, John R. The Myth of the Computer, The New York Review of Books. April 29, 1982.

6. Recommended reading

- Dietrich, Eric (ed.). Thinking Computers and Virtual Persons. San Diego: Academic Press, 1994.

- Preston, John and Mark Bishop (eds.). Views into the Chinese Room: New Essays on Searle and Artificial Intelligence. Oxford/New York: Oxford University Press, 2002.

- Searle, John R. Minds, Brains, and Programs, The Behavioral and Brain Sciences, Vol. 3, 1980

- Searle, John R. Minds, Brains and Science. London: BBC Publications, 1984; Penguin Books, 1989. Cambridge, MA: Harvard University Press, 1984.

- Searle, John R. Is the Brain's Mind a Computer Program?, The Scientific American, January 1990.

- Searle, John R. The Myth of the Computer, The New York Review of Books. April 29, 1982.

CHAPTER THIRTEEN

What's Best For Us

Asher Seidel

Those philosophers most known to the general public often suggested the best sort of life. In this, they have been joined by theologians, various novelists, and various politicians, among others. Restricted to philosophers, the suggestions have been many. Included among these are: the moral life; the life of pleasure; the life of qualified pleasure; the life of knowledge; the life of historical self-realization; the life of unalienated labor; the life of freedom; the life of creativity; the life of existential suffering. Modulations and combinations of these lives, along with other sorts of lives, have received at least some philosopher's blessing.

There is the further consideration that any such normative recommendation of optimum lifestyle invites the response, "For whom and when and where?" People are variously situated. It is only counterfactually appropriate to recommend the life of a theoretical physicist to someone laboring in neo-feudal circumstances. It is not simply the "best" part of the phrase "what's best for us" that's problematic. Who are the "us" being addressed? In surveying the historical and contemporary responses to this latter question, it is generally assumed that the "us" being addressed are human. This assumption will be initially honored here. There is the further implicit assumption that, however the transformation from the unrecommended life to the recommended life, the transition is from human status to human status. This assumption is so naturally made, as to constitute some of the background of virtually all such discussions. It should be recalled, however, that those theologies valuing a life after biological death are considering a transition that can be termed human-to-human only if the latter use of the term "human" is given some leeway.

In contrast, I am considering lives that are not human. Initially, these are lives that were once human, but have transitioned to something other than human. From this nonhuman, or posthuman, position, continued transitions will be considered that start from a nonhuman state and proceed to a further altered nonhuman state. The general proposition I seek to defend is that what's best for us humans, more or less all humans, is that we become something other than human. Among the complications to be addressed in elaborating this proposition, there is the likelihood that this evolution will not be a one-stage affair, and consequently "we" will refer to humans as presently understood, humans in transition to posthumans, and posthumans in continued transition.

Put briefly, I recommend that we successively evolve into entities other than human, at each stage enhancing cognitive and creative abilities as we acquire further understanding of, and means to implement, such enhancements.

This recommendation is offered in all seriousness and with the utmost caution. I recognize the possibility that what I am suggesting might, if implemented, prove disastrous. I have no airtight argument in support of my recommendation. I am not so far gone as to be unaware of the manner in which many thoughtful individuals will regard this recommendation, even after I have elaborated it below and offered such defense of it as I am able.

There are minimally the following three reasons for the recommendation. Humanity may be approaching various crises which cannot be resolved, as long as we remain human. Of course, humanity has faced and survived many crises. The newer crises – environmental, political (of the sort having military implications), economic, biological – are some of them species-threatening. Some of them allow for the continued existence of humans, albeit at greatly reduced numbers and what many would consider harsher conditions of life. If one accepts the premises that: there are clear and present species-threatening dangers; these dangers likely cannot be avoided by us due to deficiencies in our nature; these dangers are more likely avoidable if we change our psychological and social nature to the extent that we are no longer human, then one may be open to the suggestion that some such change is desirable.

Those are a tall list of premises. There is a long history of doomsaying, and the continued existence of humanity to the present testifies to the resilience of our species. Even if it is granted that there are special features regarding our current situation (proliferation of nuclear weapons, for example), it remains unlikely that a majority of humanity will be persuaded to consider seriously a trans-species migration on the basis of present and foreseeable circumstances.

Allowing the non-persuasiveness of the imminent-doom rationale for a transhumanist migration, there is an optimistic reason for such change. A large number of humans admire and respect certain exemplars of human wisdom and behavior. Names such as Socrates, Leonardo, Einstein, Gandhi resonate with significant segments of literate humanity. Many other names could be cited. Allowing that one's socio-historical positioning, as well as one's more immediate environment, together with one's genetic endowment, profoundly shape one's character, and that all manner of luck (being at the right place at the right time, etc.) determine whether one achieves the sort of recognition granted these well-known individuals, the question remains: why aren't more of us like them? Even if few of us are located so as to achieve world-historical prominence, the haunting question is: why don't more of us possess these qualities of mind?

Of course, general answers to this question have been supplied. Most of us lack the conjunction of favorable circumstances, as well as the genetic endowment, to possess such minds. The matter is not finished, however. What if we possessed better theories of child rearing, child development, education, together with as yet unknown brain-modifying techniques which yield markedly improved cognitive abilities? The response to such what-ifs might be, "What if we could flap our arms and fly to the moon"? Agreed, we all have our utopian wishes. Many of these wishes presume the best of social circumstances, whatever constitutes such circumstances. Presently, detailing these circumstances is sufficiently difficult and controversial, without the added burden of elaborating biological means of "brain-boosting".

The above speculation seemingly has to do with creating better humans, in some intuitive sense of "better". Appearances are possibly misleading here. Were the above optimistic speculations realized, there is a chance that human biology would be altered to the point that humans, homo sapiens sapiens, are no longer being considered. Since the above-mentioned exemplars were human, one might question the supposition that alterations allowing much of humanity to possess their characteristics would transform humans to posthumans. Agreed that if the speculated alterations were limited to replication of the desired characteristics to the degree of the mentioned human paragons, the transformation would be human-to-human. The question arises, however, as to the stopping point. What if, for example, we could transform ourselves into beings that individually were cognizant of biology, chemistry, physics, and mathematics at the current level of the most advanced practitioners in these disciplines? What if, on average, our transformed selves, or descendants, could simultaneously solve advanced engineering problems while composing excellent (by some agreed upon standard) poetry and planning a gourmet meal? If it is not agreed that these are cognitive abilities beyond those of any human, then with small effort one can imagine tasks where such boundary-crossing is beyond dispute.

Doubtless more is needed here. Details regarding such cognitive enhancement are wanting. Perhaps more important, there are considerations of the accompaniments of such seeming improvements. What if, for example, beings with such abilities are either generally, or in the particular manner in which these abilities have been instantiated, ruthlessly self-centered? It lends scant comfort to believe that such beings will likely not long survive.

There will be further detailing of these enhancements below. For now, let us consider the following premises. Suppose that in the sometime future we gain considerable understanding of ourselves as psychological and social beings. This is not to deny the achievements of the various social sciences in their present forms, but rather to project large advances in the relevant fields of study. Perhaps hitherto unknown or not yet adequately developed sciences will allow for such understanding. Assume that together with techniques for enhancing our cognitive powers, we have procedures for optimizing our social characteristics, so as to live harmoniously with others. Allowing for the vagaries of these assumptions, if these conditions were realized while we are still human, they might provide incentive for many of us to transform ourselves, or our offspring, to something other than humans.

Even granting the preceding premises, it is not obvious that humans will choose to transform themselves or their children under such ideal circumstances, rather than choosing to remain human. Predicting what the majority of humanity will do is generally problematic. Yet there are circumstances where such predictions are given more confidence. One such circumstance constitutes the third reason for recommending the transformation from the human to the posthuman.

Consider the following story. Through various media outlets reports emerge of a breakthrough in retarding aging in mammals. Initially, these reports are couched in tentative and cautionary language. It appears that one of the contributing factors of aging, the limitation of the number of replications of which a strand of

DNA is capable, has been overcome in some instances. But, the more sober reports stress, findings are inconclusive and results of further investigations are awaited. Even if preliminary results are confirmed, testing on humans will not take place for at least several years.

Within a period of time, further reports tell of laboratory successes in aging research, including such matters as the purging of accumulated toxins in mature cells. At some point it is announced that human trials are to commence. Preliminary results of trials on humans indicate that aging is not simply arrested, but actually reversed to the point that older humans have been brought back to their young adult physiological constitutions with no apparent ill effects.

To this point in my story telling, many questions are raised. For instance, is it possible to return an adult to an infant's physiological status and, if so, does the adult lose or suffer diminution of various cognitive capabilities? Can the reversal process cure diseases of aging? Must one continue whatever treatment effected age retardation or reversal, and if so in what manner?

Let us leave these and most other questions unaddressed. To the best of my knowledge, none of what is presented in the story has been accomplished. However, there has been research which has direct and indirect bearing on the story.[1] There are optimistic estimates by some researchers in this area that, in a matter of decades, at least some aspects of the above story will be realized.[2] Of course, this may prove excessive optimism on the part of these researchers, as such enthusiasm has occasionally been shown to be in other scientific fields.

Granting that the more optimistic estimates of when profound age retardation and age reversal in higher animals will happen might prove incorrect, there is a strong sooner-or-later feel to this prospect. Admitting that probability does not entail actuality, let us assume that the content of the story is realized. This assumption allowed, let us consider consequences.

Assuming that these age-extending techniques, or technique, are initially tested and verified under strict, clinical conditions, it is likely that various authorities will implement control and restriction on access to these techniques. This assumption is made under the supposition that these techniques are not of an everyday sort, such as drastic restriction of caloric intake. If these techniques do not require large, specialized machinery for their administration, we may expect that many people will make appreciable efforts to secure them for themselves and those close to them. If, for example, either ingestion or injection of substances is required, many people will attempt to obtain these substances. There will be the usual black market in hijacked and independently synthesized substances. It is likely that authorities will issue cautions and warnings: cautions that side effects and long-term consequences are not yet properly understood; warnings that those unauthorized to undergo these techniques are subject to prosecution. Depending on the perceived seriousness of these cautions and warnings, they will to a greater or lesser extent go unheeded, even if media reports provide reasons supporting the cautions. One can imagine that if authorities implement widespread testing procedures to detect unwarranted administration of these techniques, a market in "masking" procedures will ease the reticence occasioned by announcements of such testing.

There are complications which would demand qualifications, if not outright rejection, of the above story. Still, there is a history of officially restricted or banned procedures and substances desired by the general populace, and that history suggests that such efforts at control are circumvented. Perhaps in this speculated case the situation is such that initially only those with wealth and influence are able to obtain these procedures or substances illegally. Those members of theocratic societies might not be quite so motivated to pursue these age-retardation techniques. Many younger members of more secular societies might not feel the need for them. It is virtually undeniable, however, that a significant percentage of middle-aged and older members of those in the developed sector would desire these techniques, and would make strong effort to secure them.

Some might object that simply extending human lifespan, including age-retardation, is not by itself a human to posthuman transformation. I do not believe it is useful at this point to argue definitions of "human". There are intuitively clear examples of the human, and of the nonhuman. As the discussion continues, there will be more focus on the latter. For now, I offer that an average life expectancy of, say, two hundred years, most of which is lived in the average biological state of current humans at thirty, strains the notion of human, at least in the biological, if not the psychological, sense.

Regarding these three reasons there are at least the following two false-sounding notes. First, there have been many assumptions made, and most of these assumptions occur within a chain of assumptions. Hence, withdrawing any of these assumptions strongly affect the reason in whose service they exist. It is possible, of course, that the reason could be supported in another manner. Second, the last reason given, the reason based on the imagined "story", seems less a reason than a cause. That is, in the sense in which "reason" has normative weight, the claim that people will attempt vigorously to secure any proven means of significantly extending their lives avoids the question as to whether they should be availing themselves of these means at the earliest possible moment. I shall argue that, in the absence of the second reason - the reason having to do with optimistic assumptions of human and posthuman improvement - the last reason is not only not a normatively positive reason, but rather as a normative reason it functions negatively.

To note that many people will likely seek available means to extend their lives is not to say that they should, morally or prudentially, seek these means. Even if such means were widely available, there are those who likely would refuse them. Some might refuse them for religious reasons. Others out of ignorance. Still others from lack of availability. To premise wide availability is not to assume complete availability. Whole segments of population; aboriginal people, for example, or those living in dire poverty, might lack access to these speculated widely available means. That many would at least initially either actively refuse extended lifespans, or not undergo such life-extension due to ignorance, poverty, or other circumstance of unavailability, in itself supports the judgment that those willing and able should refrain from extending their lifespans, other matters being as they currently stand.

A world of "extendeds" and "nonextendeds" would be a world of disparity greater than the current world of developed and underdeveloped populations. The latter world, our world, is a world of humans. Granting that there are those who believe that different geo-cultural groups have different innate cognitive capacities, I am writing for those who do not hold such beliefs (I will further grant that the matter of innate differences, whether racially distributed or randomly distributed, if they exist at all, is not a settled matter). Those who do not hold such beliefs hold that present disparities have largely historical and political causes.

But imagine a world in which some people have a life expectancy of hundreds of years, with access to advanced bio-technology that, in its progress, will afford continued extensions on the initial hundreds of years. Imagine further that this world has a substantial population of people not so endowed. There are obviously too many variations on this general scheme to examine individually. There are possible variations in the geographical distribution of extendeds/nonextendeds, as well as variations in their respective numbers. These differences would likely have varying results. But many of these variations promise mistrust, hatred, factionalism, on an unprecedented scale. It would be as though two distinct species of humans were occupying the earth, and often in competition.

There are various concerns voiced in technical and popular media: overpopulation, pollution, diminished resources, weapons proliferation, systematic economic malaise, sectarian hostilities. Yet I propose that what has been reviewed above, the possibility of radically extended lifespan, is potentially as threatening as any item in the preceding list. Of course, these items are none of them exclusive of the others. Yet of all those mentioned, I believe this one in itself has the highest probability of dramatically increasing human suffering. We currently live with the other concerns, whose realization has not to date caused a worldwide human conflagration. I worry that the possibility of radically extending human lifespan, a possibility that might be realized in the nearer term, could have horrendous consequences.

I believe it likely that means for radically extended lifespan will be available in the not too distant future, and that the less than universal application of these means (compared to such worldwide applications as polio and smallpox vaccination) will have significant repercussions. The extent of human suffering, should my belief be realized, is problematic. It may be that there are significant repercussions and they are not harmful to the degree I fear. It would probably be better for us if we universally agreed to refrain from so extending human lifespan until such time as we were better prepared to implement it in a rational manner; rationality here understood in terms of maximizing our gain and minimizing our suffering.

"Gain" and "suffering" remain to be given satisfactory elaboration. I do not believe that such elaboration has been given. This is to say, I do not believe that normative philosophy, in its present or historical instantiations, has fulfilled its mission, which is to show us how best to live. While I cannot offer a proper support for this claim here, I shall indicate my overall reasons for making this statement. I believe that a successful normative theory would fulfill two necessary and likely jointly sufficient conditions; it would have widespread acceptance roughly akin to the extent that various theories of natural science enjoy, and it would show us how to live so that an overwhelming majority of humanity could lead self-satisfied lives at cultural (in the broadest sense of "cultural") levels enjoyed by the educated elites in the developed sector.

I emphasize that what I have said needs elaboration and proper support. Although I cannot supply elaboration or support here, I will continue to make philosophically inflammatory statements. I do not believe that what I have termed a "successful normative theory" can be given for humans. Some would agree with this on the ground that humans are free in the deep, metaphysical sense of the notion, and consequently must define themselves and their values by their free choices.[3] Some maintain that normative terminology is purely or fundamentally prescriptive.[4] Inasmuch as a genuinely successful theory is an organized collection of true statements, and prescriptive statements are essentially imperatives, and imperatives are neither true nor false, there cannot be a successful normative theory.

While I agree with the general statement that a successful normative theory is impossible for humans, I do not agree with either of the above reasons for rejecting the possibility of such a theory. That is, I believe a successful normative theory is possible, but not for humans. Humans have various deep-seated characteristics which do not permit them to adopt any such theory. Put another way, we might recognize the theory - if it were ever stated, but we could not live by it.

As an example of what I am urging, I do not believe people possess freewill in the deep sense. But even if they do, it is likely that such will is manifested only occasionally, most of our choices and resultant actions happening in a manner in which there is nothing answering to an autonomous moral self as the originator. If this is correct, consider that many humans, if not most, act in manners that are inconsistent with this position. Individually and collectively, most of us act as though we are all autonomous moral agents. Criminal law is retributivist to the extent that "the punishment should fit the crime" is virtually an axiom. The wrongdoer suffers in rough proportion to the degree that the wrongdoer has criminally occasioned suffering. But such an institutionalized attitude, as contrary to the metaphysics of human action as it may be, is not the whole story. Human society might become enlightened to the point that criminal law adopts a non-retributivist stance on punishment. What seems most unlikely is that, as individual humans, possessed of human psychology, we can eliminate or significantly mitigate our more immediate reflex and near-reflex responses to perceived personal injury. These responses have likely served us well in our evolutionary history. However, we might question their current usefulness, all things considered. If it is the case that we would be better off without such responses, or at least without the degree to which we are subject to them, but that as humans we are unable to eliminate or seriously temper them, then in a significant manner we cannot adopt some part of a successful normative theory (assuming that a successful normative theory would prescribe not treating others as autonomous agents when they are not).

If the difficulty in overcoming our so-called "reactive attitudes"[5] is not sufficient reason to despair of our adopting a successful normative theory, consider the following. It is apparently in our nature to show kindness to others, to empathize with their distress, and in general support their well-being. This is not a universally distributed characteristic, nor does it manifest without qualification. But it is an aspect of many people. Similarly, many of the same people who have a tendency towards kindness, also have a tendency to outdo others. People want what those around them have, loosely speaking, but they also want somewhat more than what those around them have, be it personal characteristics, material possessions, or the reputation of oneself, one's family, or other associated group.

In the broadest sense, people seek to treat each other equally, in the manner of equality that regards helping one another, raising others up to one's own standards (and, as previously noted, people also seek equality in retribution). But people are also naturally competitive. None of this is news to anyone, nor need it be emphasized that remarks such as these are in need of refinement, elaboration, and critical scrutiny. Some believe that most such generalizations, including the concepts within these generalizations, are subject to future elimination by deeper, underlying theories.[6] I follow those who believe that many such psychological and sociological commonplaces will someday join with a more complete neuroscience, but they will retain, and perhaps strengthen, their applicability when so supported. These are matters currently under examination and debate.

Assuming some descriptive, if not explanatory, utility for such psychological and sociological commonplaces, it may come to pass that humanity will look more fully at contradictions and conflicts seemingly inherent within. These are considerations that most of us tend to avoid considering, given that there seems so little we can do regarding them. We make resolutions, institute educational and other social reforms, ingest relevant pharmaceuticals, celebrate our progress, and yet in many ways remain who we were since recorded history.

Some will think the generalities given above verge on vacuity. Since the readers of this are likely to be academics, or those of comparable cultural level, I urge all interested to attend any of the more popular venues, be they sporting events, rock concerts, firework displays; in short, low-brow gatherings. I do not pretend that what I am gesturing towards does not exist at the higher arenas, be they progressive political gatherings, scholarly conventions, art exhibits or the like. Rather, it is easier to note the all-too-human among those with whom one is less familiar. I believe many will be struck by the characteristics manifested by the attendees or participants. In the case of the "cultured" observer of those less "cultured", the former may well wish the latter had better manners, a higher set of cognitive skills, more developed aesthetic discernment, and so forth. I leave it to the reader to speculate as to what the less cultured observer might think of the characteristics of those at an avant-garde opening.

One might say that whatever is observed by anyone, on either side of the questionable cultural divide I have composed, is behavior in a comparatively artificial situation. Agreed, but does not the manifestation of such behavior indicate underlying characteristics, by whatever workable theory such characteristics are delineated? It is undeniable that we possess intuitively recognizable positive attributes, however these are encompassed theoretically. To say that we are not perfect, that we are human, is to avoid a question that many of us may soon come not to avoid; what would it be like for us to become more perfect, to resolve some of our deficiencies, even if the price were to no longer be human - at least in some of the senses in which we were and are now human? Waiving that question for the moment, consider the following. There are those among us, such as Gandhi, Mother Teresa, and other compassionate individuals, who apparently lacked traits so common to most humans, or at least controlled them to a greater degree than many are capable. One can ascribe their personalities largely to their genetic endowment, or to environmental influences, or perhaps some balanced combination of both. If and when we learn enough about ourselves, we may also have learned how to control the relevant factors so as to endow everyone with what will generally be acknowledged to be favorable characteristics. If one human is capable of the mind of Einstein, Leonardo, Plato, Shakespeare, Gauss, perhaps all humans can possess one or the other of such minds.

If I seem to be arguing against myself, in that I am considering the possibility of widespread improvement of *human* emotional, cognitive, and creative abilities, allow me to add that I do not believe this will be enough to secure what is best for us. For one, I believe this is a counterfactual situation, in that I believe that before we have become scientifically advanced to the point of enabling such profound human transformation, we will have already transitioned to beings having radically extended lifespans. As I stated above, I fear that this transition will not go smoothly. But should humanity survive this transition largely intact, the possibilities indicated in the previous paragraph will be found limiting, if and when they become available.

Living radically extended, perhaps open-ended extended lives, these beings who are our descendants will have cognitive and creative possibilities beyond those of any current humans, once the science and technology of enhancing mindfulness exists. Imagine for a moment individual minds comfortable and facile in complex mathematics, theoretical physics, chemistry, biology; such minds having expanded memory and recall, together with the ability to interconnect these various areas in manners notably beyond current human abilities. I do not wish to restrict this imaginative exercise to the domain of natural science. Imagine these same minds having expanded abilities in the realm of aesthetics, as well as in other areas open to cognitive and creative mindfulness.

The projection of enhanced cognitive and creative abilities onto these transitioned humans invites the criticism that nothing has been said regarding their moral attributes, or their affability towards each other. Indeed, as newly-evolved posthumans, they would likely maintain the various envies and ill-wills their human ancestors had, along with the intuitively positive affections and dispositions of humans.

Given the implicit assumption that much will have been learned about human mindfulness, we are permitted the hope that if these speculations are realized, our posthuman descendants will eventually have the capability to "engineer" the emotional aspects of their minds to a greater extent than is currently possible. We should not ignore that the various socialization procedures to which our own children are subject, together with other techniques at our disposal, fall under the notion of "engineering" emotional characteristics.

There are some further things I wish to say regarding this, although I am unsure that these will have the intended persuasive effect, rather than the result of turning readers away from the foregoing speculations. It seems possible to me to alter human mindfulness to the extent that our descendants will not compete with one another in areas of cognition or aesthetic creation, but rather will rejoice in each other's accomplishments. Such is possible now in limited contexts, but transformed humans, posthumans, might generally be this way. Transformed humans might be to a large extent cooperative beings. Our competitive aspects undoubtedly have had survival value for humans individually, and as a species. Many would argue that they still serve us, in fostering innovation and persistence. Yet we can imagine these aspects, and the benefits we derive from them, transformed so that our posthuman descendants desire to innovate for the benefit of the general community, rather than only for themselves, their family, or some such limited group.

Humans in the developed sector are often in competition for limited employment, especially at the more respected and remunerative levels of endeavor. At a future time in which much labor is performed by increasingly capable machinery, and in which what are then people are roughly equal in cognitive abilities - which substantially outrun the abilities of current humans; where relative deprivation does not exist, and where economics, - as the science of wealth - is not practiced for the sake of a privileged class, there is the chance that people are not in economic competition. There is the chance that these speculated descendants lead self-satisfied lives, as far as their attention to whatever answers then to "household management" and general governance. There is the chance that they work cooperatively together, using their faculties at high levels.

Thus far, what has been speculated is perhaps agreeable, or at most, occasions polite criticism. The following speculation will prove harder to accept.

If our posthuman descendants lead ever-longer lives, population levels will need be managed. Since these people are, by assumption, highly educated, intelligent, and other-regarding, such management will be easier to effect than is currently the case for various human societies. That posthumans enjoy longer average lifespans does not free them from events, both external and internal, that significantly shorten their lives. External impacts and internal maladies and malfunctions will terminate virtually any conceivable biological entity, if sufficiently severe. Population levels can be controlled initially by some form of allowance for offspring, in balance with cessations.

As these posthumans increase their lifespan, however, and as they discover techniques for either avoiding premature cessation, or restoring themselves in the event of termination, there will be less opportunity for offspring. While we cannot imagine all the possibilities of extended lifespan, one possibility is noteworthy for showing a manner in which lifespans might be indefinitely extended. Suppose a person's full range of what we currently term "psychological characteristics" are periodically up-loaded to some storage device. Assuming these posthumans have biological brains, such "back-up" might be some sort of sufficiently detailed brain scan. Assume further that all people have a DNA file, and that reconfiguring the full biological aspects of a person (cloning of some sort, perhaps) is a routine matter. As a final assumption, allow that the most recent brain scan of a deceased person can be downloaded into that person's nascent body-duplicate. These assumptions realized, the deceased person is only temporarily gone.

In time, it may prove unacceptable to produce posthuman offspring. Of course, people could still enjoy sexual intimacy, biology permitting. In the absence of children, however, and in the passage of time, the dualities of sex might come to be viewed as outmoded, a holdover from a more biologically primitive past. People might become unisexual and still practice some form of interpersonal intimacy, or they might choose to migrate to a physiological and psychological position of nonsexuality. I offer that this last is their best alternative.

Those who reject my suggestion of an evolved, sexless posthuman life, but are otherwise sympathetic to posthumanism, have recourse. These speculated posthumans might agree to some limit to lifetime. Or they might disallow person-restoration, such as was entertained above. They might so proceed due to favoring continued raising of offspring, or continued sexual intimacy, with or without sexual duality. Sex, procreation, and child rearing occupy our current adult lives to the extent that we find it difficult, if not unimaginable, to project our lives - especially our radically extended lives, absent them. This may be so for posthumans also.

We are sufficiently far into speculative terrain as to be unable to resolve this matter with any degree of confidence. That being so, I am obliged to supply reasons for my suggested radical departure from the sort of lives with which we are familiar, and from which we take much enjoyment.

I believe that if humanity transitions to posthumanity in anything like the manner I describe, including: initial instability due to conflicts between those having extended lifespans and those having currently average lifespans; a cessation of tumult due to widespread adoption of enhanced lifespan; a population generally endowed with enhanced cognitive and creative abilities, together with solid moral characteristics; indefinitely extended lifespan due to "restoration" techniques, then these posthumans will want to continue in their intellectual and aesthetic pursuits. They will want this for themselves, and all others. There will be initial regrets at loss of sexuality, a loss having origins in childlessness, but there will also be awareness of gains. Interpersonal sexuality, as more than simply a pleasurable pursuit, implies emotional involvement. With such involvement comes the likelihood of conflicts. If sexuality loses its procreative function in the sketched posthuman circumstances, our descendants might at some point resolve that, having taken its full measure, what remains is not worth keeping. A more joyful existence might emphasize mindfulness unencumbered by the sorts of demands sexuality places on those whose nature it forms a significant part.

There are responses to what I have urged here. If sexuality currently fosters competition and conflict, perhaps in a posthuman future people will satisfy sexual cravings in virtual reality, complete with virtual partners. Such a scheme would avoid actual conflicts, and would allow people to continue their sexual pleasure (perhaps leading to virtual families).[7]

I do not wish to prolong this discussion here. If humanity at some time does transition to posthumanity, the matter will be resolved in one form or another. From our present perspective, many of us cannot imagine an enjoyable life absent emotional involvement with other individuals, and excluding yet other individuals. Sexual attraction is a strong component in some of these emotional involvements. I believe that a better life awaits those sentient beings with enhanced mindfulness, and that life does not include the trials (and joys, as well as sorrows), of limited interpersonal relationships. Given that you are reading this volume, your models are likely those celebrated for their various manifestations of mindfulness, rather than their social and sexual successes. Because we are human we wish for a life including the pleasures of friends and family, as well as various of the bodily pleasures. It is difficult to imagine an indefinitely extended life with pleasures other than these; a life in which these now former pleasures create sufficient dissension so as to interfere with the pleasures that will matter most to those who we have become.

What I have said is taxing enough, yet it is perhaps only a starting point. It may come to pass that our posthuman descendants decide to abandon their biological aspects. Speculation on this possibility is sufficiently removed from our present situation as to be fantastical. The reason for doing this, in whatever manner it might be done, is the reason I have been giving. When all is settled, I believe, what humans value most is their mindfulness. Not all humans, and rarely solely their mindfulness. Further, "mindfulness" needs considerable elaboration. Rather than give what is needed, and what I can only supply at length, if I can supply it at all, I return to our models. Here again is a list, obviously incomplete: Plato, Leonardo, Galileo, Shakespeare, Beethoven, Gandhi, Einstein, von Neumann. All commonly regarded to be extraordinary humans for some aspect of their mindfulness, and not for their bodily abilities (except insofar as their mindfulness was a function of their embodiment). Perhaps what is best for us is that we consider this human mindfulness as only a beginning.

There are objections to consider:

1. The incomplete list of "models" given above exemplifies various forms of mindfulness; cognitive, creative, philosophical, mathematical. Are some of these forms to be preferred to others? By what criteria? Are there other forms? If so, what are they? If not, is this list provably complete as to forms?

Response: I cannot confidently answer these questions. These various forms of mindfulness, and other ostensible forms not exemplified on the list, might be incorporated in many, or all, speculated posthumans at some future period. In some ways, we are at the beginning of our understanding of that which is vaguely gestured at by the term "mindfulness". Perhaps modes of this something are more different and less compatible than I am supposing. Perhaps there is a unification of them at some level that yet allows such terminology as "cognitive", "creative", "affective". At present it seems to me that, in our study of mindfulness, we are in a period somewhat like that of the decades preceding Dalton's chemical theory.

2. Speaking of the "affective", what sorts of emotions, if any, do these later posthumans possess? If they are rid of emotions, what sort of enjoyment do they receive from their mindful pursuits? If they receive none, if they are unfeeling, what is the attraction of being one of them?

Response: I agree with much that is stated in this objection. Of course, I cannot confidently predict what manner of mindfulness these possible posthumans will manifest. I believe that we do not have adequate understanding of the various roles of the emotional characteristics that are distributed throughout human society to comment on their value at present. It seems that some of them, quickness to anger, for example, have largely outlived their evolutionary usefulness, but I am not confident of this. I am still less confident in any speculating I might attempt regarding the emotional characteristics of possible later posthumans. As I have considered these speculated descendants, they will have greater understanding of, and control over, such characteristics.

I agree that if there is nothing comparable to the enjoyment humans have in their various endeavors, then there would be little attraction in being posthuman. I therefore presume they enjoy their lives.

3. Our species, homo sapiens sapiens, has existed for over one hundred thousand years. Approximately 10,000 years ago our

ancestors developed agriculture and transitioned from hunter-gatherers to villagers. We acknowledge the existence of these earliest human settlers, and some few of us study their lives as far as possible, but most of us do not identify with them, in that we have little concern for, or understanding of, their thoughts. They, of course, could not imagine the details of our lives. Tracing the hominid tree further backwards, there are species biologically distinct from us, from which we have evolved, but towards which most of us feel no sense of identity. Needless to say, they had no idea of us. Assuming relations between ourselves and our speculated posthuman descendants remain similar to the relations between ourselves and our hominid predecessors, what benefit is it for humans to transition to beings with whom we cannot identify, and who are unlikely to identify with us?[8]

Response: I realize that this is not a well-formed objection, and if it were offered as a sharp criticism of what I have been urging, it would not be difficult to dismiss on grounds of looseness (and, of course, I would be open to the charge of setting up and knocking down a strawman). However, I want to consider this objection at some length because it expresses, however hazily, the deepest beliefs contrary to my speculations.

One need only survey the science-fiction genres to note the prevalence of humans; people with the biological, psychological, and sociological characteristics of one's friends and neighbors, projected into future circumstances, such futurity being constituted mainly by technological advances. There is an intuitively understandable desire on the part of all concerned; viewers, readers, directors, authors, to feel comfortable in these surroundings. We would likely feel less so were the protagonists biologically and psychologically unfamiliar to us (there are notable exceptions within science-fiction; I am speaking generally).

Perhaps this unfamiliarity with alien species is due mostly to psychological difference. Imagine for a moment a species whose physiognomy matched that of an octopus, who inhabited a watery environment, but who differed from actual octopus in their intelligence, lifespan, mating characteristics, and communicative abilities. In these divergences, they resemble humans. They can communicate with us in well-formed, thoughtful sentences, having learned our language. This fantasy is in need of further detail, which I shall not supply. It is already likely a physiological impossibility. But if these imaginary octopuses were to duplicate closely many of our psychological traits, we might not feel greatly removed from them (consider the manner in which many dog owners regard their pets).

It may well be that presumed psychological disparity accompanies perceived physiological disparity. In any case, the notion of psychological difference covers much ground. That the posthumans of my speculation have altered emotional characteristics and heightened cognitive abilities relative to humans might not imply a loss of identification, us to them and them to us, provided certain conditions are met. The requisites for such identification include metaphysical, epistemological, and psychological constraints.

A word regarding identification is in order. There is a body of philosophical literature addressing the overall problem of identity over time. For many physical objects, spatio-temporal continuity - uniqueness of orbit - is an intuitively adequate criterion of such identity. There are complicated cases, however, including: objects with exchanged parts; entities with consciousness; institutional objects (e.g., churches). There are two distinct items of concern in our investigation. Posthumans who live indefinitely extended lives might, in the course of their duration, undergo dramatic alterations, thereby leading to questions regarding their identity over time. Then, there is the looser notion of identification in which, for example, parents identify with their children as in some manner carrying on their "bloodline", or in which the members of a generation of some social group, such as a nation, identify with their national predecessors and their national descendants. It is this latter notion of identity or identification that is of concern in considering the transition of humans to posthumans.

There is one metaphysically necessary condition I wish to emphasize; realism, by which I mean the philosophical attitude that there is a world independent of mindfulness, populated by types of entities whose typifications are to some extent amenable to mindfulness, although these typifications are not determined by mindfulness. The point of this metaphysical supposition is that different instances of mindfulness exist in, and relate to, the same world, which can be understood in increasing clarity and detail by those forms of mindfulness capable of such increasing understanding. This is in contrast to those metaphysical suppositions that mindfulness, to some lesser or greater extent, creates the world, and that differently located (culturally, socially, historically) mindfulness inhabits different worlds. I believe the former metaphysical supposition allows for a higher degree of identification from human to posthuman than is permitted by the latter supposition. There is much more to be said on this topic, but here I can only indicate my preference.

The epistemological constraints are regarding the mindfulness spoken of above in its task of understanding the world. In this task, these have to do with concept formation, hypothesis generation, confirmation and disconfirmation. They have to do with dissemination of information as to investigation and results. Given my realistic metaphysical stance, it should be apparent that I take the position that beliefs, propositions, hypothesis, theories, and so forth are true to the extent that they correspond to the world and false to the extent that they fail such correspondence. Although correspondence is emphasized as an appropriate theory of truth, understanding happens from a socio-historical situatedness. Consequently, factors of coherence and pragmatics are present in belief formation. Rather than develop this necessarily brief sketch, allow me to emphasize the major epistemological constraint I have in mind. Given that I believe there is a world to discover, the appropriate epistemological attitude is that which supports such discovery. Whatever the recorded thoughts of natural scientists, I believe this is predominantly their attitude. I do not question, indeed I believe, that there are alternative conceptual schemes which can successfully correspond to reality, perhaps in some significant cases with roughly equal detail, comprehensiveness, and fecundity. I also believe that investigators are often, perhaps most often, not positioned to judge the absolute correspondence between the world and their thoughts thereof, although I believe on occasion they are so positioned.

Again, what is important in this discussion of the degree of identification of current humans with speculated posthumans is acceptance of the epistemological attitude, or range of attitudes, that the world can be known in increasing detail; that knowledge is a possibility, and an increase in knowledge, measured by some sort of absolute scale of correspondence, is an outstanding part of the identification of mindful creatures, be they human or posthuman. Given that practitioners are typically not situated so as to compare their beliefs to this absolute scale, the identification of which I speak here is with regard to prevalence of the disposition among mindful entities that the scale exists, and that mindful beings have both faith in it, and intimations of it.

In brief, the psychological requirements are as follows. Our posthuman descendants will have a stronger identification with us, and us with them, if they remain the sort of beings who are capable of enjoying their lives; living them with consciousness, feelings, rather than without such. In the spirit of Spinoza, they will be entities not enslaved by the passions, and liberated by their employment and enjoyment of their powers of cognition and creation.[9] They will be freed from the biological demands of survival via propagation. They will lose impulses to dominate, temper their reactive attitudes, and they will lose various other impediments to focused, creative thought. Their identification with us will be stronger if they retain a historical interest that extends at least as far as recorded human affairs.

4. Little has been said about the social and political context of these speculated posthumans. Are these matters irrelevant or of minor importance?

Response: In my view, the details of these matters are of lesser importance. The speculated posthumans are assumed to have greater understanding of their nature, and of the requirements for living enjoyable lives. I assume they will implement these requirements. What is of major importance is their willingness and ability to secure enjoyable lives for everyone, so that the corrosive effects of alienation are absent. How they relate to one another on a daily basis is best left to imagination, under the constraint that they support, rather than undermine or dominate, others. Power relationships are similarly left to the imagination, under the same constraint.

5. As the title of this piece indicates, the intention is to show what is best for humans. Has this been shown? Have alternatives been considered? Has any indication of what would justify an answer to the question of optimum living been presented? If these questions warrant negative answers, what has been done here that is of philosophical value?

Response: Regarding the first three of these four questions, agreed. Negative answers are warranted. At the outset I voiced my uncertainty regarding my overall suggestion that humanity transform to posthumanity. Nothing that I have said mitigates that uncertainty. I do not believe we will remain human, and I have given reasons in support of this belief, but these reasons are not conclusive. If at some point we transform ourselves to posthumans, it is imaginatively possible that our posthuman descendants will choose some form of indolent lives, supported by capable machinery. From my perspective, and hopefully that of most others who have contemplated a posthuman future, indolence, violence, xenophobia (the notion of foreigner being somehow instantiated) are examples of incorrect choices for posthumanity.

How is the ethical content of my speculations justified? I have beliefs regarding such justification, but I do not believe that these beliefs would survive critical scrutiny. As I said earlier, I do not believe we have a successful ethical theory, nor do I believe such a theory is possible for humans. As for a meta-ethical theory of justification, consider that natural science has widespread acceptance, although it is difficult to find solid meta-scientific theorizing among the scientific community. As Thomas Kuhn has emphasized, in periods of deep controversy such theorizing is brought to the fore, but with respect to more mundane scientific activity done under the aegis of entrenched theory, practicing scientists feel little need for justifying their governing theories.[10] There are various philosophical justifications for the correctness of natural scientific theories, but I believe these justifications play a negligible role in the widespread acceptance of the theories.

I am not implying that ethical theories ought somehow reduce to natural scientific theories, nor am I refusing to entertain such a possibility. I find it significant that various natural scientific theories enjoy wide acceptance, in comparison to the various ethical theories. But such a comparison is misleading if it is not also noted that social scientific theories suffer a similar comparison to successful natural scientific theories. In their behavioral aspects, humans are quite complicated. My hope is that either social scientific theories reduce in some useful manner to well-entrenched natural scientific theories, or that social scientific theories retain some degree of autonomy vis-a-vis natural scientific theories, yet have a predictive-explanatory success approximating that of natural scientific theories (this last requires further elaboration, which I cannot supply here). If either of these alternatives are realized, the likelihood of a successful ethical theory - for posthumans, if not for humans - will be increased, due to the close relation between the concerns of social science and the concern of ethics.

We can and do learn much from engaging questions regarding the prudentially and morally good life, and we learn much from considering what major philosophers have said regarding these and related questions. That their answers differ is important, but to the extent that their answers can be reworked without undue violence to their original intent, and made mutually consistent, the possibility of an enlarged, more multi-faceted ethical theory is suggested.

I believe that the handful of philosophers who are sympathetically considering a posthuman future for humanity are making a contribution to our ongoing attempts at what I have termed a "successful ethical theory". It might emerge, after all, that it is best for us, morally and prudentially, that we transform to posthumans of a certain sort. If from a human standpoint we largely agree that this is best, and if this agreement is based upon acceptance of similar reasons, then there might be implicit elements of some future successful ethical theory in our possession.

I close with the following questions. Do you suppose that if humanity survives for the next one hundred thousand years, what we are then will be physiologically and psychologically what we are now, and have been, as homo sapiens sapiens? What about the next ten thousand years? The next thousand years (think of scientific and technological progress in the past one hundred years)? The next five hundred years? Have you considered these questions at length previously, or have you tacitly assumed that we would be physiologically and psychologically as we are now?

Bibliography

Churchland, Paul. "Eliminative Materialism and the Propositional Attitudes", Journal of Philosophy 78 (1981).

De Grey, Aubrey, with Michael Rae. *Ending Aging: The Rejuvenation Breakthroughs That Could Reverse Human Aging in Our Lifetime*. New York: St. Martin's Press, 2007.
Fukuyama, Francis. *Our Posthuman Future: Consequences of the Biotechnology Revolution*. New York: Farrar, Straus and Giroux, 2002.

Hare, R. M. *The Language of Morals*.Oxford: Clarendon Press, 1952.

Kuhn, Thomas. *The Structure of Scientific Revolutions*. 2d ed. Chicago: University of Chicago Press, 1970.

Ray Kurzweil, *The Age of Spiritual Machines: When Computers Exceed Human Intelligence*. New York: Viking Penguin, 1999.

Shostak, Stanley. *Becoming Immortal: Combining Cloning and Stem-Cell Therapy*. Albany: State University of New York Press, 2002.

Spinoza, Baruch. *The Ethics and Selected Letters*, trans. Samuel Shirley. Indianapolis: Hackett, 1992.

Strawson, Peter Frederick. *Freedom and Resentment and Other Essays*. London: Methuen, 1974.

Wiggins, David. *Identity and Spatio-Temporal Continuity*. Oxford: Blackwell, 1967.

Endnotes

1. See, for example, Stanley Shostak, *Becoming Immortal: Combining Cloning and Stem-Cell Therapy* (Albany: State University of New York Press, 2002).

2. For such an example of such optimism see Aubrey De Grey and Michael Rae, *Ending Aging: The Rejuvenation Breakthroughs That Could Reverse Human Aging in Our Lifetime* (New York: St. Martin's Press, 2007).

3. I am thinking of the earlier writings of Sartre, such as *Existentialism and Human Emotions*, trans. Bernard Frechtman and Hazel Barnes (New York: Citadel Press, 1957).

4. A leading advocate of this position is R. M. Hare, as in his *The Language of Morals* (Oxford: Clarendon Press, 1952).

5. The phrase is from P. F. Strawson, "Freedom and Resentment", *Proceedings and Addresses of the British Academy* 48 (1962): 1-25.

6. Paul Churchland, "Eliminative Materialism and the Propositional Attitudes", Journal of Philosophy 78 (1981): 67-90.

7. Except for virtual families, the idea of virtual sex can be found in Ray Kurzweil, *The Age of Spiritual Machines: When Computers Exceed Human Intelligence* (New York: Viking Penguin, 1999), 147.

8. A lengthier and clearer version of this objection can be found in Francis Fukuyama, *Our Posthuman Future: Consequences of the Biotechnology Revolution* (New York: Farrar, Straus and Giroux, 2002), ch. 9.

9. Baruch Spinoza, *The Ethics and Selected Letters*, trans. Samuel Shirley (Indianapolis: Hackett, 1982).

10. Thomas Kuhn, *The Structure of Scientific Revolutions*, 2nd ed. (Chicago: University of Chicago Press, 1970).

CHAPTER FOURTEEN

Camus, Plague Literature, And The Apocalyptic Tradition

David Simpson

It was in April 1941 that Camus made the first reference in his journal to a possible new play or work of fiction about an epidemic of plague. "The liberating plague [*La Peste libératrice*], he cryptically noted. "Happy town. People live according to different systems. The plague: abolishes all systems."[1]

The word "liberating" is unexpected. How, we may wonder, can a plague be liberating? One possibility is that Camus was thinking of the last plague of Egypt (by virtue of which the Israelites were released from bondage). However, if we look ahead to the completed texts of *The Plague* and *The State of Siege,* we can see that he probably meant "liberating" more in the sense of a blaring reveille or terrible awakening – a crisis that suddenly startles a population out of its complacency and ennui into fearful consciousness. In this sense, even the Black Death may be seen as an agent of existential liberation, a dark visitor who shakes people out of their habitual somnolence and living death.[2]

Whatever Camus meant by the phrase, the journal entry shows that even before the publication of *The Stranger* and *The Myth of Sisyphus*, the twin works that would propel him onto the main stage of European letters and effectively launch his career, he was already thinking about an ambitious new project and had apparently settled on bubonic plague – the fatal return of the Black Death – as his likely subject and philosophical theme. The spread of a deadly disease must have struck him as an almost perfect metaphor for the tragic course of European politics during the second quarter of the twentieth century.

Camus and Dostoyevsky

The basic idea of using a disease epidemic as the setting for an apocalyptic novel and as a symbol for the wave of totalitarianism that swept through Europe during the period 1922-1941 – was almost certainly inspired by the dream of Raskolnikov in the epilogue of *Crime and Punishment*:

> "He dreamt that the whole world was condemned to a terrible new strange plague that had come to Europe from the depths of Asia. All were to be destroyed except a very few chosen. Some new sorts of microbes were attacking the bodies of men, but these microbes were endowed with intelligence and will. . . ."[3]

In this nightmare vision of a Russia and Europe overcome by nihilism or by some similarly alien and destructive creed, Camus recognized an uncanny prophecy of the rise of fascism, Stalinism, and National Socialism in his own era. And since he had just completed a long study of Dostoevsky while writing *The Myth of Sisyphus*, it's hardly surprising to find the early seeds for *The State of Siege* and *The Plague* in the work of his lifelong influence and philosophical rival.[4]

Whatever triggered Camus' original conception, he quickly got serious about the idea of a work of fiction or drama based on an outbreak of pestilence. Before long he began a concentrated study of the history of plague, of apocalyptic literature in general, and of plague literature in particular.

The Apocalyptic Tradition

The term "apocalypse" comes from the Greek *apokalipsis* and refers to the "pulling back of a curtain" or the "lifting of a veil" to disclose a hidden or future reality. The origin of the term in Western literature goes back of course to the New Testament and the Apocalypse of St. John the Divine, more commonly known as the Book of Revelation. With its archetypal imagery and symbolism of cosmic destruction and renewal, the text has served, along with its Old Testament predecessors, as the inspiration and model for a long tradition of literary works about the fate of the earth and the end of history.

Although today it's mainly thought of as a sub-genre of science fiction, apocalyptic literature actually has a lengthy pedigree beginning with the Old Testament Book of Daniel and continuing with the later Jewish and Christian apocalyptic writings (including titles like "The Assumption of Moses," "The Apocalypse of Baruch," etc.) that became popular during the first and second centuries AD.

Ancient apocalyptic has been defined as "a genre of revelatory literature with a narrative framework, in which a revelation is mediated by an otherworldly being to a human recipient, disclosing a transcendent reality which is both temporal, insofar as it envisages eschatological salvation, and spatial insofar as it involves another, supernatural world."[5] In contrast, modern apocalyptic literature, unless it is specifically religious in its motive and purpose, typically dispenses with whatever "transcendent reality" or "supernatural world" may await us and addresses our future history and ultimate destiny on earth instead.

Ancient apocalyptic, that is to say, is concerned to tell us what we need to do to assure our future salvation and continued existence in the next world. Modern apocalyptic, on the other hand, is more concerned to reveal our likely fate (and what we must do to secure or avoid it) in this one. Consequently, in modern apocalyptic (and especially post-apocalyptic) fiction, the narrator who reveals this dire and possibly vital information is typically a lone survivor or eyewitness to some past or current catastrophe or future disaster (and is thus an "otherworldly being" only in the sense of being the sole living representative of some remote, long-forgotten, or doomed civilization). He or she in effect makes one last desperate attempt to communicate with, and possibly save, whatever remnant of humanity might still exist.

Hence the survivor's tale, typically transmitted via a notebook or diary (or occasionally by the classic device of a manuscript in a bottle or its modern sci-fi equivalent, the electronically recorded and stored message), is a standard narrative form for modern apocalyptic and post-apocalyptic fiction. Mary Shelley's *The Last Man* (1826), one of the first specimens of the genre, makes especially fanciful use of the device: In an opening introduction, the author claims that the narrative represents her own edited and translated version of some prophetic writings discovered in a cave in Italy. Supposedly produced by no less a personage than the Cumaean Sybil, these writings envision the ordeal of the last human survivor of a plague epidemic that sweeps through Europe at the end of the 21st century.

An alternative narrative convention for apocalyptic fiction in general and plague literature in particular is to have the narrator (usually an older person) relate his story to an audience of fellow survivors who are either ignorant of or too young to remember the fatal disaster or sequence of events that destroyed their former civilization and brought about the dark, bleak, and in most cases barbarous world they now find themselves in. Jack London's story "The Scarlet Plague" (1912), which describes the aftermath of a deadly epidemic that begins in the year 2013, and Margaret Atwood's novel *Oryx and Crake* (2003), which chronicles the ghastly effects of a failed experiment in bioengineering, are two examples of post-apocalyptic fiction that employ this particular narrative technique.[6]

In *The Plague*, Camus adapts elements of both narrative traditions. On the one hand, he offers a basically straightforward survivor's account of a plague outbreak, presented in the deliberately un-romantic and un-heroic style of chronicle history or documentary journalism by an initially unidentified narrator. This figure, who is eventually revealed to be Dr. Bernard Rieux, the main protagonist of the novel, explains that his purpose is to report, as lucidly and as objectively as possible, only what he has personally seen and heard and in this way provide a true and accurate record of a plague epidemic and its devastating effects. Among the various sources and documents he draws on in assembling his account is the diary of M. Jean Tarrou, a friend and co-worker who, out of a deep personal sense of moral obligation and duty, proposes that the local citizens organize themselves into groups of volunteers – in effect, into sanitation teams or anti-death squads dedicated to fighting the plague.

The tone of realism, of straightforward reportage, that runs through *The Plague* came naturally to Camus as an experienced journalist and editorialist, but it was also something that he had worked at diligently in order to produce the stark, unemotional *ecriture blanche* or "zero-degree" writing of *The Stranger*.[7] This dry, matter-of-fact style was also something that he likely picked up from Daniel Defoe, whose *Journal of the Plague Year*, a classic work of fiction in the guise of non-fiction and a nearly pitch-perfect display of verisimilitude, served as a model and inspiration for the basic plot and technique of *The Plague*.

Defoe's novel was only one of several literary and historical accounts of pestilence, from the ancient descriptions of the plague of Athens in Thucydides (Book 2, Chapter 7) and Lucretius (*De Rerum Natura*, Book VI, lines 1138 ff.) to the chronicles, fictional records, and journal entries of Procopius, Boccaccio, and Mathieu Marais, that Camus studied or consulted while developing ideas for his own contributions to the genre. Plague, he discovered, had ravaged European cities repeatedly for nearly every generation from the mid-14th century to the end of the 17th century, beginning with the massive infestation of the period 1347-1350, during which, in Jean Froissart's words "a third of the world died."[8] If Froissart's estimate is even close to accurate, this wave of death, which overwhelmed most of France and claimed a massive toll of victims in England, Scotland, Italy, Switzerland, Sweden, and other parts of Europe (and which according to the current rumor had previously devastated the populations of China, India, Egypt, and the Middle East) would have killed 20 million.

Whatever the actual death toll may have been, the graphic descriptions of the havoc and terror caused by the plague obviously made a vivid impression on Camus since there is hardly a detail from the historical record that he does not make some use of in order to add realistic depth and texture to his own grim narrative. Here's a portion of the vivid account of a modern historian:

> "So lethal was the disease that cases were known of persons going to bed well and dying before they woke, of doctors catching the illness at a bedside and dying before the patient. So rapidly did it spread from one to another that to a French physician, Simon de Covino, it seemed as if one sick person 'could infect the whole world.' The malignity of the pestilence appeared more terrible because its victims knew no prevention and no remedy. . . .

> When graveyards filled up, bodies at Avignon were thrown into the Rhone until mass burial pits were dug for dumping the corpses. In London in such pits corpses piled up in layers until they overflowed. Everywhere reports speak of the sick dying too fast for the living to bury. Corpses were dragged out of homes and left in front of doorways. Morning light revealed new piles of bodies. In Florence the dead were gathered up by

the Compagnia della Misericordia - founded in 1244 to care for the sick - whose members wore red robes and hoods masking the face except for the eyes. When their efforts failed, the dead lay putrid in the streets for days at a time. When no coffins were to be had, the bodies were laid on boards, two or three at once, to be carried to graveyards or common pits. Families dumped their own relatives into the pits, or buried them so hastily and thinly "that dogs dragged them forth and devoured their bodies."[10]

From his reading of the earlier literary and historical accounts and from available medical resources, Camus drew elements for his own fictional, but frighteningly realistic, scenario of an epidemic of bubonic plague, tracing the course of the disease from its initial outbreak and its originally slow but thereafter rapid and seemingly unstoppable spread to its ultimate (if temporary) retreat and remission. From his sources, Camus learned details not only of the physical symptoms of the disease (in this respect the accounts of Thucydides and Lucretius are particularly harrowing), but also of its social and psychological effects: the initial incomprehension and outright denial on the part of local inhabitants and civic leaders; the gradual awakening and eventual acknowledgment that something is dangerously amiss; full recognition of the scale and implications of the infestation; mass panic and dread; a desperate search for explanations (both physical and metaphysical) and for saviors and scapegoats; an upsurge of superstition and occult belief; abandonment of traditional norms and mores, with a reversion to hedonism and barbarism. Eventually the local residents reach a kind of dismal equilibrium, with isolated instances of resistance and opposition amid a general reaction of submission, resignation, and acceptance of defeat.

What Camus adds to his sources and to the tradition of plague literature in general – and this is arguably a new and original contribution to the genre on his part – is the element of sustained political and philosophical allegory. In virtually all the earlier literature of pestilence and apocalypse, from the Bible and Lucretius to Defoe and Shelley, plague is plague, pure and simple, and not also a metaphor or symbol for something else. Of course an outbreak of disease can serve (and in plague literature nearly always does serve) as the occasion for philosophical reflection and speculation, for meditations on fate, history, and the meaning of things. It can also be interpreted as an omen or portent – for example, as a harbinger of the end times -- or as a sign of a disruption in nature or of cosmic or ecological disorder. And it can be presented, as in the Bible and the *Oedipus Rex* of Sophocles (and as in Camus's novel itself, in the form of Father Paneloux's first sermon), as a heaven-sent scourge or punishment for sin. However, in no previous literary treatment is plague, in addition to its overt and literal role as a deadly disease, and beyond whatever moral or theological interpretations may be applied to it, also used as a vehicle for a consistent and fully developed layer of allegorical meaning.

In short, unlike other specimens of plague literature, Camus's novel, like Dante's *Commedia*, can be read, and in fact needs to be read, on more than one level. For in Camus's treatment the plague is never simply a disease. Instead, as Camus himself pointed out, and as most first-edition readers already understood or immediately recognized, it is also a graphic symbol or metaphor for a complex of interconnected political, economic, and historical developments, including: (a) the Nazi occupation of France (especially during the period 1942-44 when Camus himself was trapped there and unable to return to his native Algeria);[11] (b) the rise of totalitarianism and of new systems of bureaucratic oppression and terror, both on the left and on the right; (c) twentieth-century Europe's hidden history and ugly legacy of midnight round-ups, concentration camps, and mass murder – epitomized and made forever infamous by the Jewish Holocaust (recurrent images of which, from the quarantining of citizens and the separation of families to the shoveling of lime over dead bodies heaped in mass graves, cast a haunting shadow over the pages of the novel).

Over the course of the narrative, the list of allegorical associations and meanings of plague is expanded even further. The character Tarrou, for example, extends the metaphor to include support for *any* system of institutionalized inhumanity, social injustice, or officially sanctioned death, especially support for capital punishment.[12] In a long, revelatory personal monologue recorded by Rieux (245-54), he expounds his belief that in *all* our daily moral and political decisions we are either carriers of plague or opponents of it, either pro-death or anti-death:

> "And thus I came to understand that I, anyhow, had had plague through all those long years in which, paradoxically enough, I'd believed with all my soul that I'd been fighting it. I learned that I had had an indirect hand in the deaths of thousands of people; that I'd even brought about their deaths by approving of acts and principles that could only end that way. . . ." (251)[13]

> "For many years I've been ashamed, mortally ashamed, of having been, even with the best intentions, even at many removes, a murderer in my turn. As time went on I merely learned that even those who were better than the rest could not keep themselves from killing or letting others kill, because

such is the logic by which they live; and that we can't stir a finger in this world without the risk of bringing death to somebody. Yes, I've been ashamed ever since; I have realized that we all have plague, and I have lost my peace. . . ." (252)

"I know positively . . . that each of us has the plague within him; no one, no one on earth is free from it. And I know, too, that we must keep an endless watch on ourselves lest in a careless moment we breathe in somebody's face and fasten the infection on him. . . ." (253)

"All I maintain is that on this earth there are pestilences and there are victims, and it's up to us, so far as possible, not to join forces with the pestilences." (253-54)

"I only am escaped alone to tell thee."[13] Camus and Melville

Beyond question the single greatest influence on Camus at the time he began writing *The Plague* was Herman Melville. Camus considered *Moby Dick* the ultimate philosophical novel, and in Melville's use of apocalyptic themes and symbolism he recognized an artistic and conceptual template that he could use for his own attempt to combine socio-political allegory, personal philosophy, metaphysical speculation, meditations on life, death, and the human condition, along with elements of personal experience and autobiography all within a realistic fictional narrative. The difference would be that instead of following Melville's use of high rhetoric and a style in the vein of the King James Bible and Shakespearian tragedy, Camus would present his own apocalyptic novel, as noted earlier, in the plain, unadorned style of documentary realism.

It was Camus himself who pointed out that *The Plague* "may be read in three different ways. It is at the same time a tale about an epidemic, a symbol of Nazi occupation (and incidentally the prefiguration of any totalitarian regime, no matter where), and, thirdly, the concrete illustration of a metaphysical problem, that of evil . . . which is what Melville tried to do with *Moby Dick*, with genius added."[14] This insight reminds us that when he first began working on the novel, he had just completed an intensive study of Melville's epic and had come away from the experience with admiration and even awe for the author's achievement. In *Moby Dick* he found not only a classic specimen of apocalyptic fiction, rich with Biblical allusions and omens of destruction, but also a survivor's tale of struggle and doomed enterprise. Here too was a work of realism and high seas adventure that could also be read as a tragic parable or allegory for the downfall of a nation (the age-old metaphor of the Ship of State is never far below Melville's surface) as well as a philosophical inquiry into the nature of the universe and the problem of evil. Camus determined early on that *The Plague* would be a novel like *Moby Dick*, though on a more modest scale.

In Melville's scheme, the Pequod (whose name is taken from an extinct Native American tribe) is not just a ship and a community of whaling men; it is also a world – a compacted globe or 19th-century ark of Noah bearing an international cast of characters.[15] Such a use of symbolism and synecdoche, of microcosm representing macrocosm, of part standing for whole, is a standard device within the apocalyptic tradition, which typically uses a relatively small or isolated community – a town or village, a ship or castle, or even a public building or single-family dwelling – as a symbol or epitome for a larger social group: a nation or empire, or even the earth and the human race as a whole. Camus's Oran is entirely conventional in this respect – a relatively small and isolated, yet nevertheless cosmopolitan city whose inhabitants offer a representative spectrum or cross-section of humanity at large. In the fate of Oran (as in the fate of occupied Paris) we can find an image or figure of the possible fate of the world.

There is at least one further respect in which *The Plague* resembles *Moby Dick*, and that is that both novels involve substantial elements of autobiography.[16] This is not to say that Melville is Ishmael or that Camus is Rieux, but simply to point out that each author draws heavily upon his personal experiences and knowingly projects his own memories, passions, philosophical ideas, and even obsessions into the minds of his characters.

In Camus's case, he pours his own political and moral theories as well as aspects of his own life and personality into nearly all of his characters. As an added autobiographical note, it's important to remember that Camus' recurrent tuberculosis, the basic symptoms of which are very similar to those of pneumonic plague, served him as a constant and awful reminder of his own and every human being's potentially fragile health and vulnerability to disease. Plague was not a condition or metaphor that he took lightly or impersonally.

Of the main characters of the novel, those whose experience reflects that of Camus himself or who resemble the author in some interesting or important way include:

Dr. Bernard Rieux. The narrator's physical description is a close fit for Camus, and he clearly embodies the author's moral philosophy. His ordeal in fighting plague in Oran parallels the author's encounter with and resistance to the Nazi occupation of France.

Joseph Grand. The hopelessly aspiring writer and would-be Latinist is a satirical self-parody of the author, who himself at one time seemed destined to a disappointing career as a struggling writer, petty bureaucrat, or grammar school teacher in Oran or Algiers. Grand's futile, never-ending effort to perfect the opening sentence of his prospective magnum opus is a parody version, both comical and pathetic, not only of novel-writing but also of the Oran community's effort to defeat the plague and of Sisyphean struggle in general.

Father Paneloux. The Jesuit priest and scholarly authority on St. Augustine is another alter ego of the author, a lapsed Catholic and atheist who nevertheless maintained a lifelong respect for and dialogue with Christianity and who wrote his college thesis on Augustine. Though Paneloux, so windy, judgmental, and self-assured, is in many ways an anti-Camus, his notion of "active fatalism" – the belief that an acceptance of divine will is no excuse for inaction – comes close to the author's own concept of Sisyphean revolt.

Robert Rambert. As a Paris journalist who is effectively imprisoned in Oran by the plague and cut off from his wife and homeland, Rambert's situation mirrors that of Camus, who owing to the Occupation of France and the war in North Africa became separated for two years from his own wife and native country and who served for most of that period of exile as a journalist and member of the resistance in Paris. (See note 11.)

Jean Tarrou. Quiet, affable, and reserved, Tarrou is a perfect stand-in for Camus. In the novel he serves as a spokesman for Camus's own visceral and lifelong opposition to the death penalty, and in his paradoxical notion of the agnostic saint – the "saint without God" (255) -- he both effectively articulates and personally embodies the author's philosophy of rebellion.

Oran. The Algerian port city (which like the plague itself is frequently personified in the text) can be considered a character in the story. Camus's love-hate relationship with Oran is well documented (particularly in his essay "The Minotaur" where he describes it, in imagery that seems to come straight out of TS Eliot's *The Waste Land*, as both culturally and physically barren – a desert wilderness of rock, sand, and pebbles as well as the capital city of ennui). Yet Oran was home to Camus during the period 1941-4 and was also the native city of his wife Francine, so he had pleasant memories and feelings associated with this "happy city"[17] as well.

The Plague and the Holocaust

One aspect of *The Plague*, often noted but seldom accurately or extensively commented on, is how the allegorical framework of the novel relates to and incorporates the Holocaust. One commentator offers the astounding interpretation that the dying rats in Part I represent "French Jews," i.e., dispensable plague victims whose mounting death toll is largely overlooked or swept quietly aside, and observes that it is only when the *human* inhabitants of Oran (i.e., "French Christians") begin dying that any kind of resistance or counter-action begins.[18]

What most commentaries on this particular facet of the novel overlook is how Camus's use of plague as an allegory and sustained metaphor effectively takes the vocabulary and imagery of Nazi rhetoric -- whereby Jews are viewed as vermin or as a pestilence to be exterminated (and Nazism is presented as a divine scourge or exterminating angel) – and stands it squarely on its head. The classic example of this style of rhetoric was Himmler's notorious Posen speech of October 4, 1943, in which he declares a program of *Judenevkuierung* ("Jewish evacuation") and Ausrottung ("Extermination"), adding that "extermination" is not only a word that is easily said but also an action that can be just as easily performed. "Fie!" a mere "*Kleinigkeit,*" he calls the project - - "a trifling thing."[19] That a cyanide-based industrial pesticide called Zyklon B became one of the principal agents used to "exterminate" the "Jewish pestilence" accorded with this rhetoric and became a source of mordant jokes within the Death Camp cadres and the Nazi high command.

That Camus was aware of the sinister resonance, the undertones of Nazi hate-speech and anti-Semitic propaganda, that the plague metaphor signified, and yet went ahead anyway with his own system of allegory (whereby, instead of European Jewry, the Reich itself is equated with the Black Death) was a bold decision and adds a layer of ironic frisson to each occurrence of the words "plague" and "pestilence" in the novel.

The irony is intensified by the fact that, as Camus discovered while doing preliminary research for his novel, the Jews themselves had been held responsible for earlier outbreaks of plague in Europe. For example, as the Black Death spread through Europe during the period 1347-50, Jews were accused of poisoning local wells and were frequently rounded up for execution or mass slaughter. Pope Clement issued two papal bulls in 1348 (July 6 and Sept 26) in which he condemned the violence and said those who blamed the plague on the Jews had been "seduced by that liar, the Devil." He urged clergy to take action to protect Jews, as he himself had done.[20] In 1391 an outbreak of plague in Spain was also followed by a massacre of Jews, and when the plague recurred in Spain a century later, the Inquisition again blamed the Jews. This led to the infamous forced conversion or expulsion of the Jews from Spain in 1492, in effect the first official European policy of "*Judenevakuierung*" and "*Ausrottung*."

In one of the grimmer and more memorable sections of *The Plague* (175-78), Camus successfully superimposes upon a realistic description of Oran's sanitary precautions and mass burial procedures shadowy traces of Europe's awful recent history. The resulting fused images of a public health disaster and the Holocaust constitute one of the great triumphs of his understated style and allegorical method:

> "The naked, somewhat contorted bodies were slid off into the pit almost side by side, then covered with a layer of quicklime and another of earth, the latter only a few inches deep, so as to leave space for subsequent consignments." (176)

When the death harvest in Oran reaches its high point, mass burial gives way to a new policy of cremation. Once again a realistic account of emergency procedures that would likely happen during an actual plague epidemic creates, via allegory, an effect of photographic double exposure: the shuttling and burning of plague victims in Oran inevitably reminds the reader, as it is meant to, of the brutally efficient system of rail cattle cars delivering their daily sacrifice to the gas chambers and crematoria (notable for their expertly engineered smokestacks and powerful burning capacity) at Dachau and Auschwitz.

"Then a municipal employee had an idea that greatly helped the harassed authorities; he advised them to employ the streetcar line running along the coastal road, which was now unused. So the interiors of streetcars and trailers were adapted to this new purpose [of transporting plague victims], and a branch line was laid down to the crematorium, which thus became a terminus.

During all the late summer and throughout the autumn there could be seen moving along the road skirting the cliffs above the sea a strange procession of passengerless streetcars swaying against the skyline . . . .Little groups of people contrived to thread their way unseen among the rocks and would toss flowers into the open trailers as the cars went by. And in the warm darkness of the summer nights the cars could be heard clanking on their way, laden with flowers and corpses.

During the first few days an oily, foul-smelling cloud of smoke hung low upon the eastern districts of the town. These effluvia, all the doctors agreed, though unpleasant, were not the least harmful. However, the residents of this part of the town threatened to migrate in a body, convinced that germs were raining down on them from the sky, with the result that an elaborate apparatus for diverting the smoke had to be installed to appease them. Thereafter, only when a strong wind was blowing did a faint, sickly odor coming from the east remind them that they were living under a new order and that the plague fires were taking their nightly toll." (178-79)

Kafka, Orwell, and *The State of Siege*

At various times in *The Plague*, the narrator personifies the disease, emphasizing its grinding monotony and describing it (in a way that anticipates Hannah Arendt's notion of the "banality of evil") as if it were a dull but highly capable public official or bureaucrat:

> "It was, above all, a shrewd, unflagging adversary; a skilled organizer, doing his work thoroughly and well." (180)

> "But it seemed the plague had settled in for good at its most virulent, and it took its daily toll of deaths with the punctual zeal of a good civil servant." (235)

This identification of the plague with oppressive civil bureaucracy and the routinization of charisma looks forward to the author's play *The State of Siege*, where plague is used once again as a symbol for totalitarianism – only this time it is personified in an almost cartoonish way as a kind of overbearing government functionary or office manager from hell.

The play has never been especially popular, and its original performance received largely negative reviews.[21] Without getting into the debate over the merits and alleged deficiencies of the work, we can simply observe that Camus' plays in general have been thought to be of a lower candescence than his best fiction and that *The State of Siege* in particular, though certainly an interesting stylistic and theatrical experiment, never achieves the same level of moral intensity and high seriousness as his novels and essays.

Like the Oran of *The Plague*, the Cadiz of *State of Siege* is depicted as a place of easy habits and drowsy contentment where the inhabitants go about their daily routines more or less aimlessly and where, as the chorus declares, "nothing will happen." In fact "nothing" seems to be the local byword, and the town spokesman is a drunken cripple and street-corner prophet aptly named Nada.

At the start of the play a wailing music -- "recalling the sound of an air-raid siren" -- pierces the silence, and a bright comet suddenly illuminates the dark stage. "It's the end of the world," a voice mutters. Cadiz is about to be jolted out its customary lethargy. Indeed a strange new variety of plague is about to descend upon the city.

Many of the familiar conventions, devices, and imagery of traditional apocalyptic writing and plague literature are here, along with, as in *The Plague*, lightly veiled references to the Holocaust. We encounter death-carts heaped with bodies, crematoria, closed gates, death knells, star-shaped symbols hung over doorways, identification badges, prophets and sorceresses, astrologers and priests, prayers and incantations, signs in the sky. However, in this case, instead of a text featuring the vocabulary and imagery of medicine and disease, we have the language and iconography of the law court, the Politburo, industrial management, and the concentration camp.

Of course the most obvious change between *The Plague* and *The State of Siege* is the alteration of literary genre and mode from a fictional narrative in the journalistic and verisimilar style of Defoe to stage drama in the form of a jazzed up and modernized Medieval morality play containing elements of farce, pantomime, and musical comedy. And instead of the primary thematic and philosophical influences being, as in *The Plague*, Dostoyevsky and Melville, in *The State of Siege* Camus's main artistic inspiration comes from two of his other favorite writers, Franz Kafka and George Orwell.

Although Camus's admiration for Kafka is well documented,[22] his respect for and debt to Orwell is less well known. Obviously the two writers shared core values and traits in common, from opposition to Fascism to a jaundiced view of bureaucracy and political language. Both men served in resistance movements, Orwell in Spain and Camus in France, and both suffered (as did Kafka) from tuberculosis -- so in a way the two activists were brothers in arms as well as fellow plague victims. Both authors were also members of the European Federalist Movement (precursor of the EU). However, there is no record that they ever actually met or corresponded.[23]

.

In the play Camus's bold attempt to blend the allegorical and dramatic conventions of *Everyman* with the social and linguistic concerns and thematic preoccupations of the authors of "In the Penal Colony" and "Politics and the English Language" produces intriguing results. To begin with, as in a Medieval mystery play, we have a personified Plague and Death. But in this case instead of a skeletal figure with hooded cowl and scythe, we get a corpulent narcissist (Plague), who wears a gaudy uniform draped with a large medal, and a prim, fussy, official-looking woman who is also in uniform and who carries a clipboard or notebook. This figure turns out to be Death, in her busy role as Plague's Secretary. With this couple serving as the principal villains and embodiments of evil in the work, *The State of Siege* emerges as not only a new form of plague literature but also as a dystopian allegory and political satire very much in the vein of Orwell's *Animal Farm* (1945), with perhaps a touch of Charlie Chaplin's *The Great Dictator* (banned in France during the Occupation) thrown in for added effect.

In *The Plague* one of Camus's recurrent themes is the very post-modern (but actually centuries-old) concern over the slipperiness and adequacy of language, especially the question of whether it's possible for writers to achieve clarity, precision, and fidelity to truth. Rieux's scrupulous effort to be objective and his promise not to heroically embellish or romanticize his report; Grand's frustration at his own inarticulateness and his obsessive effort to write a perfect sentence; Paneloux's histrionic and florid sermons; and Tarrou's remarks on rhetoric and silence (116) and insistence on "exact truth" (287) – all these form part of a web of references and allusions to the difficulties of communication. At one point, dismayed by the distortions of fact and the shrill style typical of contemporary newspaper and radio reporting, Rieux vows to steer clear of such practices and to take particular care not to include language and feelings "overcharged with emotion in the ugly manner of a stage play" (138).

Camus must have found it amusing if, at the very time he was writing that sentence about the "ugly manner of a stage play", he was already considering an eventual stage adaptation of *The Plague* or was beginning preliminary work on *The State of Siege*. Regardless, one thing that can be said about *The State of Siege* is that it's certainly *not* "overcharged with emotion." On the contrary, for all its avant-garde staginess and music-hall style, it's actually a fairly cerebral play about a possible future in which natural human emotions (especially love) are either suppressed or eliminated.

The theme of the corruption or debilitation of language and of loss of speech, especially political speech (a theme graphically depicted onstage in the form of a gagged chorus and a political ceremony performed entirely in pantomime), dominates the drama from beginning to end. Much of the language of the play, with its euphemisms, slogans, bureaucratic circumlocutions, and official jargon can truly be described as Kafkaesque or Orwellian and seems to blend the obscure legalese of *The Trial* (particularly the proceedings before the Examining Magistrate) with the infamous doublespeak of *1984*. The word "love" is forbidden, and Death and Plague prove to be as smoothly "double-tongued" as any media-savvy spin artist or modern politician. Nada, now acting as an official public relations agent for the new order, declares that the age of meaningful communication is over:

"We want to fix things up in such a way that nobody understands a word of what his neighbor says. We are steadily nearing that perfect moment when nothing anybody says will rouse the least echo in another's mind . . .well on the way to that ideal consummation, the triumph of death and silence." (186)[24]

Even critics of *The State of Siege* concede that it's a daring artistic experiment. By shifting from narrative to drama, Camus abandons the traditional forms and models of plague literature and apocalyptic yet retains certain features of each (e.g., the blazing comet that begins the play and the sea-storm and predicted tidal deluge that end it are straight out of the apocalyptic tradition.) And at the same time that he effectively reshapes the Medieval morality play (including its conventional use of allegory) and adapts it to the modern stage, he also adds a chorus in the style of ancient Greek drama along with elements from the contemporary music hall and cabaret and even a couple of features (e.g., the characters Plague and Death are given to periodic bursts of laughter and slapstick) that look back to the Spanish Golden Age and Calderon and forward to the newly emerging Theatre of the Absurd. In a way, the play is a virtual portmanteau of theatrical history from the classical stage and the Middle Ages and Baroque era to the modern avant-garde.

Conclusion: The Plague, The State of Siege, and Camus's Philosophy of Revolt

It seems appropriate to close with a few observations on the meaning of *The Plague* and *The State of Siege* along with some suggestions on how these two works align with and illustrate Camus's philosophy. The first thing to observe is that there has been a tendency on the part of critics and viewers to oversimplify the meaning of *The State of Siege* and to view the play as in some respects either a more optimistic restatement of the writer's philosophy (as articulated in his essays and non-fiction and especially in *The Myth of Sisyphus* and *The Rebel*) or even as a watered down or artificially sweetened version of it. The problem arises because Diego, the main protagonist of the play, appears to *defeat* the Plague (he does indeed send the annoyingly vain and pompous figure – an obvious stand-in for Franco – packing, usually to audience applause) by his simple act of courage and self-sacrifice – i.e., by pledging his own life for that of his lover Victoria (a reprise of the Orpheus-Eurydice theme that Camus included in *The Plague*). But this interpretation overlooks several important facts: 1. Diego dies. 2. Death (no longer in her temporary, and overworked, assignment as Plague's administrative Secretary but once again in her normal role as life's random pillager and harvester) endures. 3. The Plague isn't defeated or destroyed; he is merely in retreat. He departs laughing and sneering, confident of his eventual triumph. He is, as he says, "stubborn."

Understood in this way, the moral and political meaning of *The State of Siege* is not significantly different from that of *The Plague*. Courage and self-sacrifice count. Individual acts of rebellion, of drawing a line in the sand and saying, No further!, count, especially when these acts are performed in solidarity with and for the sake of others. But courage and resistance, as Rieux in the final pages of his chronicle makes clear, can never actually defeat the plague; they can only check it or put it temporarily to flight. Eventually, it will make its reappearance in some other venue, some other Oran or Cadiz, at some future time.

> "[H]e knew that the tale he had to tell could not be one of final victory. It could be only the record of what had to be done, and what assuredly would have to be done again in the never-ending fight against terror and its relentless onslaughts . . . .
> He knew . . . that the plague bacillus never dies or disappears for good; that it can be dormant for years and years in furniture and linen-chests; that it bides its time in bedrooms, cellars, trunks, and bookshelves; and that perhaps the day would come when, for the bane and enlightening of men, it would rouse up its rats again and send them forth to die in a happy city." (308)

The fact that the plague (in its broadest symbolic signification as *any* agent or source of human oppression) is essentially undefeatable and thus capable of returning at any time with renewed or even greater potency is the main reason why Rieux (and by extension Camus) refuses to romantically embellish his account or turn it into a saga of heroic victory or martyrdom.[25] For to romanticize Oran's ordeal would be to deny its fundamental absurdity, to obscure the fact that it is a true Sisyphean struggle ("the same thing over and over and over again" – p. 161) and doomed to fail.[26] In addition, an account in the style of Hollywood melodrama or heroic adventure fiction would also be false to Camus's basic philosophy and sense of reality. For it would tend to color over or even negate the fact that, from his strict philosophical perspective, Oran's (and Cadiz's) temporary reprieve from plague, and for that matter France's liberation from and eventual triumph over Nazism, are basically fortuitous – since there is no *metaphysical* reason why rebellion should prevail. That is, there is nothing in the structure of history or reality, no dialectical process or cosmic order, that guarantees the success of human struggle or that values human life over the existence of the plague bacillus; no divine *Nous* or transcendent moral system that assures the victory of liberty over terror or prevents the triumph of evil over good.

And yet despite this recognition and full acceptance of the fact that we live in an absurd universe, an immense void (as Pascal said) of which we are ignorant and which knows us not, Camus nevertheless clings to his personal belief (and this is very much an article of faith rather than a reasoned conclusion on his part) that there is an intrinsic value and essential dignity to human life. And that dignity, his narrator reminds us, is never more visible or pronounced than in time of plague, a time when we learn again a simple and modest truth about human beings: "that there are more things to admire in men than to despise" (308).

After-note: In 1946, while Camus was in France and completing work on his novel, five cases of pneumonic plague were reported by a medical team at a hospital in Oran. A further small outbreak of bubonic plague (18 cases) occurred, virtually without warning, in the city in 2003.[27]

Notes:

1. Camus, Albert. *Notebooks, 1935-42*. Philip Thody, tran. New York: Modern Library, 1965, p. 193.

2. It's noteworthy, for example, that at the beginning of *The Plague* Oran is described as "ugly," and banal, its townsfolk as bored and boring. They are basically presented as automata performing familiar rituals and sterile routines. Even their love-making is described as essentially half-hearted and mechanical. "Treeless, glamourless, soulless, the town of Oran ends by seeming restful, and, after a while, you go complacently to sleep there," complains the narrator (p.6). In such a tired, lifeless environment, a sudden eruption of plague might indeed seem "liberating," a dramatic wake-up call. Herbert R. Lottman has conjectured that the phrase *la peste libératrice* may refer to a 1934 essay by Antonin Artaud, who compares live theatrical performance to plague. See *Albert Camus: A Biography* (Corte Madera, CA: Gingko Press, 1997), p. 271.

3. Fyodor Dostoyevsky, *Crime and Punishment*, Constance Garnett, tr. (New York: Bantam Books, 1987), p.459.

4. For an in-depth discussion of Camus's debt to and rivalry with Dostoevsky, see Ray Davison, *Camus: The Challenge of Dostoyevsky* (Exeter, UK: University of Exeter Press, 1997).

5. J J Collins, ed., *Apocalypse: The Morphology of a Genre. Semeia* 14. (Missoula, MT: Scholars Press, 1979), p. 9.

6. Job 1: 15. Melville used this verse as the epigraph for the final chapter of *Moby Dick*.

7. The title of the French edition of Atwood's novel, *Le Dernier Homme*, is a likely homage to the identically titled early plague novel (1805) by Jean-Baptiste Cousin de Grainville.

8. Roland Barthes, *Writing Degree Zero*, Annette Lavers and Colin Smith, tr. (New York: Hill and Wang, 1977), p. 77. Jean-Paul Sartre, "Camus' *The Outsider*" (1943), in *Literary and Philosophical Essays*, (New York: Collier Books, 1955), p. 26-44.

9. Jean Froissart, *Chronicles*, Geoffrey Brereton, trans. (London: Penguin Books, 1968), p. 111.

10. Barbara W. Tuchman, *A Distant Mirror* (New York: Random House, 1978), p. 92-94.

11. In November, 1942, Camus was in Lyons and temporarily separated from his wife Francine, who had returned to Algeria to assume a teaching position. Camus was about to return to Algeria himself when he was prevented from doing so by Operation Torch, a naval invasion in which Allied forces took control of the North African coastline, and by the nearly simultaneous German invasion and occupation of formerly Vichy-controlled southern France. As a result of these coincidental developments, Camus and Francine were forced to live apart for two years, she in newly liberated Algiers, he in self-described "exile" in occupied France.

12. Camus, as is well documented, was an outspoken opponent of the death penalty and called repeatedly for its abolition. See especially "Reflections on the Guillotine," in *Resistance, Rebellion, and Death*, Justin O'Brien, trans., (New York: Vintage International Books, 1995), pp. 175-234.

13. Albert Camus, *The Plague*, Stuart Gilbert, trans. (1948, rpt. New York: Vintage Books, 1991). All page citations refer to this edition.

14. Olivier Todd, *Albert Camus: A Life*, Benjamin Ivry, trans. (New York: Alfred A Knopf, 1997), p. 168.

15. Queequeg, for example, is a South Sea Islander; Tashtego, a Native American. The crew also includes Daggoo, and African, and Fedallah, a Persian. The cabin boy Pip and the ship's cook Fleece are both African-American.

16. Of course to a certain extent all novelists do this. But Melville, who began his career with an autobiographical novel (*Typee*) and Camus, who died while he was working on one (*The First Man*) probably more so than usual.

17. Camus uses the phrase "happy city" in an early notebook reference to a plague novel (see note 1 above) and also uses it in the last sentence of *The Plague* (p. 308). He seems to have meant it not only as an ironic characterization of the ennui and complacency that he considered typical of urban bourgeois culture (note 2 above), but also as a sarcastic refutation of the optimistic and progressive visions of Marx and his followers and of the overly positive views of popular utopian and apocalyptic writers like Edward Bellamy and HG Wells.

18. See: <http://www.cclapcenter.com/2010/05/the_cclap_100_th e_plague_by_al.html>. For a more detailed and thoughtful discussion of Camus in relation to the Holocaust, see Sal Tessler, *Levinas and Camus: Humanism for the Twenty-first Century* (London: Continuum Books, 2008).

19. Full text versions, translations, and commentaries on Himmler's speech are available at multiple sites online. For example, see: <http://www.holocaust-history.org/himmler-poznan/speech-text.shtml>.

20. Tuchman, p. 113.

21. Audiences and critics who expected to see a dramatization of *The Plague* were undoubtedly disappointed by *The State of Siege*. The shift from tragic realism (Camus's normal mode) to satire (not an obvious Camus forte) probably also caught many viewers off guard. Whatever the reason, as Camus remarked, "few plays have ever enjoyed such a unanimous slashing." Albert Camus, "Preface" to *Caligula and Three Other Plays* (New York: Vintage, 1958), p.viii.

22. By the time he began working on *The Plague* and *The State of Siege* Camus had already written an important essay on Kafka that was to have been included in *The Myth of Sisyphus* (1942). Unfortunately, the Nazi censors prohibited Gallimard, Camus's French publisher, from including any material promoting the works of a Czech Jew. The essay, now bearing the title "Hope and the Absurd in the Work of Franz Kafka" was eventually added as an Appendix to the first American edition in 1955.

23. Camus's play, inspired in part by *Animal Farm*, was first performed in October, 1948, and so preceded by several months the publication of Orwell's *1984*, which came out in June of 1949. *The Plague* was published in 1947. Yet Camus apparently regarded both his play and his novel as in certain respects an answer or alternative vision to Orwell's dark narrative of a totalitarian dystopia sustained by overpowering technologies of behavioral engineering, "re-education," brain-washing, and media control. For an illuminating comparison of the thought of Camus and Orwell, see Miho Takashima, "George Orwell and Albert Camus: A Comparative Study – Their Views and Dilemmas in the Politics of the 30s and 40." *The International Journal of the Humanities*. Volume 2, Issue 3; and "Revolt and Equilibrium: A Comparative Study of *Nineteen Eighty-Four* and *L'Homme Révolté*; the Views and Struggles of Orwell and Camus." *The International Journal of the Humanities*. Volume 4, Issue 4, pp. 105-116.

24. Albert Camus, *The State of Siege*. In *Caligula and Three Other Plays*., Stuart Gilbert trans. (New York: Random House, 1958). All page citations refer to this edition.

25. Cf. Nada's proclamation (apropos of every change of regime) of a new era of myth-making, revisionist history, and self-congratulation. "Look! The writers of history are coming back and we shall soon be reading all about our heroes under the plague. . . . Look! Do you see what they're up to? Conferring decorations on each other!" (230).

26. The theme of plague resistance as an absurd, never-ending, ever-enduring Sisyphean task is indicated throughout the novel – e.g., in Grand's ceaseless effort to write his opening sentence, in the Oran movie theatres showing the same films over and over, in the opera's repeated performances of *Orpheus and Eurydice*, and in Rambert's continual replaying of the same musical recording (which, fittingly enough, is "The St. James Infirmary Blues").

27. Eric Bertherat, et.al., "Plague Reappearance in Algeria after 50 years, 2003," *Emerging Infectious Diseases*, vol. 13, no. 7 (October, 2007).

Bibliography:

Barthes, Roland. *Writing Degree Zero*. (Originally published as *Le Degre Zero de L'Ecriture*, 1953.) Annette Lavers and Colin Smith, tr. New York: Hill and Wang, 1977.

Bertherat E, Bekhoucha S, Chougrani S, Razik F, Duchemin JB, Houti L, et al. "Plague Reappearance in Algeria after 50 years, 2003." *Emerging Infectious Diseases*. 13:7. (October, 2007). Available from <http://www.cdc.gov/EID/content/13/10/1459.htm>.

Camus, Albert. *The Myth of Sisyphus*. Justin O'Brien, tr. New York: Vintage books, 1955.

_____. *The Rebel. An Essay on Man in Revolt*. Anthony Bower, tr. New York: Vintage Books, 1956.

_____. *Caligula and Three Other Plays*. Stuart Gilbert, tr. New York: Vintage Books, 1958.

_____. *Notebooks, 1935-42*. Philip Thody, tr. New York: Modern Library, 1965.

_____. *The Plague*. Stuart Gilbert, tr. New York: Vintage Books, 1991.

_____. *Resistance, Rebellion, and Death*. Justin O'Brien, tr. New York: Vintage International Books, 1995.

Collins, J. J., ed. Apocalypse: The Morphology of a Genre. *Semeia* 14. Missoula, MT: Scholars Press, 1979. P.9.

Davison, Ray. *Camus: The Challenge of Dostoyevsky*. Exeter, UK: University of Exeter Press, 1997.

Lottman, Herbert R. *Albert Camus: A Biography*. Corte Madera, CA: Gingko Press, 1997.

Sartre, Jean-Paul. "Camus' *The Outsider*" (1943). Reprinted in *Literary and Philosophical Essays*. New York: Collier, 1955.

Sessler, Tal. *Levinas and Camus: Humanism for the Twenty-first Century*. London: Continuum Books, 2008.

Takashima, Miho. "George Orwell and Albert Camus: A Comparative Study – Their Views and Dilemmas in the Politics of the 30s and 40." *The International Journal of the Humanities*. Volume 2, Issue 3.

_____. "Revolt and Equilibrium: A Comparative Study of *Nineteen Eighty-Four* and *L'Homme Révolté*; the Views and Struggles of Orwell and Camus." *The International Journal of the Humanities*. Volume 4, Issue 4, pp. 105-116.

Todd, Olivier. *Albert Camus: A Life*. Benjamin Ivry, tr. New York: Alfred A Knopf, 1997.

CHAPTER FIFTEEN

The Absurd Walls Of Albert Camus

Charles Taliaferro

In "Absurd Walls," an important section of *The Myth of Sisyphus and Other Essays*, Albert Camus offers a succinct overview of many of the main topics in his work: the absurdity of life, suicide, the limits of reason, time, anxiety, personal identity. Some of his claims are difficult to assess, such as "All great deeds and all great thoughts have a ridiculous beginning" and "mere 'anxiety' . . . is at the source of everything!"[1] But I shall offer a modest defense of Camus' view that there is a sense in which life in general, and personal identity in particular, is absurd, given secular naturalism. I shall also suggest that, under the same circumstances, the feeling that life is absurd can count as a modest, *prima facie* reason for thinking life is not absurd.

Why Life Might Be Absurd

In "Absurd Walls" the sense of life's absurdity seems to emerge due to three theses: the uniqueness and unfathomable nature of free persons; the impersonal, non-teleological nature of the cosmos ("the unreasonable silence of the world"); and the one-directional nature of time. Let's consider each thesis.

The uniqueness and unfathomable nature of free persons:

Camus recognizes that in practical contexts we may claim to know the minds of others, but he also contends that any certainty about others is elusive. We can practically and gradually build up our beliefs about the minds of others but this is not a matter of complete certitude.

> It is probably true that a man remains forever unknown to us and that there is in him something irreducible that escapes us. But *practically*, I know men and recognize them by their behavior, by the totality of their deeds, by the consciousness caused

in life by their presence, and so on. Likewise, all those irrational feelings which offer no purchase to analysis. I can define them *practically*, appreciate them *practically*, by gathering together the sum of their consequences in the domain of the intelligence, by seizing and noting all their aspects, and by outlining their universe.[2]

This seems right, especially if you grant the reality of libertarian freedom. Such freedom guarantees that behavioral predictions of voluntary acts cannot amount to incorrigible, infallible knowledge; we must instead live with probability.

It is certain that apparently though I have seen one actor a hundred times, I shall not for that reason know him any better personally. Yet if I add up the heroes he has personified and if I say that I know him a little better at the hundredth character counted off, this will be felt to contain an element of truth. For this apparent paradox is also an apologue. There is a moral to it. It teaches that a man defines himself by his make-believe as well as by his sincere impulses. There is thus a lower key of feelings, inaccessible in the heart but partially disclosed by the acts they imply and the attitudes of mind they assume. It is clear that in this way I am defining a method, but it is also evident that that method is one of analysis and not of knowledge.[3]

Our inescapable, and irreducible freedom sets us apart from things that do not have such freedom and it also fuels a longing that Camus identifies as nostalgia that results from our desire for happiness and reason.

The non-teleological nature of the cosmos:

The sense of the absurd is born in our desire for reason and happiness and the realization that the cosmos is utterly indifferent to such a desire. Our longing desire for significant happiness hits a wall.

At this point of his effort man stands face to face with the irrational. He feels within him his longing for happiness and reason. The absurd is born of this confrontation between the human need and the unreasonable silence of the world. This must not be forgotten. This must be clung to because the whole consequence of a life can depend on it. The irrational, the human nostalgia, and the absurdity that is born of this encounter -- these are the three characters in the drama that must necessarily end with all the logic of which an existence is capable.[4]

From the standpoint of Christian theism -- a standpoint that interests Camus greatly but which he decisively rejects -- the existence and value of humans, and the cosmos as a whole, is created and conserved in existence by God by virtue of the goodness of creation. In such a framework, the desire for happiness and reason is carried out in the context of a loving personal reality. But in secular naturalism, there is no undergirding purposive reality. This is perhaps made especially vivid when one takes into account the prediction that our galaxy (and perhaps all galaxies) will become extinguished.

According to contemporary astrophysics, in roughly 4.5 billion years the sun will run out of hydrogen, collapse, and then (using contemporary jargon) it will become a red giant and then a black dwarf. The earth will be vaporized, and then what is left of our former solar system will drift (along with the Milky Way) on its collision course with the neighboring Andromeda galaxy. In a famous essay in "A Free Man's Worship," Bertrand Russell writes:

> The world which Science presents for our belief [is even more purposeless, more void of meaning than a world in which God is malevolent]. Amid such a world, if anywhere, our ideals henceforward must find a home. That Man is the product of causes which had no prevision of the end they were achieving; that his origin, his growth, his hopes and fears, his loves and his beliefs, are but the outcome of accidental collocations of atoms; that no fire, no heroism, no

intensity of thought and feeling, can preserve an individual life beyond the grave; that all the labors of the ages, all the devotion, all the inspiration, all the noonday brightness of human genius, are destined to extinction in the vast death of the solar system, and the whole temple of Man's achievement must inevitably be buried beneath the debris of a universe in ruins -- all these things, if not quite beyond dispute, are yet so nearly certain that no philosophy which rejects them can hope to stand. Only within the scaffolding of these truths, only on the firm foundation of unyielding despair, can the soul's habitation henceforth be safely built.[5]

In *The First Three Minutes* Steven Weinberg offers a similar portrait:

It is almost irresistible for humans to believe that we have some special relation to the universe, that human life is not just a more-or-less farcical outcome of a chain of accidents reaching back to the first three minutes, but that we were somehow built in from the beginning . . . It is hard to realize that this all [e.g., life on Earth] is just a tiny part of an overwhelmingly hostile universe. It is even harder to realize that this present universe has evolved from an unspeakably unfamiliar early condition, and faces a future extinction of endless cold or intolerable heat. The more the universe seems comprehensible, the more it also seems pointless . . . The effort to understand the universe is one of the very few things that lifts human life a little above the level of farce, and gives it some of the grace of tragedy.[6]

Actually, it is interesting that classical Christian theology also submits that the cosmos will be destroyed (New Testament, Revelation 22:1-5), though such theology also posits a re-creation whereas there is no space for such redemption in the world of Camus, Russell, and Weinberg.

Neither Camus nor I are arguing that life is of no value because life will end. But I do find in Camus' work a sense of absurdity (perhaps related to a sense of vanity as in Ecclesiastes 1:2) in the face of the fact that all of human life, with its own teleology or purposes and value, will perish into oblivion. We will, as it were, eventually run into an absurd wall or radical termination.

The one-directional nature of time:

Camus leads us to see that the weariness of the world from a secular, naturalist perspective becomes sharpened when one takes into account the one-directional nature of time.

During every day of an unillustrious life, time carries us. But a moment always comes when we have to carry it. We live on the future: we consistently procrastinate, using phrases such as "I'll do it tomorrow," "later on," "when you have made your way," and "you will understand when you are old enough." Such irrelevancies are wonderful, for, after all, it's a matter of dying. A day comes when a man notices or says that he is thirty. Thus he asserts his youth. But simultaneously he situates himself in relation to time. He takes his place in it. He admits that he stands at a certain point on a curve that he acknowledges having to travel to its end. He belongs to time, and by the horror that seizes him, he recognizes his worst enemy. This now-thirty year old man always longed for tomorrow, but everything in him ought to reject it. That revolt of the flesh is the absurd.[7]

Why think of time as an enemy? Because, in secular naturalism, it appears that all achievements (or failures or simply life itself) is subordinate to time or, more specifically, all achievements lead to irreversible, unredeemable, and ultimately unrecoverable loss. In Christian theism, temporality is subordinate to eternity -- eternal or everlasting love, knowledge, and power -- but in secular naturalism all contingent events are fragile, bordering on oblivion.

In terms of the passage of time, Camus is an existentialist and not an adherent of four-dimensionalism, according to which all times are equally real. On the later view, it is always true (for example) that you are reading this at 11:00 am in September 2012. For Camus, when an event is no longer present, it has ceased to be. Certainly Camus and other secular naturalists can recognize, honor, and learn from great historical moments (such as the Battle of Salamis, the assassination of Julius Caesar, the defeat of the Spanish Armada, and the first human being walking on the moon) but it is no longer what we refer to as September 28, 480 BCE; March 15, 30 CE; July 29, 1588; and July 21, 1969. Eventually it is highly likely that no intelligent being, if secular naturalism is true, will survive either to be uplifted or horrified by these events. Once all memory of even the possibility of remembering human life has been erased, what is the difference between a universe in which we once existed and one in which there was no such life? This is where the old Epicurean argument comes into play: if something is good, doesn't it have to be good for something or someone? But once life has ceased to be, it is no longer bad for life that it no longer exists, and as for the universe itself, presumably there can be no good or bad for it if Camus, Russell, and Weinberg are right. It is the final void that we seem destined to perish in that Camus thinks renders our current quest absurd.

In certain situations, replying "nothing" when asked what one is thinking about may be pretense in a man. Those who are loved are well aware of this. But if that reply is sincere, if it symbolizes that odd state of soul in which the void becomes eloquent, in which the chain of daily gestures is broken, in which the heart vainly seeks the link that will connect it again, then it is, as it were, the first sign of absurdity.[8]

No matter how funny or heroic or romantic or loving you are, all is destined to the void.

Why Life Might Not be Absurd
"Absurd" comes from the Latin term *absurdus,* meaning out of tune. For Camus, there is a sense in which we are attuned to a purposive reality (God and a relationship with God) that is not possible (for there is no God). Arguably, the sense of absurdity as

articulated by Camus may be simply a testimony that the one who senses such absurdity is doing so because she has entertained a theistic worldview, and perhaps even longs for it to be true, but then rejects it. But sometimes a sense of absence can be evidence of the reality one longs for.

First, consider a commonplace observation about feeling an absence. We typically do not feel that something is absent unless we have some idea of what it would be like to feel the thing's presence. Moreover, we usually don't feel something is absent unless we either have an experience of its presence or we think the thing ought to be or may be present. I have never felt the absence of a lion or elephant in a classroom, but I have felt the absence of a student or two (much as Sartre describes the haunting absence of Pierre in the café in *Being and Nothingness*) from time to time.

Consider the following three cases. One, you are in charge of a group of students on a trip and you appear to have everyone together on a bus (students have been counted) but you have a keen sense that someone is missing. Two, you have completed your shopping and have even checked your shopping list twice, but you feel sure something is missing. Three, you are trying to make a life and death medical decision; you have been assured that all the facts you need have been presented to you, but you feel that some more information is required.

Granted, the feeling of absence is not a sufficient ground to be certain a student is missing or to decide you should purchase a vital ingredient or to believe there is some available information you could get, but wouldn't the feeling of absence be a reason to look for another student or review your plans or pursue further information? I submit that it would, especially given the conceivability of what you are missing or longing for and the persistence of a feeling that something is missing.

In the above cases, we are considering incidents of when the sense of absence or the missing of a thing or person can provide a reason to believe that the thing or person is elsewhere, or is somehow available or it was present in the past. Of course, there are other cases of when the sense of absence or missing is indicative of something not existing any more (I miss my father who is deceased, or a friendship that has ended) or indicative that some expected outcome is not to be or was based on false premises (American inspectors may have felt the absence of the weapons of mass destruction in Iraq). One way to distinguish these cases is by proposing that in the first set of cases, the sense of absence may be best explained by the reality of what or who is missed. One of the reasons why you feel a student is missing may be because you have a vague recollection that an additional student came along and was not part of the original counting. A reason why you have a feeling that you haven't completed shopping may be because you have forgotten an upcoming party and a guest's special need. And you may feel that more information is needed to make the medical decision is partly because you vaguely recall a friend in a similar situation who found further research vital. This leads me to propose the following principle:

If a subject feels the absence of X (e.g., that a student, a product, or some information is missing), X is possible (or not highly implausible) and the explanation for the feeling of absence may be explained by the existence of X (e.g., a student is missing, there is a product that should be purchased, there is further information required), then feeling the absence of X is a *prima facie* reason either to believe X exists or it is a *prima facie* reason to inquire into the existence of X.

This principle may need to be tightened up to distinguish legitimate cases from mere wish-fulfillment. So, lets consider an expanded principle:

If a subject feels the absence of X (e.g. feels that a student is missing, a product is missed, or some information is missing), X is possible (or not highly implausible) and the explanation for the feeling of absence may be explained by the existence of X (e.g. a student is missing, there is a product that should be purchased, there is information needed), and that explanation is better than accounting for the feeling of absence based on wish-fulfillment.

How might one apply this to a sense that life is absurd?

If the case for finding (and feeling that) life is absurd comes from well-grounded, reasonable beliefs in the non-teleological nature of the cosmos, the certainty of all life perishing, the one-directional nature of time, and the non-existence of any ultimate teleological reality (God), then a sense of life's absurdity (the absence of purpose) seems fitting and reasonable. But what if we have a sense of absurdity in terms of the world *as understood and experienced from the standpoint of secular naturalism*? And what if the truth of an alternative framework (e.g., theism) would explain a person's feeling of absence (e.g., something is being left out of secular naturalism) better than wish-fulfillment, secular naturalism, or some other non-theistic alternative? Let's add a further condition: imagine that the sense that secular naturalism is missing. Something that ought to be recognized (God) is accompanied by the realization that theism or some other teleological alternative is coherent and not implausible. Under those conditions, wouldn't the persistent experience of absurdity or that something is missing be grounds for thinking, or at least inquiring, into the reality of what is missed? C. S. Lewis proposed that a desire for transcendent fulfillment (a desire for what Camus described as a happiness and reason) can be an indication that there is an object of such desire" "Creatures are not born with desires unless satisfaction for these desires exists. A baby feels hunger; well, there is such a thing as food. A duckling wants to swim; well, there is such a thing as water. Men feel sexual desire; well there is such a thing as sex. If I find myself a desire which no experience in this world can satisfy, the most probably explanation is that I was made for another world."[9] In one of C. S. Lewis's charming Narnian chronicles, one of its characters finds he longs for a better world rather than the world of shadows he is facing. The character Puddleglum, is not content with a world missing the divine, represented in the story by Aslan.

> "One word, Ma'am," he said, coming back from
> the fire; limping, because of the pain. "One word.
> All you've been saying is quite right, I shouldn't
> wonder. I'm a chap who always liked to know the
> worst and then put the best face I can on it. So I
> won't deny any of what you said. But there's one
> thing more to be said, even so. Suppose we *have*

only dreamed, or made up, all those things -- trees and grass and sun and moon and stars and Aslan himself. Suppose we have. Then all I can say is that, in that case, the made-up things *seem a good deal more important* that the real ones. Suppose this black pit of kingdom of yours *is* the only world. Well, it strikes me as a pretty poor one. And that's a funny thing, when you come to think of it. We're just babies making up a game, if you're right. But four babies playing a game can make a play-world which licks your real world hollow. That's why I'm going to stand by the play-world. I'm on Aslan's side even if there isn't any Aslan to lead it. I'm going to live as like a Narnian as I can even if there isn't any Narnia."[10]

If Narnia and Aslan exist, life is not absurd, and because he longs for a meaningful life Puddleglum refuses to accept his current circumstances (he is in the dark realm of the evil Emerald Witch) as the whole story.

Consider an objection that is specific to theism: in Classical theism, God is omnipresent. On this view, there is no place where God is absent. God is therefore quite different from the missing student, missing products, and missing information. If God is absent *anywhere*, there is no God because if God exists, God is *everywhere*.

The above objection passes over an important distinction between metaphysics and experience. Divine omnipresence in classical theism is understood in terms of God's creative power and knowledge. To claim that God is everywhere means there is no place that is unknown to God and no place that would exist without God's creative power. More positively, every place exists by virtue of God's knowing, creative power. When people testify to a sense of God's absence they often testify to not feeling God's presence, or of feeling that God's will is not evident (profound evil occurs that God does not prevent), and the like. As a number of philosophers have agreed, a sense of God's absence would only be evidence that God is absent if it is plausible to believe you would feel God's presence if God is omnipresent. The literature on this matter is too massive to engage in here, except to offer this proposal: if one feels God is absent this might be explained on the grounds that God does not exist, or by appeal to the suspicion that that any feeling of divine presence is the result of wish-fulfillment. On the other hand, it might be explained on the grounds that God does exist, and that being experientially aware of and living in accord with God's will is good and that the sense of absence (or, more positively, the longing for God) is a natural response given a theistic understanding of the cosmos.

One can accept the above principle and soundly reject any line of reasoning that would connect Camus' feeling of absurdity with evidence of the truth of theism. Perhaps theism is incoherent or a hypothesis of self-delusion or perhaps of wish-fulfillment; Ockham's razor can preserve secular naturalism in the presence of a sense of the absence of God or a longing for some meaning beyond "the silence of the world." But if you have reason believe theism is coherent, and that naturalistic accounts of religious beliefs and Ockham's razor are overrated, matters can change. Allow me to fill this out in connection with another philosopher who has written admirable about the absurdity of life: Thomas Nagel.

The Interesting Case of Thomas Nagel
Thomas Nagel is probably not in the position of missing God's existence! In *The Last Word* Nagel writes: "It isn't that I don't believe in God and, naturally, hope I'm right in my belief. It's that I hope there is no God! I don't want there to be a God; I don't want to universe to be like that."[11] In a more recent publication, *Secular Philosophy and the Religious Temperament*, Nagel seems less sanguine about the prospects of a secular philosophy addressing the "question of making sense not merely of our lives, but of everything." Nagel contends that without theism, it is difficult to understand human's ability to achieve harmony with the cosmos:

> Without God, it is unclear what we should aspire
> to harmony with. But still, the aspiration can
> remain, to live not merely the life of the creature

one is, but in some sense to participate through it in the life of the universe as a whole. To be gripped by this desire is what I mean by the religious temperament. Having amazingly, burst into existence, one is a representative of existence itself -- of the whole of it -- not just because one is part of it but because it is present to one's consciousness. In each of us, the universe has come to consciousness, and therefore our existence is not merely our own.

To live not merely one's own life is also a demand of those forms of morality that take up a universal standpoint as part of their foundation. And something of the kind will very likely form part of a secular response to the religious question. But it is only a part, dealing specifically with recognition of the existence of other people. There is more to question than this. The extrahuman world that contains and generates all these people also has a claim on us -- a claim to be made part of our life. Existence is something tremendous, and day-to-day life, however indispensable, seems an insufficient response to it, a failure of consciousness. Outrageous as it sounds, the religious temperament regards a merely human life as insufficient, as a partial blindness to or rejection of the terms of our existence. It asks for something more encompassing, without knowing what that might be.[12]

Nagel envisions three alternatives: the religious option, a platonic vision in union our lives make sense in terms of some great good, or a secular atheism in which life is experienced as absurd. I cite him at length:

But does it really make any difference whether we are the products of natural teleology or of pure chance? Without an intentional designer, perhaps there is no sense to be made of our lives from the larger perspective in either case: We just have to start from what we contingently are and make what sense we can of our lives from there.

If the question is about whether our lives have a cosmic purpose, I would agree. But that is not the only possibility. The Platonic sense of the world is that its intelligibility and development of beings to whom it is intelligible are nonaccidental, so our awareness and its expansion as part of the history of life and of our species are part of the natural evolution of the cosmos. This expands our sense of what a human life is. It seems in that case at least somewhat less plausible to say that all sense begins with the contingent desires and choices of the particular individual-that existence precedes essence, in the existentialist formula.

In the Platonic conception, even the biological and cultural evolution that has led to the starting point at which each of us arrives on Earth and reaches consciousness is embedded in something larger, something that makes that entire history less arbitrary than it is on the reductive view. But if the Platonic alternative is rejected along with the religious one, we must go back to the choice between hardhearded atheism, humanism, and the absurd. In that case, since the cosmic question won't go away and humanism is too limited an answer, a sense of the absurd may be what we are left with.[13]

Let's imagine that Nagel or someone like Nagel finds hardhearted atheism and humanism problematic, and because he cannot find a Platonic secular account plausible he concludes that life is absurd. But imagine further he regards theism to be a live philosophical option and he believes that if theism is true, life is not absurd and a theistic worldview and practice would satisfy his deep aspiration to be in harmony with the center of reality.

Perhaps then his feeling that life is absurd, given secular naturalism, would itself function as a reason at least to inquire into the truth of theism and perhaps even function as a reason to believe we are not made to be secular naturalists. Admittedly this would only be the beginning of further inquiry, but it might be enough for us to think there is something good on the other side of what Camus saw as an absurd wall.

Bibliography

Camus, Albert. "Absurd Walls" in *The Myth of Sisyphus and Other Essays*, trans. Justin O'Brian.   New York: Vintage Books, 1955.

Lewis, C.S. *The Silver Chair* in *The Chronicles of Narnia*. New York: Harper Collins, 2004.

----------------. *Mere Christianity*. New York: HarperCollins, 2001.

Nagel, Thomas. *The Last Word*. New York: Oxford University Press, 2007.

--------------------. *Secular Philosophy and the Religious Temperament: Essays 2002-2008*. New York: Oxford University Press, 2010.

Russell, Bertrand. "A Free Man's Worship" in *Mysticism and Logic*. Nottingham: Russell House, Ltd.,   2007.

Weinberg, Steven. *The First Three Minutes*. New York: Basic Books, 1993.

Endnotes

[1] Albert Camus, "Absurd Walls" in *The Myth of Sisyphus and Other Essays*. trans. Justin O'Brian (New York: Vintage Books, 1955).

[2] *Ibid.*, p. 9.

[3] *Ibid.*

[4] *Ibid.*, p. 21.

[5] Bertrand Russell, "A Free Man's Worship" in *Mysticism and Logic* (Nottingham: Russell House, Ltd., 2007), p. 51.

[6] Steven Weinberg, *The First Three Minutes* (New York: Basic Books, 1993), pp. 154-155.

[7] Camus, pp. 10-11.

[8] *Ibid.*, p. 10.

[9] C. S. Lewis, *Mere Christianity* (New York: HarperCollins, 2001), pp. 136-137.

[10] C.S. Lewis, *The Silver Chair* in *The Chronicles of Narnia* (New York: Harper Collins, 2004), p. 633.

[11] Thomas Nagel, *The Last Word* (New York: Oxford University Press, 2007), p. 130.

[12] Thomas Nagel, *Secular Philosophy and the Religious Temperament: Essays 2002-2008* (New York: Oxford University Press, 2010), p. 6.

[13] *Ibid.*, p. 17.

[14] I thank Therese Cotter for her assistance in composing this paper.

CHAPTER SIXTEEN

Camusian Thoughts About The Ultimate Question Of Life

Charles Tandy

§1. Introduction
§2. Our Absurd Time
§3. Camus And Absurdity: *The Myth Of Sisyphus*
§4. Camus And Anti-Absurdity: *The Rebel*
§5. Camus As The First Man

§1. Introduction

In *The Hitchhiker's Guide to the Galaxy* by Douglas Adams, Deep Thought (a supercomputer) spends some time in deep thought about the Ultimate Question of Life, the Universe, and Everything. Indeed, after millions of years, Deep Thought finally provides the answer. The "answer" is 42. (This is reminiscent of Stephen Hawking's "answer": M-theory.)

Albert Camus (1913-1960) claims not to have definitive answers to the ultimate questions of life but instead wants to engage in genuine dialogue with the reader. (I use the present tense because his time, "an age of absurdity", remains our time, as I will explain below.) In his view, one may choose to attempt to learn and advance – instead of dangerously leaping to premature certainty. (In an age of absurdity, leaping to premature certainty or engaging in other irrational behavior is a great temptation.) He says that for him the reasonable approach is to admit uncertainty while simultaneously attempting to live his life with integrity (authenticity). Camus admits uncertainty about ultimate questions – but also notes that at a given point in time some things will seem to him more reasonable or less reasonable than other things. Camus argues for, or paints pictures of, what seems reasonable to him; the reader may look at the pictures and agree or disagree.

Although Camus saw himself primarily as a creative artist in the form of writer, rather than professional philosopher, I will in this paper focus primarily on what are generally recognized as his two major philosophical works: *The Myth of Sisyphus* and *The Rebel*. According to Camus, these two works may be identified with the two stages of his intellectual development. Unfortunately, he died ("absurdly") at the age of 46 (after only two years as a young Nobel Laureate), so we have no stage three in his development. At the age of 17, however, he had already encountered one of his numerous bouts of tuberculosis; at the time there was little treatment available and he thought he was then going to die.

§2. Our Absurd Time [1]

Camus said he lived in a time of absurdity. Above I claimed we still live there. Before exploring Camus further, I will now defend the notion that our age – the 20th century and beyond – is (continues to be) a time of absurdity. (Indeed, our attempts at problem solving sometimes cause absurdity.) As explained below, the following sets of historical events suggest to me that our age is both absurd and unprecedented:

- World Wars One And Two
- Events Between The Wars And After The Wars
- 20th Century Developments In Science
- 20th Century Developments In Mathematics
- 20th Century Developments In Psychology
- The Early 20th Century Modernist Movement
- 20th Century Existential Literature And Absurd Drama
- The 20th And 21st Century Postmodernist Movements

World Wars One And Two

Many of the late 19th century believed that human progress was inevitable. This included, for example, belief that, over time, wars would decrease in intensity, scope, and duration. Indeed, perhaps western civilization had already devised a   continental system   guaranteeing nothing more severe than limited battles or small wars between European nations.

[World War One (1914-1918)]

The war that began in 1914 only gradually turned into a great "World War." The war started in an almost careless mood with full-dress parades. But World War One proved to be larger than any previous war in European history.

Approximately 74 million were mobilized by all sides in the "Great War." The war was also unprecedented in another respect: This war involved entire societies, not merely soldiers. The concept of "total war" was born.

World War One resulted in the collapse of traditional empires, and, from the wreckage, new nations arose. Moreover, to Europe's further embarrassment, it appeared Europe had been unable to settle its own affairs (end the long war) without intervention from the New World (the USA). Europe and the west no longer seemed worthy as the leading edge of world progress, leadership, and perfection.

[World War Two (1939-1945)]

"The war to end all wars" failed too in that a Second World War began in 1939. The war of 1939, far more than the war of 1914, was a **world** war. (Thus, it seems progress is **not** inevitable.)

Events Between The Wars And After The Wars

The "**Roaring Twenties**" (1920s) were called "roaring" because of the exuberant, freewheeling popular culture of the decade following World War One. The Roaring Twenties were a time when many people indulged in new or illegal styles of dancing, dressing, and behavior, and rejected many traditional moral standards. This decade is also known in the United States as the "**Jazz Age**," marked by increased popularity of ("wild") jazz (music), and by attacks on convention in many areas of American life. Many Americans defied Prohibition (the outlawing of alcoholic beverages nationwide from 1920 to 1933). The nickname "flappers" was given to young women in the 1920s who defied convention by their dress and by such behavior as drinking and smoking in public.

Throughout the 1930s there was a **"Great Depression**," a worldwide economic crisis that continued into World War Two. Indeed, World War Two and its aftermath left the world in an unprecedented crisis situation. The attempted genocide of the Jews, the large scale death camps (gas chambers), and the use of the atomic bomb in World War Two raised terrifying questions about the extreme possibilities of human inhumanity.

There were Germans who listened to their Bach or read their Goethe after a day's work at the gas chambers (death camps). The Americans not only initiated development of the Manhattan Project, but actually authorized use of the atomic bomb – using atomic mass death and destruction not against military targets, but against cities of civilians. Indeed, the possibility of instant extermination through nuclear war or weapons of mass death and destruction makes many traditional values and conventions seem obsolete.

20th Century Developments In Science

In addition to nuclear weapons, additional science related developments radically altered our views of the universe and of ourselves. The atomic bomb was a kind of "test" (so to speak) of some of the ideas of Albert Einstein. With Einstein's new physics the dividing line between mass and energy was now far from clear. His theory of relativity says that space, time, and motion are not absolute (as Newton had assumed), but relative to the observer.

Previously, the intelligent layperson could read and understand the great thinkers like Newton and Darwin. But the thoughts of Einstein and other scientists were now only available to specialized experts. Moreover, the new physics seemed to postulate a world without continuity or absolutes, a world in which nothing was certain. A world in which scientific laws are given names like "relativity" (Einstein) or "indeterminacy"/"uncertainty"

(Werner Heisenberg) is a world in which nothing is as it appears to be to the human senses.

20th Century Developments In Mathematics

If physical reality itself is inherently "relative" and "indeterminate" or "uncertain," then surely we can at least depend on mathematics ("the language of the sciences") to give us nonabsurdity and certainty? But the work of Kurt Gödel and others in the 20th century showed that even mathematics is less secure and dependable than had been assumed. Gödel, often cited as the greatest logician of the 20th century, is credited with proving two theorems that must be bad news for anyone wanting to construct a theory that will tell the whole truth.

Gödel's First Incompleteness Theorem states that for any consistent logical system able to express arithmetic, there must be true sentences within the system that are undecidable (cannot be proved true) within the system. Gödel's Second Incompleteness Theorem states that no such (consistent) system can prove its own consistency. The moral of the two incompleteness theorems would seem to be that truth inevitably outstrips formal provability.

20th Century Developments In Psychology

Foremost in upsetting traditional notions about human nature was the work of Sigmund Freud. The new psychology proclaimed that it is not unusual for humans to engage in **self-deception**. Freud's theories that our actions and behavior are rooted in our **unconscious**, rather than in our conscious mind, seemed to many a highly **pessimistic** view of human nature. Freudian influences, including the idea of sexual repression, can be found throughout 20th century art and literature. Human nature, reason, and logic are apparently far more dark and fallible than had been thought possible.

The early 20th century west experienced a radical questioning of past traditions. For example, the western cultural roots of Greek-Roman "Reason", Hebrew-Christian "Religion", and Modern-Progressive "Science" were no longer believable traditions or were of highly uncertain value. It seems neither Reason nor Religion nor Science provided a sure path out of the absurd wilderness.

The Early 20th Century Modernist Movement

Even in the very early 20th century, it seemed to a number of western artists and intellectuals that the 20th century had broken, in a new and radical way, with previous tradition. A radical break in the western arts became evident with a series of new styles that can all be loosely grouped under the name "**modernism**." (Many "isms" may be grouped under "modernism." Modernism is the philosophy and practices of "modern art" – especially a self-conscious break with the past and a search for new forms of expression.)

In the early 20th century west we find:

- Desire for sexual freedom
- Demand for greater freedom for women
- Motion pictures
- Jazz music expressed the fast, frantic, free way of life of the "Roaring Twenties" (1920s)
- Life was speeding up and changing
- The city of Paris (France) was the place to go to exchange ideas and create the new

Almost every serious artist of the 20th century felt the necessity either to react against modernism or to build on its innovations. Most of the great "modernist" writers of the early 20th century spent some time living in Paris (France). A few of the "modernist" literary/artistic "isms" include:

- **Dadaism** is based on deliberate irrationality and negation of traditional artistic values; seeks the fantastic and absurd; life is random and uncontrolled.
- **Surrealism** is related to dadaism; heightened awareness of the conflict between the rational and irrational; produces dreamlike, fantastic, or incongruous imagery or effects. (Two

artistic examples: Salvador Dali, painter; Jean Cocteau, filmmaker.)

- **Expressionism** seeks to depict the subjective emotions and responses that objects and events arouse in the artist; German expressionists expressed the deep, hidden drives of human beings – modern society alienates the individual.
- **Futurism**, initiated in Italy about 1909, sought to give formal expression to the dynamic energy and movement of mechanical processes.

In some ways the popular American writer Ernest Hemingway was an "atypical" modernist. In general, modernist writers and artists experienced a rift between artist and public. The lack of contact with a public, except for a small group of fellow intellectuals, gives the writer or artist greater freedom to experiment but also encourages esotericism.

20th Century Existential Literature And Absurd Drama

The temper of our time is a mixture of anticipation and anxiety, optimism and pessimism. The philosophy of existentialism illustrates this blend of hope and despair. "Existentialism" is said to have been "founded" in the 19th century – but its popularity dates from the period of the 20th century's World Wars. The Theater of the Absurd represents an extension of existentialist philosophy and literature into drama in the 20th century. (Also note that we identify utopian literature with the 19th century and earlier, back to Thomas More – while we identify dystopian literature and dystopian movies with the 20th century and beyond.)

[Existentialism]

For the existentialist, it is the individual person, not the abstract concept of person or humanity, that constitutes true reality. The individual is a **stranger** or an **outsider** in an alien, hostile world. Such loneliness is seen as a call to action and free choice. Two of the 20th century's leading existentialists were Jean-Paul Sartre (1905-1980) and Albert Camus (1913-1960). (Please note that many or most such intellectuals we call existentialists today claimed not to be existentialists. This "no" includes Camus but not the existentialist Sartre.)

In his novel **The Stranger** (or **The Outsider**), and in his philosophic essay **The Myth of Sisyphus**, Camus demonstrates his concept of the "absurd" – the fundamental meaninglessness of human life and traditional beliefs. As Camus said in **The Myth of Sisyphus**: "A world that can be explained even with bad reasons is a familiar world. But, on the other hand, in a universe suddenly divested of illusions and lights, man feels an alien, a stranger. His exile is without remedy since he is deprived of the memory of a lost home or the hope of a promised land. This divorce between man and his life, the actor and his setting, is properly the feeling of absurdity." [pp.4-5] Thus individuals must create their own morality, their own way of resisting or rebelling against the "absurd."

[The Theater of the Absurd]

Beginning in the 1950s, some dramatists came to experience, as had the dramatist Albert Camus, a profound sense of absurdity. For example, Samuel Beckett, Harold Pinter, and others seemed to share a pessimistic vision of humanity struggling to find a purpose and to control its fate. Absurdist playwrights did away with the logical structures of traditional theater. The busyness of the characters underscores the fact that nothing happens to change their existence. The ridiculous purposeless behavior and talk of the characters give the plays a sometimes dazzling comic surface, but there is an underlying serious message of metaphysical distress.

The 20th And 21st Century Postmodernist Movements

The contemporary historical period I have called "an age of absurdity" is sometimes characterized as "a time of discontinuities, pluralities/diversities, absurdities, and uncertainties". Sometimes it is called "post-modernist" or "postmodern." But there are at least two additional ways these terms ("post-modernist"; "postmodern") are sometimes used: In the late 20th century, two drastically different sets of artistic-literary movements were each given the same name: "post-modernist" or "postmodern." One of the two sets of movements may be characterized as a reaction against the Modernist artistic-literary movement of the 20th century. Post-modernist or postmodern in this case means revival of traditional artistic-literary elements and techniques.

But the second set of artistic-literary movements may be seen as an intensification or yet further evolution of Modernism. In this case the post-modernist or postmodern perspective goes beyond Modernism's aspiration to (a new) unity. The search for unity has been abandoned in favor of pluralities/diversities of styles and interpretations.

Thus no single cultural tradition or mode of thought can serve as a **metanarrative** (a universal voice, or totalizing story, for all human experience). Jean-Francois Lyotard defines **The Postmodern Condition** as "incredulity toward metanarratives." Incredulity does not mean disbelief – it means **inability** to believe.

The Modernists of the 20th century felt exiled, like outsiders or strangers, and sought a new unity that they, and perhaps all persons, could believe in. But today many postmodernists **celebrate** incredulity as a liberating, humanizing force. They hear a new key: For them, pluralities of voice are beautiful, not terrifying.

Does this great celebration celebrate uncertainty? According to my Camusian-inspired interpretation, it **does**. Does this great celebration celebrate absurdity? According to my Camusian-inspired interpretation, it does **not**. Does the great celebration embrace anti-absurdity? According to my Camusian-inspired interpretation, it embraces absurdity for the purpose of rebelling against it and giving each person their due. A life of celebration and multiple flourishing is not without its values. Its values are those associated with celebration or flourishing and with resisting or rebelling against the "absurdity" of anti-celebration and anti-flourishing. Arguably, the celebratory kind of incredulity involved here is neither totalizing, absolute, nor nihilistic.

§3. Camus And Absurdity: *The Myth Of Sisyphus*
("Is my life worth living?")

So how do we make the transition from an absurd here and an uncertain now – to a vital world of celebration and multiple flourishing? Perhaps nourishing a certain kind of uncertainty and of celebration may help transform an absurd uncertain universe into a flourishing uncertain multiverse. Voltaire wrote to Frederick the Great in 1767: "Doubt is not a pleasant condition, but certainty is an absurd one."

In our history of learning to advance toward dialogue and friendship, we should know by now that one's fundamental beliefs should be held tentatively rather that absolutely. Socrates is said to have been the world's wisest person because he knew that he did not know with certainty, whereas everyone else was certain. Yet Socrates was committed to dialogue and friendship even at the risk of his very life.

Let us now proceed to dialogue with our friend Albert Camus. We will begin at the beginning by asking with Camus, in his *The Myth of Sisyphus*, "Is my life worth living?" If one convincingly feels that neither Reason nor Religion nor Science are traditions that provide a sure path out of the absurd wilderness, then must one conclude that one's life has no meaning and is not worth living?

In a 1955 Preface to *The Myth of Sisyphus*, Camus tells us that he wrote it in 1940 (during World War Two) and that he has "progressed beyond several of the positions which are set down here;" [perhaps referring to *The Rebel*?]. He continues: "but I have remained faithful, it seems to me, to the exigency [absurdity?] which prompted them."

Part One Of Five Parts Is Entitled: An Absurd Reasoning

I gather that Camus is suggesting that Reason (as in philosophy), Religion (as in Christianity), and Science (as in physics or psychology) are but games compared to the more serious question: "Judging whether life is or not worth living."

[p.1] Accordingly, judging "whether or not the world has three dimensions, whether the mind has nine or twelve categories – comes afterwards." [p.2]

The man who committed suicide had lost his daughter years earlier and had never gotten over it. Her loss had "'undermined' him. A more exact word cannot be imagined. Beginning to think is beginning to be undermined." [p.3] On the other hand, the non-suicides "continue making gestures commanded by existence for many reasons, the first of which is habit." [p.4] (Camus notes that humans learn to breathe before they learn to think.)

Can one live without the eternal or certain values sponsored by Tradition/Authority (Reason; Religion; Science)? Contrary to what authority/tradition (the philosopher or theologian or scientist) may say, an autonomous Camus is able to begin with absurdity without ending there. Camus says that choosing for or against suicide forgets the third option: One can choose to continue questioning instead of ending it with a premature YES or NO. Instead of taking an "either-or" or "all or nothing" approach, why not choose the middle ground of autonomy? Camus fights passionately against the absurdity of death from the ("Mediterranean") position of cool moderation instead of extreme certainty. (Camus' Algeria, as had Golden Greece, bordered on the Mediterranean sea.)

Thus perpetual questioning leads one [Camus] to say NO to mortality. Many of those who say NO to suicide, however, act as if they said YES. Instead of choosing the anti-death moderation of autonomy, they choose the supposed immortality or certainty of a Tradition/Authority (Reason or Religion or Science as an ideology or way of life). They are the living dead.

So there are the dead and there are the living dead. Camus takes the third way, that of individual autonomy. One does not have to know that The Ultimate Answer to Life, the Universe, and Everything is "63" or "42" (here the living dead have their lively, sometimes deadly, disagreements) in order to value autonomous living. (As for the literal dead, it seems that those of the dead who took an autonomous or anti-death stance were not entirely successful in that they are dead. [2]) The living dead includes those who have not given much thought to the life-death question, including those who consciously or unconsciously avoid the topic. The living dead also includes those who have made a leap of faith to some eternal or absolutist ideology (religious or secular).

If I may speak in a technical rather than everyday way, we can say that Camus is a moderate in that he is both anti-death and anti-immortality. Unless one joins the living dead and gives up one's autonomy, it would seem to take an infinite amount of time to achieve immortality. Instead of fantasizing, one can attempt (in one's finite presentness in an age of absurdity) to take steps against both literal death and living death. Here is the book's epigram:

> O my soul, do not aspire to immortal life, but exhaust the limits of the possible.
> -- PINDAR, Pythian iii

Anti-death autonomy is the middle way between the extremes of literal suicide and the immortal life supposed by the living dead. Recovering the certainty of knowledge is a fantasy of the living dead. Camus is unwilling to give up his anti-death autonomy in favor of such dangerous nostalgia.

I can feel a heart within me and I can touch a world: "There ends all my knowledge, and the rest is construction." [p.17] "In psychology as in logic, there are truths but no truth." [p.18] There are three characters in the mortal drama: the irrationality of the world; the human heart's desire for meaning; and the encounter between the two (absurdity). First comes the feeling of absurdity or meaninglessness, then comes the notion or concept. The perpetual struggle has meaning (!) "only in so far as it [the absurd] is not agreed to." [p.30]

So what are the consequences of the absurd? Some pretend to forget or ignore it. Some leap to premature suicide. Others leap to unevidenced certainty in Tradition/Authority (one or another ideology of Reason or Religion or Science). Thus there are many ways to escape from the Sisyphusian struggle. Some, however, freely choose autonomy (integrity; authenticity) and continue to continue to continue the struggle against death.

But the Camusian rejection of Tradition or Authority or Ideology is a partial one. Sisyphus finds it natural to use experience or tools associated with reason or religion or science to the extent they are helpful in his unending battle against literal death and living death. For example, one can use reason in a piecemeal way to counter both Ideology and Mortality. Camusian moderation maneuvers between the extreme paths of triumphal reason and humiliated reason.

"The laws of nature may be operative up to a certain limit [prior to individual autonomy], beyond which they turn against themselves to give birth to the absurd [individual autonomy]. Or else, they may justify themselves on the level of description without for that reason being true on the level of explanation." [p.35] But whatever the case may be, Sisyphus will not give up his autonomous anti-death struggle against the universe.

Sisyphus chooses integrity rather subterfuge. Without appealing to eternal or absolute values, he removes Reason and Religion and Science from their pedestals but does not commit them to annihilation. This is the middle way between the extremes of hubris and humiliation. Autonomous rebellion against death and absurdity is a meaningful and moderate way to perpetually search for meaning.

"By the mere activity of consciousness I transform into a rule of life what was an invitation to death." [p.62] Reasoning about the absurd leads Camus to infer three natural consequences for Sisyphus:

1. **My Life** (my rebellion): "It is essential to die unreconciled and not of one's own free will." [p.53]
2. **My Liberty** (my autonomy): inner authenticity (free self-legislation with integrity) instead of external rules (eternal or absolute values) with respect to thought and action.
3. **My Pursuit of Happiness** (my passion): diversity of experience, "But the point is to live." [p.63]

Part Two Of Five Parts Is Entitled: The Absurd Man

Camus then attempts to present a few cameos of persons (the seducer-lover; the actor; the adventurer-conqueror) as each might live their very own individual lives of my-rebellion, my-freedom, and my-passion. The Absurd Man does not claim to know about eternal or absolute values; indeed, he has no need of rules to justify his behavior. The Absurd Man finds it difficult to believe that anyone deserves a death sentence, whether from the universe or from other men. Such Absurd Persons live their lives rather than obey someone else's rules.

Part Three Of Five Parts Is Entitled: Absurd Creation

According to Camus, the creator-artist is "the most absurd character." [p.89] Camus may be referring to an artist-writer such as himself. If the world is clear and certain, we get description rather than (absurd) art. Artists prefer images over arguments. Camus wants to create without appeal to eternal or absolute values and thus "liberate my [literary] universe of its phantoms and to people it solely with flesh and blood truths whose presence I cannot deny." Even the great writers such as Dostoievsky and Kafka do not altogether succeed in this respect. One of the truly absurd works, Camus says, is Melville's *Moby Dick*.

Part Four Of Five Parts Is Entitled: The Myth Of Sisyphus

Sisyphus was a wise and clever mortal accused of lacking proper respect for the gods. Indeed, he succeeded in putting Death in chains. However the gods then unchained Death and condemned Sisyphus. His punishment was severe: Forever rolling a stone up a mountain only to have the boulder fall back again and again and again. His situation is tragic because he is without hope, forever doomed to failure. YET: "If the descent is thus sometimes performed in sorrow, it can also take place in joy. The word is not too much." [p.117] "The struggle itself towards the heights is enough to fill a man's heart. One must imagine Sisyphus happy." [p.119]

Kafka writes in such a way that it forces the reader to re-read and re-interpret. We find that Samsa has "a 'slight annoyance'" – "his boss will be angry at his absence." [p.125] Samsa otherwise seems unconcerned – that his body has metamorphosed into that of a huge insect!

Camus opines that truth is contrary to conventional morality. Sometimes Kafka seems to almost realize that it is fatal to give God what does not belong to him. But then Kafka will sneak in hope, unaware that hope "is not his business. His business is to turn away from subterfuge." [p.134]

§4. Camus And Anti-Absurdity: *The Rebel*
("How do I live a meaningful life?")

Camus, in his *The Myth of Sisyphus*, had asked "Is my life worth living?" We have seen that his conclusion was affirmative. Now we will ask with Camus, in his *The Rebel*, "How do I live a meaningful life?" Here we will find some evolution in his thought.

Indeed, if you have read only *The Myth of Sisyphus*, and no other work by Camus, you may declare my interpretation above severely distorted. You may be correct, for I read *The Rebel* before reading *The Myth of Sisyphus*. I have read the two volumes as if they were one. Moreover, in my anachronistic fusion of the two, I am **not** attempting to articulate a definitive Camus interpretation of Camus. That is why this chapter's title says (**not** "The Thoughts Of Camus", but) "Camusian Thoughts" – thoughts inspired "About The Ultimate Question Of Life" upon reading these two works by Camus. With this caveat in mind, I now present to you my Camusian Thoughts about *The Rebel*.

In the introduction, Camus says that in our peculiar time, good-will ("innocence") is suspect. It seems these days we have to explain why we want to do good. Camus wants to do good simply because it is the good thing to do. Yet in our complex world, is it even possible to do good without directly or indirectly killing or harming someone? In the face of the absurd, Camus chose to live. But given this, and given our age of ideologies (Reason; Religion; Science), is it possible to live without killing others? Simply by living in an age of ideologies, do we not participate in the literal death or living death of others?

Yet is it not the case that: "From the moment that life is recognized as good, it becomes good for all men"? [p.6] Not only should one not kill oneself, but likewise one should not kill others. My autonomous or natural rights imply your rights too. My rights and your rights "must be accepted or rejected together." [p.6]

Choosing to live is a value judgment, a standard or limit with implications for one's relation to others. When I rebel against absurdity, my action is on behalf of all. My individualistic act is an act of solidarity or unity.

Part One Of Five Parts Is Entitled: The Rebel

A rebel is one "who says no, but ... who [also] says yes. ... there are limits." [p.13] By rebelling, we are saying that there are limits or standards or values or rights which we all should respect. Thus there is a sense in which we are all naturally equal and "a human nature does exist." [p.16] Humans have metaphysical solidarity – they are a natural community: My rebellion "is for the sake of everyone in the world." [p.16]

The spirit of rebellion tends not to be expressed in societies of either extreme inequality or extreme equality. But over time we humans experience absurdity, and humanity's self-awareness grows. Yet we are tempted to forget either the basis of rebellion or that there are limits. We tend to ideologically leap into a living death – either groveling before God or intoxicating on power.

I, Camus, rebel against both servitude and tyranny. I, Camus, refuse to give in to either metaphysical other-worldliness or to Caesaristic historicism. In order to exist, we humans must rebel: "Man's solidarity is founded upon rebellion, and rebellion, in its turn, can only find its justification in this solidarity." [p.22]

Thus we have gone beyond individualistic absurdity. Descartes had said: "**I think – therefore I am.**" But in rebellious solidarity, "suffering is [now] seen as a collective experience. ... **I rebel – therefore we exist.**" [p.22]

Part Two Of Five Parts Is Entitled: Metaphysical Rebellion

The Man refuses to give in to metaphysical other-worldliness. The Man refuses to ideologically leap into a living death, groveling before God. To the "I rebel, therefore we exist," the Man adds: "**And we are alone.**"

Part Three Of Five Parts Is Entitled: Historical Rebellion

The Man refuses to give in to Caesaristic historicism. The Man refuses to ideologically leap into a living death, intoxicating on tyrannical power. To the "I rebel, therefore we exist," and the "We are alone," the Man adds: "**Live and let live.**" "Instead of killing and dying in order to produce the being that we are not, we have to live and let live in order to create what we are." [p.252] Thus the key importance of creativity or art:

Part Four Of Five Parts Is Entitled: Rebellion And Art

Creation or art is pure rebellion – it demands unity while partially rejecting the world. The world is used by the artist in order to attempt to create a better world that meaningfully unites everyone and everything. The Protestant Reformation, the French Revolution, Russian nihilism, and German ideology are all examples of artistic banishment. Should we not choose morality or usefulness or progress (take your pick) over beauty?

The artist rebels against these stark, narrow worlds – by constructing alternatives to such prisons. The artist believes in life and living, not death and chains. "Rebellion ... is a fabricator of universes." [p.255] The artist finds both absurdity and beauty in the world. This suggests construction of a less absurd, more beautiful universe. Thus the artist becomes part of the process of evolution.

The modern novel "competes with creation and, provisionally, conquers death." [p.264] Proust's *Time Regained* "appears to be one of the most ambitious and most significant of man's enterprises against his mortal condition. ... this art consists in choosing the creature in preference to his creator." [p.267] But even more, it supports "the beauty of the world and its inhabitants against the powers of death and oblivion. It is in this way that his rebellion is creative." [p.268]

Real creation attempts neither to escape from reality nor to accept it as it is. The moderation and passion of the Meridian Rebel or true artist "simply adds something that transfigures reality." [p.269] Thus both "formal' art and "realist" art must be seen as extremes. Much "modern" art unwisely attempts to replace one totalitarian "unity" with another. The Meridian artist intervenes to make the world more beautiful, between the extremes of leaving the world as it is either by pure description or by pure escapism.

Real literary creation is not to be identified with commentary or criticism. Terror and tyranny contradict creativity and art. Perhaps a renaissance of creativity and civilization is possible. Often movements proclaiming a new world are actually the extreme opposite or climaxing contradiction of the old one. Today this extreme, in one form or another, is bent on industrial production. But: "The society based on production is only productive, not creative." [p.273]

Every creative act of love denies the world of master-slave. Yet today it seems our leaders have no time for love. "But the fact that creation is necessary does not perforce imply that it is possible." [p.274] Of every ten potential artists, maybe one or none will become artists if all that counts is industrial competition in a hellish world.

Even if history has an end, it is not our task to end it. Those who choose to ignore nature or the sea or the stars or beauty are constructing a world devoid of freedom and dignity. We, the Meridian rebels, must uphold beauty and creativity if we are to live in a beautiful and creative world with freedom and dignity.

Part Five Of Five Parts Is Entitled: Thought At The Meridian

Forgetting its Meridian roots (I rebel – therefore, we are), rebellion too easily oscillates between murder and sacrifice. If a "single human being is missing in the irreplaceable world of fraternity, then this world is immediately depopulated. ... On the level of history, as in individual life, murder is thus a desperate exception or it is nothing." [p.282]   Killing and rebellion are contradictions; death and life are contradictions.

"Rebellion is in no way the demand for total freedom. On the contrary, ... the rebel wants it to be recognized that freedom has its limits everywhere that a human being is to be found ... The more aware rebellion is of demanding a just limit, the more inflexible it becomes. ... The freedom he claims, he claims for all; the freedom he refuses, he forbids everyone to enjoy. He is not only the slave against the master, but also man against the world of master and slave." [p.284]   Indeed, "rebellion, in principle, is a protest against death." [p.285]

But in the real, absurd world in which we live, the rebel is confronted with hard choices. Violence versus non-violence. Justice versus freedom. And: Not to choose is itself a choice or risk. "In so far as it is a risk it cannot be used to justify any excess or any ruthless and absolutist position." [p.289]

Rebellion with limits changes everything: We all have a common nature and individual rights. But technology, without proper guidance from the rebel, does not know this. Science has forgotten that it originated in rebellion with limits. Science and technology may yet return from deadly extremes to their origins and serve "individual rebellion. This terrible necessity [against terrorism, destruction, and enslavement] will mark the decisive turning-point" in history. [p.295]

The value that gives historical development meaning is not unknown: I rebel – therefore, we are. Thus: "Virtue cannot separate itself from reality without becoming a principle of evil. Nor can it identify itself completely with reality without denying itself." [p.296]   We have here a new form of virtue and a new kind of individualism. "I have need of others who have need of me and of each other. ... the individual, without this discipline, is only a stranger ... I alone, in one sense, support the common dignity that I cannot allow either myself or others to debase. This individualism is in no sense pleasure; it is perpetual struggle, and, sometimes, unparalleled joy when it reaches the heights of proud compassion." [p.297]

The moderation of the rebel, Meridian Man, is a perpetual tension and never-ending task. Contradictions ensue if we try to exist either above or below the meridian. History and the future must be viewed as opportunities, not as objects of worship. Passionate rebellion with limits is Meridian Man's approach to such opportunities. "Even by his greatest effort man can only propose to diminish arithmetically the sufferings of the world. But ... no matter how limited they are, they will not cease to be an outrage. ... confronted with death, man from the very depths of his soul cries out for justice." [p.303] This applies not only to literal death, but also to living death: "Thus Catholic prisoners, in the prison cells of Spain, refuse communion today because the priests of the regime have made it obligatory in certain prisons." [p.304] Thus moderation and life perpetually struggle against extremity and death.

Now to the final paragraph of *The Rebel* [p.306]: Meridian Man refuses the temptation of Caesaristic intoxication "in order to share in the struggles and destiny of all men," birthing a strange joy; happily "we shall remake the soul of our time ... which will exclude nothing." Following the intoxicating "pride of a contemptible period": "All may indeed live again, ... but on condition that it is understood that they correct one another, and that a limit, under the sun, shall curb them all. ... it is time to forsake our age and its adolescent furies." At "this moment ... at last ... [the first] man is born."

§5. Camus As The First Man

Camus, age 46, died absurdly as a passenger in an automobile accident in 1960. With him was his uncompleted manuscript, *The First Man*. I suggest that it is only by **each one of us learning to**

become the first man that each one of us can become the first man – neither slave nor tyrant. With Camus, we may yet learn to become Meridians – passionate, moderate rebels. If Camus is right, our common task – the struggle against death and the birthing of Mankind or Meridiankind – is an occasion of strange joy.

Bibliography

Asimov, Isaac (1991). *Asimov's Chronology of the World*. (HarperCollins Publishers). 1991. [See the Epilog.]

Camus, Albert (2005/1942). *The Myth of Sisyphus*. (Penguin Books). 2005. [Originally published in French in 1942.]

Camus, Albert (1991/1954). *The Rebel: An Essay on Man in Revolt*. (Vintage Books). 1991. [Originally published in French in 1954.]

Lyotard, Jean-Francois (1984). *The Postmodern Condition: A Report on Knowledge*. (University of Minnesota Press). 1984.

Tandy, Charles (2004). "Earthlings Get Off Your Ass Now!: Becoming Person, Learning Community," In Tandy, Charles (Editor), *Death And Anti-Death, Volume 2: Two Hundred Years After Kant, Fifty Years After Turing*, Ria University Press: Palo Alto, California (USA). (ISBN 0974347221). (Pages 373-391).

Endnotes

[1] This section is based on pages 374-381 of Tandy (2004).

[2] But see below: "**§4. Camus And Anti-Absurdity:** *The Rebel*" (see specifically the subsection: "Part Five Of Five Parts Is Entitled: Thought At The Meridian").

CHAPTER SEVENTEEN

The UP-TO Project:
How To Achieve World Peace, Freedom, And Prosperity

Charles Tandy

§0. Preliminary Note on Human Nature
§1. Preliminary Note on Terminology
§2. The "UP": Union of Well-ordered Peoples
§3. The "TO": Treaty Organization Acting for a Better Cosmos
§4. Conclusion

§0. Preliminary Note on Human Nature

For many centuries it was often said that due to nature or human nature, stable world peace was not possible and perhaps not desirable. For many centuries it was often said that due to nature or human nature, human heavier-than-air flight was not possible and perhaps not desirable. For many centuries it was often said that due to nature or human nature, the banning of human slavery was not possible and perhaps not desirable.

According to the third-century-BCE Chinese prince and scholar, Han Fei Tzu: "King Yen practiced benevolence and righteousness and the state of Hsü was wiped out; Tsu-kung employed eloquence and wisdom and Lu lost territory. So it is obvious that benevolence and righteousness, eloquence and wisdom, are not the means by which to maintain the state." (Watson, 1964).

The so-called "Parable of the Tribes" or "Prisoner's Dilemma" helps explain why things sometimes go from bad to worse. Understanding the Parable or Dilemma may help us, individually and collectively, engender a future that goes from good to better instead of from bad to worse. Our apparent inability – in many past centuries – to ban human slavery, to engage in heavier-than-air flight, or to achieve stable world peace **cannot** be reduced to nature or human nature.

Schmookler (1984) explains the **error** of reducing all human problems to nature or human nature; Schmookler explains the Parable of the Tribes this way: "In an anarchic situation ... no one can choose that the struggle for power shall cease. ... **no one is free to choose peace, but anyone can impose upon all the necessity for power**. [And humans, unlike nonhuman animals, have the ability to modify their anarchic situation.]"

Evans (1995) also explains the **error** of reducing all human problems to nature or human nature; Evans explains the Prisoner's Dilemma this way: "The prisoner's dilemma describes a possible situation in which prisoners are offered various deals and prospects of punishment. The options and outcomes are so constructed that it is rational for each person, when deciding in isolation, to pursue a course which each finds [in terms of actual results] to be against his interest and therefore [in terms of actual results] irrational. ... Such a scenario postulates a lack of enforced cooperation; and to avoid the undesirable outcome, the actors in the drama need to be forced into cooperation by a system of [enforced] rules."

§1. Preliminary Note on Terminology

My "UP-TO" proposal is meant to be a general outline or approach to achieve world peace, freedom, and prosperity. If "the devil is in the details," I leave such details to those more qualified than I. Nevertheless I have to use words even to suggest the general proposal such as it is. These words may be easily misunderstood. Thus herewith below a preliminary note on terminology. (I suggest consulting *The Law of Peoples* by John Rawls if one wishes to attempt to improve on the terminology – i.e., the meaning of "peoples" and "well-ordered peoples".)

I have chosen to use the word "peoples." As a first approximation: By "peoples" I mean to include "states" (e.g., members of the United Nations) and "quasi-states" (e.g., Taiwan/ ROC/ Chinese Taipei). Strategic ambiguity and strategic clarity each have their proper roles to play in human life and world betterment. Such decisions of ambiguity or clarity I leave to those actually involved in negotiating to achieve a better world. Given such a concrete context, they (not I) may decide, for example, whether "quasi-states" should be included or whether only "states" should be counted as "peoples."

I have chosen to use the term "well-ordered peoples." As a first approximation: By "well-ordered peoples" I mean to include "democratic peoples" and "decent peoples" (somewhat along the lines suggested by John Rawls in his *The Law of Peoples*, but so as to be consistent with the previous paragraph). Here again I leave it to those actually negotiating the agreement or treaty to decide whether "decent peoples" should be included or whether only "democratic peoples" should be counted as "well-ordered." **The UP part of the UP-TO proposal will be identified primarily with the term "well-ordered peoples"** (as distinguished from the broader term "peoples"). **The TO part of the UP-TO proposal will be identified primarily with the term "peoples"** (as distinguished from the narrower term "well-ordered peoples").

Below, I do NOT say that the European Union MUST take the lead to achieve world peace, freedom, and prosperity. Again, I am not the most qualified person to decide this pragmatic question (of leadership in the present context). Nevertheless on this issue I wish now to express my opinion. My **opinion** is that the European Union **should** take the lead in achieving world peace, freedom, and prosperity. A detailed defense of this tentative opinion ("that the European Union should take the lead") is beyond the purview of the present paper. However below I outline an "UP-TO" approach which I believe is doable at the level of international politics. For it to prove successful, it will require assistance from the United States. But I see the project as having a better chance of success if the EU (European Union) takes the leadership role. My belief or opinion is that it is in the interest of the United States and of the world for the United States to encourage the EU to take the leadership role in this unprecedented endeavor. So what exactly is the UP-TO proposal I am so excited about and want the EU and others to implement? First I will describe UP, then I will describe TO. [1]

§2. The "UP":
Union of Well-ordered Peoples

Immanuel Kant's *Perpetual Peace*, published in 1795, is a remarkable piece of social science foresight. [2] In 1795, very few republics existed and no liberal democracies existed (e.g. consider civil rights issues related to slavery and women). Kant argued for republicanism and for an expanding concert of peaceful republics. He believed this approach (as distinguished from the universal membership approach) would eventually lead (in the 21st century?) to a global stable peace.

The UP proposal is a proposal for a terrestrial Union of Peoples Well-ordered. On today's Earth, the People of China and the People of Russia are NOT well-ordered – but the EU Peoples and the North America Peoples, **among many others**, do have the actual substance of cultural norms and working mechanisms (rather than the mere form of words or documents or elections) that considerably support the rights of persons and that considerably tend to prevent war among such like-minded well-ordered peoples. Reminiscent of Kant, this is sometimes referred to by today's political scientists as the world's **"zone of peace."**

Let me point out that although the UP and the TO are separate entities with separate terrestrial-extraterrestrial functions, they are nevertheless expected, with help from the EU and others, to be birthed on the same date and to be perpetually on speaking terms with each other. Signatories of peoples to the (terrestrial) UP are members of the UP **Voting** Council; signatories of peoples to the (extraterrestrial) TO are members of the UP **Advisory** Council. Likewise, signatories of peoples to the (extraterrestrial) TO are members of the TO **Voting** Council; signatories of peoples to the (terrestrial) UP are members of the TO **Advisory** Council.

The UP (Union of Well-ordered Peoples) idea is inspired in part by the Daalder and Lindsay proposal for a CD (Concert of Democracies). (The article by Ivo Daalder and James Lindsay is entitled "Democracies of the World, Unite".) [3] While the UP and the CD concepts are not altogether identical, either realization may be expected to provide a number of benefits to world betterment, including: (1) Strengthening and expanding the positive relationship zone of peace and peaceful activities among UP or CD societies; and, (2) Weakening the negative temptations of UP or CD societies, such as (A) crusading imperialism; (B) imprudent appeasement; and, (C) moralistic isolationism.

§3. The "TO":
Treaty Organization Acting for a Better Cosmos

The TO proposal is a proposal for an extraterrestrial **Treaty Organization Acting for a Better Cosmos**. Before I explain the TO proposal in more detail, let me again point out that although the UP and the TO are separate entities with separate terrestrial-extraterrestrial functions, they are nevertheless expected, with help from the EU and others, to be birthed on the same date and to be perpetually on speaking terms with each other. Signatories of peoples to the (terrestrial) UP are members of the UP **Voting** Council; signatories of peoples to the (extraterrestrial) TO are members of the UP **Advisory** Council. Likewise, signatories of peoples to the (extraterrestrial) TO are members of the TO **Voting** Council; signatories of peoples to the (terrestrial) UP are members of the TO **Advisory** Council. **Note that the People of China and the People of Russia may indeed be founding (voting) members of TO but not of UP; on the one hand, they are peoples – on the other hand, they are not yet well-ordered.** (At some point in time in the 21st century, perhaps China and Russia will become well-ordered.)

Previously above I have explained the "Union of Peoples Well-ordered" (UP) idea. Now I present the TO ("Treaty Organization Acting for a Better Cosmos") part of the UP-TO proposal. [4] I believe that both the UP and TO ideas are desirable and feasible for today's world, even more so if implemented together (UP-TO). These two Concerts, acting more or less in concert, may have historically unusual ("doubly-synergistic") abilities to transmute our outmoded world into a "transcivilization" of peaceful prosperous societies.

We may not know the actual or secret (classified) policies of the United States and others with respect to extraterrestrial space.

It is nevertheless true that over 100 nations (including all of the "major" ones) publicly claim to support the 1967 Outer Space Treaty (or have at least signed it). The purpose of the 1967 Outer Space Treaty is to promote the peaceful uses of outer space. But today, as we approach its half-century mark, the 1967 peace treaty continues to have no enforcement mechanism.

The TO would serve two functions: (1) as gateway between planet earth and peaceful space – this includes **enforceably** banning weapons and weapons-making from extraterrestrial space; and, (2) as midwife to an evolving Extraterrestrial Society.

As I proceed to discuss these two functions, I want to attempt to offer a **realistic vision** of future extraterrestrial technology. This should help us better understand how to think about TO (and UP-TO). **But realistic visions sometimes change. Accordingly, we will want TO to be flexible enough to change in case our "realistic vision" changes!**

It has been said that if your prediction of far-future technological capacities does not sound like science fiction, **then your prediction is wrong.** Accordingly, please momentarily assume that the following vision (which may sound like science fiction) is indeed realistic. You can then later critically examine the notes and bibliography to independently decide for yourself.

The astounding capacity of future technology can be glimpsed at by taking a **non-controversial** look at the future of extraterrestrial O'Neill Habitats and of molecular Drexler Technology (and their eventual melding together). [5,6] I say **non-controversial** because the controversy in each case is over when, not if. Thus for present purposes we can overcome this dispute by simply talking non-controversially about these kinds of capacities in the far future (bypassing timeline predictions of near or far). Abundant extraterrestrial resources will be used to construct greener-than-earth, self-sufficient, self-replicating O'Neill Habitats in orbit around planets and suns. Accordingly, barring catastrophe, it seems highly likely that in the long run almost all of our multitudinous offspring will be permanently living and working somewhere in the universe other than on planet Earth.

Indeed, molecular Drexler Technology is not required for construction and development of extraterrestrial O'Neill Habitats ("SEG communities") – it just makes the task easier.

Historically one of the reasons that Terrestrial civilizations of old engaged in wars against each other was to gain more territory, and the power and glory that came with empire. (Human-caused global warming, human-made weapons of mass death and destruction, and a human-crowded global village did not yet exist.) As life grew denser on planet earth, the environment on which each organism depended increasingly consisted of other living things. But the development of advanced O'Neill Habitats (Sustainable Extraterrestrial Greener-than-earth communities, or SEGs) in orbit around a planet or a sun will mean multiple, self-reproducing biospheres; "unlimited free land" (freely available territory); and the realistic possibility of **intentional** (i.e., voluntary) communities for all persons. Instead of remaining in the community or culture of one's birth, one will be realistically free to experiment living in one kind of community or another. New kinds of cultures and communities will be enabled by the new extraterrestrial technology.

According to Carter and Dale (1974): "Most of the progressive and dynamic civilizations of mankind started on new land – on land that had not been the center of a former civilization." The following metaphorical insights have been widely quoted by SEG experts: "The Earth was our cradle, but we will not live in the cradle forever." "Space habitats [SEGs] are the children of Mother Earth." According to Carl Sagan, our long-term survival is a matter of "spaceflight or extinction"; all civilizations become either space-faring or extinct. According to the "mass extinction" article in *The Columbia Encyclopedia* (6th ed.): "The extinctions, however, did not conform to the usual evolutionary rules regarding who survives; the only factor that appears to have improved a family of organisms' chance of survival was widespread geographic colonization." [7]

[3.1 Gateway Between Planet Earth And Peaceful Space]

The TO would serve as gateway between planet earth and peaceful space – this includes **enforceably** banning weapons and weapons-making from extraterrestrial space.

As explained above, eventually there will be many Extraterrestrials, few Terrestrials. We can understand the practical or special interests that might prevent us from banning weapons and their manufacture from today's Earth. **Indeed, someday there might be analogous practical or special interests in extraterrestrial space unless we engage in foresight today to proactively and enforceably ban weapons and their manufacture from extraterrestrial space.**

On the one hand, our political interests today may constrain us in our present time and place. But, on the other hand, our political interests today may free us with respect to future times and places (e.g. our extraterrestrial future). **What this means is that today we have a realistic prospect of proactively establishing the legal structure and enforcement powers needed for a world at stable peace in extraterrestrial space.**

If we wait until later, we may not be so free to "do the right thing" and establish stable peace in extraterrestrial space. Extraterrestrial space is immense; it is all of the universe except for a single small planet. Eventually it might even become feasible to extend stable peace to planet Earth and thus the entire universe.

It is my belief that the suggested Extraterrestrial Space Treaty Organization (TO) will make a fine gift to our offspring and, by the way, help present Earthlings as well. But if we want a good world at stable peace (whether that world be Terrestrial Civilization or Extraterrestrial Transcivilization), it would seem we must be willing to unblinkingly face up to the following questions: Is stable peace possible if each person or each people is passionately convinced that their worldview is basically good and correct – and that other worldviews are evil or bad or incorrect? If we could enforceably prevent each and every person from killing any person over a conflict (say, a conflict of worldviews), would we do so? If so, how would we resolve our conflicts fairly or justly?

One advantage we have in facing up to these difficult questions is that we can use our imaginations to futuristically view ourselves as Extraterrestrials living in intentional communities (SEGs or O'Neill Habitats). We can further assume that a political structure there and then exists that we describe as a good world at stable peace. These Extraterrestrials of the future have liberties and technologies that Terrestrials do not have today. Yet humans today have the ability and perhaps the practical political will – via the TO proposal – to help insure humanity's **non-extinction** and promote human **flourishing** in a free and prosperous world at stable peace in extraterrestrial space (**almost** all of the universe).

I will assume that it is a fact that if today's Terrestrials are to produce such an extraterrestrial Treaty Organization (including effective enforcement provisions), it will require agreement from a number of Peoples. How many persons or peoples would accept or endorse a Space Treaty that effectively and enforceably bans weapons and their manufacture from extraterrestrial space? In this context (a good and practical legacy to our offspring), I should think we should be diligent enough to rally enough supporters. For example, TO might be signed originally by, say, twenty Peoples (including most of the "major" ones). **But the Treaty would be strongly effectively enforced by TO's Agency for a Better Cosmos (ABC) – NOT by Peoples/States – against ALL and EVERYONE, whether or not they sign the Treaty.** Once in force, I would expect many others to sign on – since the Treaty applies to them even if they do not sign it. Sooner or later the Treaty really would have to be strongly effectively enforced by the ABC against all and everyone, because eventually persons and communities will permanently settle in extraterrestrial space. (Such a Treaty also offers hope and inspiration to those of us of the present.)

Okay, you may say, this is a reasonable enough start, but what other liberties, responsibilities, and political structures would be appropriate for the Extraterrestrial World? So far, what we presumably have is a partial prototype for an Extraterrestrial

World at stable peace. But what about conflicts and the plurality of deeply held religious and philosophic worldviews?

[3.2 Midwife To An Evolving Extraterrestrial Society]

The TO would serve as midwife to an evolving Extraterrestrial Society.

What seems to me both practical and fair in this context is to think in terms of an Extraterrestrial Society of Intentional Communities. Each person is free to found new (intentional) communities. Each Community would determine its own membership requirements. Each Community would have **its own** culture of liberties and responsibilities; a member would generally be free to leave the community. A mechanism or set of mechanisms would be established to insure that each member is fully and properly informed of their liberty to leave the (intentional) community. (I suppose some communities might still allow their members the possibility of experiencing serious physical pain – but they would also allow a member to voluntarily leave their community. Too, I suppose banning animal cruelty and serious animal pain would be desirable and feasible. At least at first, this might mean with respect to animals that only domesticated or farm animals would be allowed permanent residence in extraterrestrial space?)

Note that some ("hermit") communities would consist of only one person. On old Terra, it was often difficult or impossible to leave one's community – sometimes expulsion effectively meant the individual's death. The sustainable prosperous context of the Extraterrestrial Society of Intentional Communities is radically different. If in the unlikely event this turns out not to be the case, then TO may not work exactly as envisioned by me. But whatever the case may be, the TO would have to be realistically flexible and convincingly knowledgeable of all relevant technology including on the cutting-edge and perhaps too the merely imaginable.

So at the level of the **Society** (of Communities) we have: (1) **Peace**: Weapons, weapons-making, and violence (including animal cruelty and serious animal pain) are strongly effectively enforceably banned (and so-called "research" would not be permitted as a way to get around the ban); and, (2) **Freedom**: Every individual person is fully aware of and fully informed of their general liberty to leave their community. This too is strongly effectively enforced. The Society and the communities necessarily work closely together to fully insure the liberties and responsibilities associated with both **Peace** and **Freedom**. Also note that since there is "unlimited free land," this fact will additionally help prevent some old terra-style conflicts and resolve or manage others (this would include some old-style civil conflicts).

At the level of **Communities** (in the Society) we have: (1) **Intentionality** (voluntariness): Within the good-faith transparent enforcement of Society's basic principles of peace and freedom, each Community has wide latitude for experimentation. Although there is a general liberty of members to leave the (intentional) Community, this does not necessarily relieve such persons from certain good-faith responsibilities to the Community; and, (2) **Transparency**: Each Community must strongly, effectively, and transparently help enforce the Society's basic principles of peace and freedom.

I believe the political theory or moral-political approach I have invented above with respect to TO is unique and original. It differs from the "Law of Peoples" conception of John Rawls in that TO is meant to be structured so as to necessarily ultimately generate an Extraterrestrial Law of Persons. Yet TO takes seriously the distinction Rawls makes between a "political conception" and "comprehensive doctrines." (A political conception or model addresses persons only with respect to, or at the level of, citizenship. Comprehensive doctrines or worldviews, whether religious or secular, address the full range, or deep levels, of one's personhood and relationships.) In my "Society of Communities" theory, **Society** corresponds to a political conception or model, and **Communities** represent comprehensive doctrines or worldviews.

"Is stable peace possible if each person or each people is passionately convinced that their worldview is basically good and correct – and that other worldviews are evil or bad or incorrect?" If you can sincerely and in good faith agree to my TO political conception (my approach above), the answer to this question appears to be YES, such stable peace is possible. If you can at most only agree to my TO approach as a temporary compromise, then the answer may be NO.

"If we could enforceably prevent each and every person from killing any person over a conflict (say, a conflict of worldviews) would we do so? If so, how would we resolve our conflicts?" If you can sincerely and in good faith (instead of merely as a temporary compromise) agree to my TO approach above, then stable peace in extraterrestrial space seems both possible and desirable. This approach, so I believe, realistically outlines a structure of stable peace for World Society and local Communities in extraterrestrial space – pointing toward conflict management in the new framework and encouraging subsequent projects to invent needed specifics.

According to TO, the architectures of extraterrestrial settlements will have to be PFIT (Peaceful, Free, Intentional, Transparent). Indeed, the architectures of all extraterrestrial structures will have to be congruent with PFIT. TO, via TO's ABC (Agency for a Better Cosmos), will proactively enforce the PFIT requirements. TO and ABC will have to be on the cutting edge of such changing technologies if they are to successfully fulfill their missions. PFIT **preplanning** and PFIT **retrofitting** of PFIT extraterrestrial settlements and structures will be an ongoing task.

§4. Conclusion

The political structure of island Earth, which is neither a Law of Peoples nor a Law of Persons, is unworkable. But at this unique point in history it is both desirable and feasible to establish a Terrestrial Law of Peoples (via the UP proposal). The political structure of extraterrestrial Space, which is neither a Law of Peoples nor a Law of Persons, is unworkable. But at this unique point in history it is both desirable and feasible to establish a Treaty Organization of peoples structured so as to necessarily ultimately generate an Extraterrestrial Law of Persons (via the TO proposal).

At this doubly-unique terrestrial-extraterrestrial point in world history, UP-TO is politically doable. World posterity would be eternally grateful to the EU and others for playing leading roles in birthing the joint UP-TO project. A window of opportunity has opened to the world. It is an open invitation for the world to achieve peace, freedom, and prosperity for the first time in history.

Bibliography

Carter, Vernon, and Tom Dale (1974). *Topsoil and Civilization*. University of Oklahoma Press: Norman, OK. (Page 12).

Daalder, Ivo, and James Lindsay (2007). "Democracies of the World, Unite," *The American Interest* (January-February 2007). Available free on the internet: <http://www.the-american-interest.com/article.cfm?piece=220>.

Diamond, Jared (2005). *Collapse: How Societies Choose to Fail or Succeed*. Viking: New York.

Drexler, K. Eric (1986). *Engines of Creation*. Anchor Press: New York. Available free on the internet.

Evans, J. D. G. (1995). "Prisoner's Dilemma," in *The Oxford Companion to Philosophy* (Ted Honderich, Editor), Oxford University Press: Oxford, UK. (Page 719).

Feather, Judith Light, and Miguel F. Aznar (2010). *Nanoscience Education, Workforce Training, and K-12 Resources*. CRC Press: Boca Raton, FL.

Ford, Brian J. (2009). "Culturing Meat For The Future: Anti-Death Versus Anti-Life," In Tandy, Charles [Editor] (2009). *Death And Anti-Death, Volume 7: Nine Hundred Years After St. Anselm (1033-1109)*, Ria University Press: Palo Alto, California (USA). (Chapter 2: pages 55-80).

Gore (2006). Al Gore. *An Inconvenient Truth: The Planetary Emergency of Global Warming and What We Can Do About It*. Rodale Books: Emmaus, Pennsylvania. [This is the first book in history produced to offset 100% of the CO_2 emissions generated from production activities with renewable energy; this publication is carbon-neutral.]

Hall, J. Storrs (2005). *Nanofuture: What's Next for Nanotechnology*. Prometheus Books: New York.

Heppenheimer, T. A. (1977). *Colonies in Space*. Stackpole Books: Harrisburg, Pennsylvania.

Hornyak, Gabor L., et al. (2008). *Introduction to Nanoscience*. CRC Press: Boca Raton, FL.

Kowal, C. T. (1988). *Asteroids, Their Nature and Utilization*. Ellis: Chichester.

Lewis, J. S. (1997). *Mining the Sky*. Helix Books, Addison-Wesley: Reading, MA.

O'Neill, Gerard K. (2000). *The High Frontier: Human Colonies in Space*. Apogee Books: Burlington, Ontario, Canada. (3rd edition). Also see: <http://www.nss.org/resources/books/non_fiction/review_008_highfrontier.html>.

Ratner, Mark A., and Daniel Ratner (2002). *Nanotechnology: A Gentle Introduction to the Next Big Idea*. Prentice Hall: New York.

Rawls, John (1971). *A Theory Of Justice*. The Belknap Press Of Harvard University Press: Cambridge, MA. Revised Edition, 1999. (Original Edition, 1971).

Rawls, John (1999). *The Law of Peoples: with "The Idea of Public Reason Revisited"*. Harvard University Press: Cambridge, MA. This edition, 2001.

Rosen, Carol (2009). The Institute for Cooperation in Space (website): <www.peaceinspace.com>.

Schmookler, Andrew Bard (1984). *The Parable of the Tribes*. Houghton Mifflin Company: Boston, MA. (Page 21).

Seg-communities (2008). <www.ria.edu/seg-communities>. [Seg-communities = Self-sufficient Extra-terrestrial Green-habitats, or O'Neill communities.]

Sen, Amartya (1999). *Development as Freedom*. Anchor Books: New York. This edition, 2000.

Spaceflight or Extinction (2010). <www.spaext.com>.

Space Quotes to Ponder (2010). <www.spacequotes.com>.

Spaceset (2010). [Bibliography.] <http://spaceset.org/p.bib.mm>.

Stein, G. Harry (1979). *The Third Industrial Revolution*. Ace Books: New York.

Tandy, Charles (2010a). *21st Century Clues: Essays in Ethics, Ontology, and Time Travel*. Ria University Press: Palo Alto, California.

Tandy, Charles (2010b). "'Wild-West' Versus 'Space-Age' Systems Science: An Extraterrestrial Prisoner's Dilemma?" <http://www.ria.edu/papers/wildwest/index.htm>.

Ulmschneider, P. (2006). *Intelligent Life in the Universe, Principles and Requirements Behind Its Emergence*, 2nd Ed. Springer Verlag: Heidelberg, Berlin.

Ulmschneider, P. (2009). "O'Neill-Type Space Habitats and the Industrial Conquest of Space," In Tandy, Charles [Editor] (2009). *Death And Anti-Death, Volume 7: Nine Hundred Years After St. Anselm (1033-1109)*, Ria University Press: Palo Alto, California (USA). (Chapter 16: pages 469-494).

Watson, Burton (1964). *Han Fei Tzu: Basic Writings*. Columbia University Press: New York. (Page 100).

Williams, Linda, and Wade Adams (2006). *Nanotechnology Demystified*. McGraw-Hill Professional: New York.

Wilson, Mick, et al. (2002). *Nanotechnology: Basic Science and Emerging Technologies*. CRC Press: Boca Raton, FL.

Endnotes

1. With reference to my UP-TO related research, I would like to thank the philosophy department of National Chung Cheng University (Taiwan) for their "Visiting Scholar" assistance. I am also grateful to Giorgio Baruchello, Al Globus, Jack Lee, and Asher Seidel for their comments on earlier drafts or partial drafts. For previous UP-TO related work by me, see for example the reprint of my 2007 article – Chapter 13 (pages 177-188) of Charles Tandy, **21st Century Clues: Essays in Ethics, Ontology, and Time Travel** (Palo Alto, California: Ria University Press, 2010). Also see: <http://www.segits.com>.

2. Much of the UP section of this paper (§2) is based in part on John Rawls, *The Law of Peoples: with "The Idea of Public Reason Revisited"* (Cambridge, Mass.: Harvard University Press, 2001).

3. See *The American Interest* (January-February 2007). The article by Ivo Daalder and James Lindsay ("Democracies of the World, Unite") is available at <http://www.the-american-interest.com/article.cfm?piece=220>.

4. Part of my inspiration for the TO idea comes from Dr. Carol Rosin and her website at <www.peaceinspace.com>.

5. "Extraterrestrial O'Neill Habitats" (or "SEG communities") – See, for example: (A) P. Ulmschneider, "O'Neill-Type Space Habitats and the Industrial Conquest of Space," Chapter 16 in Charles Tandy (editor), *Death And Anti-Death, Volume 7* (Palo Alto, California: Ria University Press, 2009). (B) *Seg-*

communities, <www.ria.edu/seg-communities> (2008). (C) P. Ulmschneider, *Intelligent Life in the Universe, Principles and Requirements Behind Its Emergence* (2nd ed.: Heidelberg, Berlin: Springer Verlag, 2006). (D) Gerard K. O'Neill, *The High Frontier: Human Colonies in Space* (3rd ed.: Burlington, Ontario, Canada: Apogee Books, 2000). (E) J. S. Lewis, *Mining the Sky* (Reading, Massachusetts: Helix Books, Addison-Wesley, 1997). (F) C. T. Kowal, *Asteroids, Their Nature and Utilization* (Chichester: Ellis, 1988). (G) G. Harry Stein, *The Third Industrial Revolution* (New York: Ace Books, 1979). (H) T. A. Heppenheimer, *Colonies in Space* (Harrisburg, Pennsylvania: Stackpole Books, 1977). **(For a more complete bibliography, see: <http://spaceset.org/p.bib.mm>.)**

6. "Molecular Drexler Technology" – Molecular nanotechnology's founder and founding book is K. Eric Drexler, *Engines of Creation* (New York: Anchor Press, 1987). This book is available free on the internet. Also see, for example: (A) Judith Light Feather and Miguel F. Aznar, *Nanoscience Education, Workforce Training, and K-12 Resources.* (Boca Raton, FL: CRC Press, 2010). (B) Gabor L. Hornyak et al., *Introduction to Nanoscience.* (Boca Raton, FL: CRC Press, 2008). (C) Linda Williams and Wade Adams, *Nanotechnology Demystified.* (New York: McGraw-Hill Professional, 2006). (D) J. Storrs Hall, *Nanofuture: What's Next for Nanotechnology.* (New York: Prometheus Books, 2005). (E) Mark A. Ratner and Daniel Ratner, *Nanotechnology: A Gentle Introduction to the Next Big Idea.* (New York: Prentice Hall, 2002). (F) Mick Wilson et al., *Nanotechnology: Basic Science and Emerging Technologies.* (Boca Raton, FL: CRC Press, 2002).

7. For such quotations as in this paragraph, see the websites *Spaceflight or Extinction* <www.spaext.com> and *Space Quotes to Ponder* <www.spacequotes.com>.

CHAPTER EIGHTEEN

Life And Death, And The Identity Problem

James Yount

Introduction

As technological advances make the distinction between life and death ever more obscure, the question of *"what determines personal identity?"* becomes more and more important. People who opt for cryonic preparation after their deaths are particularly interested in this question and have been since, in 1964, Robert C.W. Ettinger first presented the cryonics option in his book, *The Prospect of Immortality.*

The modern evolving ways of distinguishing life from death is very different from the notion of the two as non-overlapping polar opposites. It would seem that determining the nature and bounds of personal identity is central to current ideas on life and death.

Technological developments, particularly nanotechnology and both *information theory,* and the greatly increased availability of information will be discussed as promising techniques to reconstitute memory loss in a human subject who is to be revived or reconstituted after a frozen hiatus. We will later question whether a subject so treated can be considered as the same person he was before being frozen and then treated by these new techniques. ["Frozen" herein means "frozen/vitrified"–see below.]

Robert C.W. Ettinger can rightly be considered the father of cryonics having written the book that started the cryonics movement, *The Prospect of Immortality.* He has been a leading advocate of cryonics and has contributed much to the consideration of the "identity problem" as it relates to human consciousness and cryonics.

We will examine Ettinger's treatment of the identity question as he originally posed it in his book where he presented twenty

"Experiments," some that can be done, others simply "what if?" mind experiments.

We also pose and examine some other mind experiments that, it seems to the author, flow logically from those Ettinger originally posed and the underlying problem of determining the nature of personal identity. In 2005, in *Youniverse: Toward a Self Centered Philosophy*, Ettinger again considers the question of *"what constitutes personal identity?"* He presents his twenty Experiments again and suggests the "self-circuit" as that which best defines personal identity. We will consider this later body of work.

Life as a Vital Force

Our language reflects that outlook. The word "vitality" as defined by Merriam Webster is "1. a: the characteristic distinguishing the living from the nonliving." (Merriam Webster II: 769)

While the word *vitality* is an integral part of our language, and hence, rarely examined closely, the word *vitalism* is not but the same philosophical assumptions give rise to both words:

Vitalism:

1. A doctrine that the functions of a living organism are due to a vital principle distinct from biochemical reactions.
2. A doctrine that the processes of life are not explicable by the laws of physics and chemistry alone and that life is in some part self-determined. (wikipedia.com)

It is easy to understand how the idea that life is a vital force came about. An archer shoots a deer that then falls to the ground bleeding. Although fallen, the animal breaths, its heart beats, eyelids flutter, and muscles quiver. Then it stops moving, with no pulse, no breath, no muscle contractions. This "no move" state, once attained, cannot be reversed. It is as if a life force has just gone out of the animal. The *thing* that animates has left and the deer goes from the state of being alive (vital) to the state of death.

The New Death

Increased medical knowledge and the techniques that have arisen from such knowledge have made the distinction between life and death less acute. It is well known and generally accepted, for example, that cardiopulmonary resuscitation (CPR) can successfully treat a drowning victim. In the not so distant past, this victim would have been viewed as having irrecoverably lost the vital force.

The Persistence Question of Personal Identity

The "identity problem," especially as applied to living beings, takes on new importance with the possibilities for keeping one alive by improved medical means. What amount of change is tolerable in a being for him/her to still remain the same person? Or put another way: *"The question is what is necessary and sufficient for a past or future being to be you."* (plato.stanford.edu).

Cryonics and the Identity Problem

People who expect to undergo cryonic suspension, through freezing or vitrification, have a deep personal interest in "necessary and sufficient." They have no motivation to volunteer for a procedure, and pay their own hard-earned dollars for that procedure, if they believe that the *person* revived in the future will be someone *other than* the person who is frozen.

Please note that throughout this article we shall use the terms: "cryonic suspension," "cryonic preparation," and "freezing," (applied to humans) interchangeably. Such cryonic suspension may also include preparation through vitrification. (Current cryonic preparation often involves use of vitrification, a procedure where some of the water in the body and/or brain is replaced with a chemical solution that remains unfrozen at liquid nitrogen temperatures.)

The Identity Problem as Presented by Robert Ettinger

As previously noted, the beginning of the cryonics movement can properly be traced to the book *The Prospect of Immortality*, by Robert C.W. Ettinger, first published in 1964. Ettinger recognized the importance of survival, through the freezing processes, of the individual who was originally frozen ("suspended" in cryonics literature jargon). He devoted "Chapter VIII: The Problem of Identity" to a discussion of this question.

Ettinger begins the chapter thus:

"In considering the chances of reviving, curing, rejuvenating, and improving a frozen man, we have to envisage the possibility of some very extensive repairs and alterations. This leads to a number of very perplexing puzzles." (Ettinger, Prospect of Immortality, 129)

He then goes on to pose the question of identity in lay terms:

"What characterizes an individual? What is the soul, or essence, or ego? This seemingly abstruse question will shortly be seen to have ramifications in almost every area of practical affairs; it will be the subject of countless newspaper editorials and Congressional investigations, and will reach the Supreme Court of the United States." (Ettinger, Prospect of Immortality, 130)

Ettinger's Experiments

Ettinger then presents the reader with Experiments 1 through 20, in which some portion of a human subject's brain undergoes a degree of modification. Many of the experiments are mind experiments of "what if?" However, some of these experiments can be performed now, and some perhaps may be performed in the future.

It is somewhat surprising, given that the answer to the identity question seems so central to the premise of cryonics, that Ettinger devotes so few pages to it in his book. Perhaps that fact simply speaks to the difficulty of answering the identity question for a frozen person: further discussion may *not* provide any better illumination.

In his book, *Youniverse*, Ettinger devotes much more book-space to the topic. He writes: "The most important of all scientific questions—and possibly the profoundest—is the nature of self or of personal identity." (Ettinger, Youniverse, 174)

Information Originating Outside the Subject's Brain

Since the 1964 publication of Ettinger's *The Prospect of Immortality*, there have been many developments that make "outside" sources of information about an individual person readily available. By *outside*, we mean information that originates outside the person's own brain.

At the publication date of Ettinger's book, it was not known yet that so much information from these outside sources would emerge as soon as they have. These rich information sources and exacting details have been made possible by such technologies as enhanced computer memory, and the development of the World Wide Web.

The Importance of Memory

The outside sources of information might make it possible to replace or supplement memories lost because of disease, accident, or memory loss from the very imperfect preparation and freezing now available from cryonic suspension.

However, the importance and centrality of memories in defining "self" can be challenged. The author made this argument in *Being Human: the technological extensions of the body*, chapter entitled "An Advocation for Immortality":

> Implanting False Memories:
>
> "Just how important are your memories anyway? I have fond childhood memories of our little dog, Tippie. How

she would bark gleefully when she made a bee-line to get inside whenever the door was open. How she would howl in pain when someone slammed the door with the dog still making that mad rush. But what if I lost that memory and a technician implanted a false memory? I would remember instead, my little dog Bowser, how Bowser would catch the boomerang in his teeth as Lola and I played on the Australian desert. Come to think of it, I might actually PREFER memories of playful romps through the Outback with Lola and Bowser! Programmed Memories from Someone Else's Memories...." (Yount 434)

Ettinger touches on this same question in *Youniverse* under the topic title "ACCIDENTAL YOU!" After suggesting that someone might not want to remember an unfortunate background of child abuse or neglect he states: "Personally, I would just as soon be rid of my bad memories and traits and habits instantly, and 'know' the history only as an archive that I can look up if I wish." (Ettinger, Youniverse, 196)

Nanotechnology

One of the significant contributions to the theory of how a frozen person might be repaired and rejuvenated came about with the 1986 publication of *Engines of Creation* by K. Eric Drexler. Drexler suggested that self-replicating nano-sized repair machines could be created and used to repair human bodies that are badly damaged.

As Drexler envisioned, the tiny nanotechnology machines or repair robots could go into an individual human cell, analyze the damage, and repair it by moving individual molecules and even atoms back into proper place and alignment. This process could be applied to damaged nerve paths as well as arteries and veins, for example, to then reconstitute the individual as a healthy human being, perhaps even with complete memories.

Information Theory

The study of *information theory* may allow us to establish apparent limits to how much augmentation of personal memory from outside sources is possible as well as limits on repair. However that study is beyond the scope of this paper and this detail may not be needed to approach the basic question of personal identity as it applies to cryonics "patients."

Cyborgs and Stream of Consciousness

Information storage devices have become smaller in size and more mobile. These storage devices may be coupled with mobile and high fidelity video recorders and cameras. With this configuration, the ability of individuals to easily record events in their lives has increased significantly.

For some years Kevin Warwick has been engaged in a lifestyle that may well fashion him as a cyborg or perhaps cyborg precursor. During his waking hours he records all of his activities through miniature video cameras mounted on his head. He has continual access to the WWW. He endeavors to store the information stream coming from his camera and his computers for future reference. Other individuals have participated in similar activities, though apparently for not as extensive a period of time as has Warwick. (kevinwarick.com)

Devices, largely experimental, now exist to allow the brain to directly interact with computers making the keyboard and mouse or voice commands unnecessary.

A stream of consciousness, both from sources such as a video camera and the record of any brain-computer interface, could thus be recorded. This information is subject to classification and analysis by whatever information processing devices are available to us, sentient or otherwise. This brings forth the possibility of a "second brain," that may not even have a physical location, strictly speaking, but may reside in the internet "cloud."

From that second brain the possibility of multiple bodies and even the death of our "meat" brain with the second brain continuing are possibilities, but the development of those ideas goes beyond the scope of this paper.

For our purposes we simply recognize the fact of this increasingly rich source of information and the fact that it is available to supplement, replace, or jog our natural memories.

The Sherlock Holmes Computer

In an essay entitled "An Avocation for Immortality", published in 1999 in the book *Being Human: the technological extensions of the body*, the author introduced the Sherlock Holmes Computer.

The fictional detective Sherlock Holmes uses deductive reasoning to reach correct conclusions from imperfect information. The Sherlock Holmes Computer makes use of incomplete and imperfect information to reconstruct or reprogram (if you will) brains that have been badly damaged, including those from frozen people.

Using both physical information from the damaged brain and the rich information sources from "everything available in the world" a powerful computer could make logical inferences of the memories the human subject once had and put together either a virtual person or instruct nanites to kick the atoms and molecules back where they belong.

The Sherlock Holmes computer may be a precursor to the *Technology Singularity* after which we have no idea if Sherlock will follow the program of reanimating humans or will even have an interest in humans.

Replacement of Body Parts & Essence of the Essence

Some of the Experiments presented here involve replacement of living tissue with other living tissue and/or mechanical parts. In this article it is assumed that most of the body parts of a human will be replaceable in one way or another.

If it becomes medically necessary for a surgeon to remove a person's legs, this process does not create a new person. It is the original subject, but with his legs missing.

We may remove most of the body leaving only the brain. In fact we could discard the brain stem and perhaps part of the cerebellum that controls the autonomic nervous system. The connective tissue that hold the brain together as well as those brain centers that register sensation could be replaced.

It seems reasonable that *all tissue* not involved with the storage and processing of memory could go, to later be replaced. If this is true then the amount of biological material that makes an individual unique ("I think, therefore I am") is rather small, perhaps just the cerebral cortex and the hippocampus.

When we have arrived at that critical "part," the *essence of the essence*, we get into the questions raised in this paper and by other people on the nature of identity. What operations can be performed on this critical remaining part of the brain with the subject retaining the *identity* he came in with?

Examination of Robert C.W. Ettinger's Experiments

We will now look at a number of Ettinger's proposed Experiments as he presents them in *The Prospect of Immortality* and examine several in detail.

The brief explanatory text is taken directly from the book.

We will be referring back to some of these Experiments as we extend the discussion

> ***Experiment 1.*** *We allow a Man to Grow Older*
>
> Is this the same man? Ettinger notes that *"most of his body has been replaced and changed; his memories change and some are lost; his outlook and personality change."* (Ettinger, Prospect of Immortality, 139)

We could define the man as the same or we could define him as different. For example, Locke defined a person as "a thinking intelligent being that has reason and reflection, and can consider itself as itself, the same thinking thing, in different times and places." (Burtt, 315)

Apart from noting that the changes have occurred gradually over much time, there is little in this question that gives us guidance on the larger question of whether the man who is thawed and restored to life is the same one who was frozen.

> *Experiment 5.* *By super-surgical techniques (which may not be far in the future) we lift the brains from the skulls of two men, and interchange them. This experiment might seem trivial to some. Most of us, after thinking it over, will agree it is the brain which is important, and not the arms, nor the legs, nor even the face.* (Ettinger, Prospect of Immortality, 133)

This observation is similar to the author's own conclusions presented in this paper under the topic title: "Replacement of Body Parts & Essence of Essence". However, Ettinger does not discuss the fact that only a rather small part of the brain mass could be thus transplanted, and in essence, would thus transplant one man's "self" into the other and vice versa.

> *Experiment 6.* *By super-surgical techniques (not yet available) we divide a man's brain in two, separating the left and right halves, and transplant one half into another skull (whose owner has been evicted). Similar, but less drastic, experiments have been performed. Working with split-brain monkeys, Dr. C. B. Trevarthen has reported that " . . . the surgically separated brain halves may learn side by side at the normal rate, as if they were quite independent." This is most intriguing, even though the brains were not split all the way down to the brain stem, and even though monkeys are not men.* (Ettinger, Prospect of Immortality, 134)

There are occasions when a cryonic procedure is performed that *only the brain* is subject to preservation and storage. This may occur when a Medical Examiner has removed the brain during an autopsy before the "suspension team" is notified.

Under those circumstances the brain might be split, and the left brain subjected to one particular experimental preservation and cold storage regiment, and the right brain to another. Either half might later be used to partly "reprogram" a new companion left or right half. It must be added that as far as the author knows, this splitting of the brain during a cryonic procedure has not been done.

> *Experiment 8. A man dies, and lies unattended for a couple of days, passing through biological death and cellular death. But now a marvel occurs; a space ship arrives from a planet of the star Arcturus, carrying a supersurgeon of an elder race, who applies his arts and cures the man of death and decay, as well as his lesser ailments. (It is not, of course, suggested that any such elder race exists; the experiment is purely hypothetical, but as far as we know today it is not impossible in principle).* (Ettinger, Prospect of Immortality, 136)

It appears we need not wait for the ship from Arcturus to arrive. Something similar to the above might well be in our future with nanotechnology machines ("nanites") coupled with the Sherlock Holmes Computer taking over the task of the Arcturian supersurgeon. (See discussion in this essay under topic titles The Sherlock Holmes Computer and Nanotechnology).

> *Experiment 9. A man dies, and decays, and his components are scattered. But after a long time a super-being somehow collects his atoms and reassembles them, and the man is recreated.* (Ettinger, Prospect of Immortality, 136)

> *Experiment 10. We repeat the previous experiment, but with a less faithful reproduction, involving perhaps only*

some of the original atoms and only a moderately good copy. Is it still the same man?

Again, perhaps, we wonder if there is really any such thing as an individual in any clear-cut and fundamental sense. (Ettinger, Prospect of Immortality, 136-137)

Experiment 11. *We repeat experiment 10, making a moderately good reconstruction of a man, but this time without trying to use salvaged material.*

Now, according to the generally accepted interpretation of quantum theory, there is in principle as well as in practice no way to "tag" individual particles, e.g. the atoms or molecules of a man's brain; equivalent particles are completely indistinguishable, and in general it does not even make sense to ask whether the atoms of the reconstructed body are the "same" atoms that were in the original body. Those unfamiliar with the theory, who find this notion hard to stomach, may consult any of the standard texts.

If we accept this view, then a test of individuality becomes still more difficult, because the criteria of identity of material substance and continuity of material substance become difficult or impossible to apply. (Ettinger, Prospect of Immortality, 137)

The question of what constitutes the individual identity of the man as he was before death is at the heart of the discussion. Experiment 9, 10, and 11 highlight the problem.

The question of whether or not anything is to be gained by collecting and using the original, atoms in the identical places as the originals is one that is still debated and there has been some lively discussion of this subject on the cryonics discussion forum CryoNet. (cryonet.org)

The transporter envisioned by the writers of *Star Trek* works by breaking down the atoms and transmitting an atom stream so that both information and "same atoms" are used to reconstruct Captain Kirk. The view that the original atoms put back in the original place (in relationship to other atoms) is necessary for *you to be you or Kirk to be Kirk* appears to be commonly held.

As noted elsewhere, a cryonics repair strategy might involve the Sherlock Holmes Computer, coupled with nanite repair machines to put the *same molecules and even atoms* back to their original place. A better argument can be made for this possibility in cases where the subject is treated by cryonics preparation shortly after death and the displacement of physical material is relatively small. Treatment for badly damaged individuals may be statistically beyond any presently envisioned "best guess" strategy of such computers and nanite robots.

> ***Experiment 12.*** *We discover how to grow or to construct functional replicas of the parts of the brain - possibly biological in nature, possibly mechanical, but at any rate distinguishable from natural units by special tests, although not distinguishable in function. The units might be cells, or they might be larger or smaller components. Now we operate on our subject from time to time, in each operation substituting some artificial brain parts for the natural ones. The subject notices no change in himself, yet when the experiment is finally over, we have in effect a "robot"! Does the "robot" have the same identity as the original man?* (Ettinger, Prospect of Immortality, 137)

> ***Experiment 13.*** *We perform the same experiment as 12, but more quickly. In a single, long operation, we keep replacing natural brain components with artificial ones (and the rest of the body likewise) until all the original bodily material is in the garbage disposal, and a "robot" lies on the operating table, an artificial man whose memories and personality closely duplicate those of the original.* (Ettinger, Prospect of Immortality, 137-138)

> ***Experiment 14.*** *We assume, as in the previous two experiments, that we can make synthetic body and brain components. We also assume that somehow we can make*

sufficiently accurate nondestructive analyses of individuals. (Ettinger, Prospect of Immortality, 138-139)

Experiments 15, 16, and 17. *We repeat experiments 12, 13, and 14 respectively, but instead of using artificial parts we use ordinary biological material, perhaps obtained by culturing the subject's own cells and conditioning the resultant units appropriately. Does this make any difference?*

We draw the obvious but useful conclusion that, from the standpoint of present serenity, it is merely the prospect of immortality that is important. (Ettinger, Prospect of Immortality, 139)

Experiment 20. *We pull out all the stops, and assume we can make a synthetic chemical electronic mechanical brain which can, among other things, duplicate all the functions of a particular human brain, and possesses the same personality and memory as the human brain. We also assume that there is complete but controlled interconnection between the human brain and the machine brain: that is, we can, at will, remove any segments or functions of the human brain from the joint circuit and replace them by machine components, or vice versa.* (Ettinger, Prospect of Immortality, 140)

Experiments 12 through 21 raise the possibility of repair and replacement by natural or synthetic cells either through gradual repair/replacement or "all at once."

A close variation on Experiment 21 may now be underway. This is discussed briefly in this paper under the subject topic Cyborgs. As humans are better able to access and "become one" with computers, artificial intelligence of computer programs could closely interface with human intelligence, memories, and even emotional reactions. One may find he "lives" in the information cloud on the web as much as he does in his own head. When the "meat"-you dies the information-you world wide web ("WWWU") continues. The WWWU might even recreate or assist in the eventual repair of the frozen you, though as Ettinger notes:

The subject "may even prefer to 'inhabit' the machine. He may even view with equanimity the prospect of remaining permanently 'in' the machine and having his original body destroyed." (Ettinger, Prospect of Immortality, 141).

Identity from the Standpoint of the Observer

As mentioned earlier, we will bring some of Ettinger's experiments back into the discussion as we explore the problem of personal identity further.

Using Ettinger's form of presentation, here are several additional Experiments that seek to make the point that the question of "the unique quality of personal identity" is answered differently for the "inside" and "outside" observer.

"Beam Me Up Scotty" Experiments A through K

Many people have viewed episodes of *Star Trek* and the transporter that is used to seemingly magically make people disappear and then reappear someplace else. In discussing the topic of *human identity as information*, the transporter is referred to as a way to introduce paradoxes and anomalies of identity that would be apparent if such technology were possible.

We will avoid using *Star Trek* references since the "atom stream" in the Starship Enterprise scanning process may not fit our intended purpose. However, *Star Trek* fans may readily understand and relate to Experiments A through K.

A human (call him "James") is *scanned* head to foot. The location of every atom in his body, in its relationship to the other atoms that make up his body, is recorded. This information can then be used to create a perfect likeness of the individual anytime in the future and at any place where the "reader" is set up.

The scanning is of true fidelity so no scanning mistakes are made or information lost because of lack of the ability of the scanner to read-in the proper detail. The individual thus scanned is not even aware that the information that makes up his body has been scanned and recorded.

Experiment A. A human is scanned, then destroyed, and then reconstituted

After scanning and saving the recorded scanned data, the subject, call him "James," is painlessly reduced to his constituent atoms. A few minutes after the scanning/destruction the information from the scanning is used to reconstruct the individual. The same atoms are used and are placed in the same location in relation to other atoms in reconstructing the individual.

Experiment B. Same as A, but different atoms are used

Different atoms are used to replace those that are collected after James' body is broken down into its constituent atoms.

In The Prospect of Immortality, Ettinger correctly observes that:

> Now, according to the generally accepted interpretation of quantum theory, there is in principle as well as in practice no way to "tag" individual particles, e.g. the atoms or molecules of a man's brain; equivalent particles are completely indistinguishable, and in general it does not even make sense to ask whether the atoms of the reconstructed body are the "same" atoms that were in the original body. Those unfamiliar with the theory, who find this notion hard to stomach, may consult any of the standard texts. (Ettinger, Prospect of Immortality, 137)

Experiment C. Nano machines aid in assembly

Nano sized machines, in conjunction with a scanner, are used to map the relationship of atoms to each other and/or molecules to each other. We disrupt the function of James body enough so that he is clinically dead. After a few minutes the nano machines move any atoms that are out of place back to where they were originally and James is good as new.

If the generally accepted theory that all like atoms are interchangeable is correct, then using new atoms as in Experiment F will make no difference. If this theory is incorrect (as many cryonics advocates seem to believe on an intuitive basis) then having the nanites "tag" the individual atoms and use these same atoms to reproduce James overcomes this objection.

If and when nanotechnology attains this degree of sophistication there are apt to be practical limits on the tagging capacity of the nanites. That statement needs some further explanation:

Envision a military reconnaissance photo of a road leading up to a river where there is another road directly across from the first on the other side of the river. It is a reasonable inference that there was once a bridge connecting the two roads that was destroyed. If a giant where to come by who is very good at puzzles he might be able to find many of the parts that made up the bridge and reassemble them in their proper order.

In the neural weave in James brain, envision a neuron part that appears to line up with another neuron part with a gap in between. It is a reasonable inference that the two were once connected and a nanite that is very good at puzzles might look around and find the molecules and even atoms that made up the missing neuron part and put it back in place.

However, it does not seem likely that if James brain is badly decayed, the nanites will be sufficiently good at puzzles to tag all atoms and then put them back exactly where they were before. So in some (perhaps most) cases, for the person revived to be "identical" to the person before he was frozen, we would likely need to depend upon the premise that atoms are interchangeable.

Experiment D. We destroy the original manuscript of *Moby Dick*

Melville's classic work of fiction *Moby Dick* has been widely published with millions of individual copies now scattered around the world. If the original manuscript were to be destroyed would *Moby Dick* be destroyed?

Since there are so many faithful copies, the destruction of any one copy or even the original manuscript would have no effect on the other copies in existence, any one of which could be used as the "new" original to make more copies.

Experiment E. Destroy a single copy of *Moby Dick* where it is the only copy available to an isolated individual

A shipwrecked sailor is lost to the world on a desert island. He has no way to get off; he will be there the rest of his life. He has managed to rescue a copy of *Moby Dick* which he reads over and over. One night the book falls into the fire and is destroyed. Has *Moby Dick* been destroyed?

In the big wide world there are many copies of the book. The destruction of the copy on the island would have no effect on the other copies any of which could be used as an information source to make more copies in the future.

However, for the shipwrecked sailor, the destruction of *his* copy *is* the destruction of *Moby Dick*.

It can be said that the answer to the question: *Has Moby Dick been destroyed* depends upon the *observer*, or perhaps the location of the observer. To people outside the Island, *Moby Dick* was not destroyed. To the sailor, *Moby Dick* has indeed been lost forever.

Again, quoting Ettinger: "Consider every tiny bit of dead matter. Two hydrogen atoms are identical in most respects, but if they differ in location, and perhaps in momentum, then, even without regard to gravitational aspects, they cannot substitute one for the other." (Ettinger, Youniverse, 194)

In the Experiment where James was scanned when the original was not destroyed, the original James can be thought of as the *observer*. For observers outside the brain of original James any of the duplicates are as good as the original since these observers can't tell the duplicates from the original.

James himself wasn't even aware he had been duplicated. If we told him that we were now going to destroy him because he had been copied, and explained to him that James (*Moby Dick*) has a continued existence, the explanation would likely not be accepted so cheerfully.

Experiment F. Replace component parts of George Washington's Hatchet

This Experiment has some of the same considerations as those in Ettinger's Experiment 12.

A variation on the Theseus' Paradox is known as *George Washington's Hatchet*. In presenting the paradox, the author will exercise a bit of literary license:

An apocryphal story, often told to children to teach them to be honest, is that as a boy, George Washington chopped down his father's cheery tree. When his father asked who was responsible for the dastardly deed, young Washington replied: "I cannot tell a lie, Father. I did it."

[So the author will now exercise a bit of literary license, as follows:] A family in New Hampshire owns the original hatchet that George Washington used to chop down the cherry tree. Except that the handle has been replaced three times and the hatchet head replaced twice!

In spite of the replaced parts, we could trace the history of the hatchet and see where it was located and when and where it was put to use. There would be an unbroken time line from when it left Washington's hands until the present.

Experiment G. Replace both component parts of George Washington's Hatchet at once

This includes some of the same considerations as those in Ettinger's Experiment 13.

The farmer is using the hatchet to chop down a tree and gets frustrated with how poorly it is performing. He goes to the hardware store with the hatchet. He picks out another similar hatchet from the rack and throws the old hatchet in the trash. He goes home and continues to chop down the tree.

There could be the same unbroken line of the existence and use of a hatchet as with the Experiments above. However *no part* of the hatchet survives when the farmer makes the new purchase.

Experiment H. Component parts of Washington's Hatchet are used to fashion two hatchets

When the New Hampshire farmer replaces the hatchet head he tosses the old head in the trash. His hired hand sees the discard and uses it to make his own hatchet by fastening on a new handle.

Each of the two newly assembled hatchets could be called "the original" in that there is an unbroken time line from when the hatchet left Washington's hands until the present.

Experiment I. Same as F, but with information storage in hatchet handle and head

Consider now that both hatchet head and handle can record information about where they have been and what use they were put to. When the handle is replaced, the old handle copies the information from itself onto the new handle as part of the replacement process. If for any reason that is not possible, the hatchet head copies and transfers information *it* has in storage to the new handle.

The claim the New Hampshire family have for owning the original hatchet is strengthened substantially. Not only is there an unbroken time line of the hatchet's existence between the time of Washington through to the present but there is an unbroken information stream from the time of Washington to the present.

Experiment J. information storage in hatchet handle and head but time elapses before information is transfered between component parts

The farmer takes apart the hatchet head and handle, throws away the handle then deactivates the head and puts it in storage for several years. He then brings the head out of storage and puts on a new handle. The head then transfers its information to the new handle.

This experiment is fairly close to what may happen in the cryonic suspension process.

Further Discussion of the George Washington's Hatchet Paradox

We could define "the same hatchet" as being the continued existence of most of the original hatchet.

In that case both the original hatchet and head would have to be intact (we will excuse minor loss of wood and metal). No gross replacements would be permitted.

We could define "the same hatchet" as the continued existence of the time stream of the history of the hatchet and allow for replacement of parts.

We could define "the same hatchet" as the continued existence of any part of the hatchet that had at any time been attached to any of the component parts that made up the head and handle Washington used to chop the Cherry tree.

We could define "the same hatchet" as the continued existence of the information of where the hatchet had been and what it has been used for.

For the purpose of trying to better understand the nature of human identity and conditions where it terminates or continues, the experiment where the hatchet handle and head recorded and contained information on where they had been and what they were used for is the most useful. The condition where there is a break in time does not seem to nullify the farmer's claim if it is based on the hatchet as "information". We did not go to the extreme of making the hatchet head and handle sentient but for purposes of our analysis of identity of the hatchet there is no apparent reason to believe it would matter.

Physical continuity (continuity of place) and temporal continuity

As this is being written, a large garden rock can be viewed through the window. Most people would agree that the author can rightly regard his own personal identity as the information, mass, activity of his self. Most people would agree that the rock is not the author. The person who sits at the keyboard has physical and temporal continuity with all of his past instances of identity. The rock has neither physical nor temporal continuity with the person at the computer. The person is the author's identity; the rock is not. Between these two extremes is where the difficulty of establishing identity lies.

In almost 50 years of wresting with the problem of determining what is personal identity, and its bounds, Robert Ettinger has not been able arrive at definitive answers, though he claims progress in identifying what he calls *"self-circuits"* and *"overlapping successors"* "which may just possibly solve the problem." (Ettinger, Youniverse, 185)

Generally, though, Ettinger punts the ball on down field, or to the future:

"...the only reasonable working posture I can see is to seek as much continuity as possible both in information and physical substance. That way we keep our options open." (Ettinger, Youniverse, 192)

"At present, no clear-cut answer presents itself to me—nor, credibly, to anyone else, as far as I know. But on a pragmatic, common-sense basis, about all we can do for now is try to avoid extreme or sudden change, try to maintain the maximum physical integrity and psychological continuity, and leave this problem, or its more extreme variations, for the future." (Ettinger, Youniverse, 185)

Final Discussion and Summary

With increased sophistication in medical technology the definition of death has changed to recognize this fact. A suggested radical procedure that would make use of present and future manipulation of the human body is that of cryonic suspension.

The answer to the question *"what amount of change is tolerable in a being for him/her to still remain the same person?"* is central to the cryonic premise. If the freezing/reanimation process so changes the subject that he is not the same person as he was when he was living then most people would not be interested in volunteering for the experiment, let alone pay for it.

The man widely considered as the "father of cryonics," Robert C.W. Ettinger, recognized the importance of the personal identity question in his book *Prospect of Immortality* first published in 1964. In an effort to come to grips with the identity problem, Ettinger presented twenty "Experiments," some that could be performed, others just thought experiments. In his 2005 book *Youniverse*, Ettinger again presented these Experiments and expanded greatly on his original discussion of the identity problem.

In this paper we considered Ettinger's Experiments and proposed a number of related new thought experiments. We discussed some of the changes that have come about since the Experiments were first proposed.

Information originating outside the subject's brain could be used to replace or supplement information lost or destroyed in a cryonics subject. The importance of memory is called into question and it is suggested that implanting false memories would not necessarily negate the identity of a reanimated subject to being that of the person before freezing.

The possible role of nanotechnology to reanimate a subject was discussed. It was pointed out that nanites (self-replicating nano-sized robots) might be capable of placing the individual molecules and even atoms of a damaged brain back in their proper place so the question of whether or not the original atoms or replacement atoms need be used to retain the identity of the individual may not be a problem.

A computer with programs to use deductive reasoning to make use of incomplete information to "fill in the gaps" to attain true information based on "best guesses" can be a powerful tool in reconstructing badly damaged brains and lost personal information. We call such a computer the Sherlock Holmes Computer. Sherlock may be able to instruct nano repair mechanisms to accomplish such reconstruction.

We note that it appears that the information storage and higher level processing centers in the brain consist of the cerebral cortex and the hippocampus and repairing those centers involves much less mass than that which makes up the whole brain. It is postulated that most of the other body parts could be replaced with no loss of identity.

Some of Ettinger's original Experiments ([1964] Prospect of Immortality) that the author believes are most critical to the identity question are discussed. Experiment 20 where the brain is replaced by an artificial one appears to be most critical to the personal identity question.

We suggest that Ettinger's Experiment 20, or something rather similar may be starting to happen now. Closer interface between computers, the World Wide Web, and the human brain may create cyborgs where the ultimate death of the "meat" part of the combination will not mean the destruction of the personal identity.

A number of additional thought experiments are proposed (which are identified by letters). "Beam Me Up Scotty" Experiments, suggested by the transporter in the *Star Trek* TV episodes, are used to illustrate that if one uses identity of information alone as the criteria for personal identity many anomalies are presented that do not seem to fit our current concept of *identity* as that of an object continuing in time and space. However, a scanner capable of discerning nano sized details, along with the Sherlock Holmes Computer and fully developed nano repair robots, might reconstruct an individual who is a very close match to the original but who lacked continuity in time and space with his predecessor.

An experiment where a shipwrecked sailor has only one copy of the book *Moby Dick* was used to illustrate the fact that the question of "has *Moby Dick* been destroyed" depends upon the observer, or the location of the observer. To people *outside* the island the burning of one copy of the book makes little difference; to them *Moby Dick* is not destroyed. To the sailor (the inside observer), his loss of the book means that *Moby Dick* has been destroyed.

To people who observe the cryonic suspension of a loved one (the outside observer) who is later reconstructed the question of "is this the same person" may make little difference since the subject would act the same and be biologically the same and the subject's brain would have the same information. However, to the *inside observer* the subject herself, it would make a lot of difference. *Moby Dick* might indeed have been destroyed.

The subject of continuity of material and information was explored using the George Washington's Hatchet paradox (a variation on The Ship of Theseus' paradox). In the discussion of the paradox, it was questioned whether a hatchet could be regarded as the same hatchet after all its component parts had been replaced. It was concluded that we could define "same hatchet" in ways that it was either the same or was different. It was observed that the continuation through time of the hatchet could be established, even if presently it is missing all of its original component parts. When we empowered both the hatchet handle and head with information recording and restoring capacities we observed that there could be an unbroken information stream through the existence of the hatchet even though all component parts were replaced several times.

The author agrees with Ettinger's conclusion that: "At present, no clear-cut answer presents itself to me—nor, credibly, to anyone else, as far as I know." (Ettinger, Youniverse, 185) and "…the only reasonable working posture I can see is to seek as much continuity as possible both in information and physical substance. That way we keep our options open." (Ettinger, Youniverse, 192)

Bibliography

- Alcor Life Extension Foundation, <u>Cryonics</u>. Monthly Publication, Published monthly since 1981 through present.
- American Cryonics Society. *Board of Governors Meeting Minutes,* 1968-Present.
- Asimov, I. *The Chemicals of Life,* New American Library, New York, 1954.
- Basri, Saul A. *A Deductive Theory of Space and Time.* North-Holland, Amsterdam, 1966.
- Behan, D., 'Locke on persons and personal identity', *Canadian Journal of Philosophy* 9: 53–75, 1979.
- Burtt, Edwin A., "John Locke: Concerning Human Understanding," *The English Philosophers from Bacon to Mill*, The Modern Library, New York, 1939
- Campbell, S., 'The Conception of a Person as a Series of Mental Events', *Philosophy and Phenomenological Research* 73: 339–358, 2006.
- Chalmers, David J. *The Conscious Mind,* Oxford 1996.
- Chamberlin, Fred & Linda, *Life Quest,* Create Space, California, 2009.
- Changeux, J-P & Connes, A. *Conversations on Mind, Matter, & Mathematics,* Princeton 1983.
- Charme, Wesley M. Du, *Becoming Immortal: Nanotechonology, You, and the Demise of Death.* Blue Creek Ventures, Colorado, 1995.

- Cole, Dandridge and Roy G. Scarfo: *Beyond Tomorrow: The Next 50 Years in Space.* Amherst Press, Wisconsin, 1965.

- Donaldson, Thomas. <u>Periastron: A Newsletter Fact, Hypothesis, and Speculation</u>. Published 1994 through 2005.

- Cooper, Evan (writing as Nathan Duhring*): Immortality: Physically, Scientifically, Now.* Privately published, Washington, D.C., 1962.

- Drexler, K. E. *Engines of Creation.* Anchor Press/Doubleday, Garden City, New York, 1986.

- Ettinger, R.C.W. *Man into Superman,* St. Martin's Press, 1972.

- Ettinger, R.C.W. *The Prospect of Immortality,* Doubleday 1964.

- Ettinger, R.C.W. *Youniverse,* Immortalist Society, 2005.

- Freeman, Eugene, and Wilfrid Sellars, *Basic Issues in the Philosophy of Time.* Open Court, Illinois, 1887.

- Friedman, B. J. Frozen Guys. *Playboy*, August 1978, pp. 102+

- Friedman, David M., *The Immoralists: Charles Lindbergh, Dr. Alexis Carrel, and Their Daring Quest to Live Forever.* HarperCollins Publisher, New York, 2007.

- Garrett, B., *Personal Identity and Self-Consciousness*, London: Routledge, 1998.

- Harrington, A. *The Immortalist: An Approach to the Engineering of Man's Divinity.* Random House, New York, 1969.

- Hayflic, L. On the facts of life. *Executive Health*, Vol. 14, no. 9, June 1978.

- Heller, M., *The Ontology of Physical Objects: Four-Dimensional Hunks of Matter*, Cambridge University Press, 1990.

- Hero. A clinically doggone beagle, medical miracle Mile is a former chilly dog back from the beyond. *People,* Vol. 27, no. 16, April 20, 1987, p. 85.

- Hirsch, E., *The Concept of Identity*, Oxford University Press, 1982.

- Honderich, Ted. *Mind and Brain,* Oxford/Clarendon 1988.
- Hudson, H., 2001, *A Materialist Metaphysics of the Human Person*, Cornell University Press.
- Hume, D., *Treatise of Human Nature*, Oxford: Clarendon Press (original work 1739); partly reprinted in Perry 1975, 1978.
- Immortalist Society, Long Life: longevity through technology, bimonthly, Since 1968 with name changes several times.
- Kent, S. *The Life Extension Revolution.* Morrow, New York, 1980.
- Kelly, Kevin. "The Singularity is Always Near." Kevin Kelly. 15 February 2006. 24 November 2010. <http://www.kk.org/thetechnium/archives/2006/02/the_sin gularity.php>.
- Klebanoff, G., R. G. Armstrong, R. E. Cline et al. Resuscitation of a patient in stage IV hepatic coma using total body washout. *Journal of Surgical Research,* Vol. 13, 159-165, 1972.
- Kosko, Bart. *Fuzzy Thinking,* Hyperion 1993.
- Kurtzweil, Ray. *The Age of Intelligent Machine.* Massachusetts Institute of Technology, Massachusetts, 1999.
- Lamb, Lawrence E., *Get Ready for Immortality: A Doctor's Guide.* Harper & Row, New York, 1974.
- Martin, R. and J. Barresi (eds.), *Personal Identity*, Oxford: Blackwell, 2003.
- McDowell, J., 'Reductionism and the First Person', in *Reading Parfit*, J. Dancy (ed.), Oxford: Blackwell, 1997.
- Merricks, T., 'There Are No Criteria of Identity Over Time', *Noûs* 32: 106–124, 1998.
- Meryman, H. T. *"Mechanics of Freezing in Living Cells and Tissues."* Science, v. 124, 1956, p. 515.
- Meryman, H. T. *"Physical Limitations of the Rapid Freezing Method."* Proceedings of the Royal Society B, v. 147, 1957.

- Meryman, H. T. *"The Mechanisms of Freezing in Biological Systems."* Recent Research in Freezing and Drying, eds. A. S. Parkes.
- Minsky, Marvin, *"Steps Toward Artificial Intelligence."* Proceedings of the IRE, v. 49, 1961, p. 8.
- Minsky, Marvin. *The Society of Mind,* Simon & Schuster 1988.
- Nelson, Robert F. and Sandra Stanley: *We Froze the First Man.* Dell, New York, 1968.
- Noonan, H., *Personal Identity,* Second Edition, London: Routledge, 2003.
- Parfit, D., 'Personal Identity', *Philosophical Review* 80: 3–27, and reprinted in Perry 1975, 1971.
- Penelhum, T., *Survival and Disembodied Existence,* London: Routledge, 1970.
- Penrose, L. S. *"Self-Reproducing Machines."* Scientific American, v. 200, June, 1959, p. 105.
- Penrose, Roger & Isham, C.J. (ed.) *Quantum Concepts in Space and Time,* Oxford 1986.
- Penrose, Roger. *The Emperor's New Mind,* Oxford 1989.
- Perry, RM. *Forever for All,* Universal Press 2000.
- Prehoda, Robert W., *Suspended Animation: the research Possibility That May Allow Man to Conquer the Limiting Chains of Time.* Chilton Book Company, Philadelphia, 1969.
- Regis, Ed. *Great Mambo Chicken & the Transhuman Condition,* Addison Wesley 1990.
- Reichenbach, Hans. *The Philosophy of Space & Time,* Dover 1958.
- Rorty, A.O. (ed) *The Identities of Persons,* U. California Press 1976.
- Rosenfeld, Albert. *Prolongevity II,* Knopf 1985.
- Schechtman, M., *The Constitution of Selves,* Cornell University Press, 1996.
- Segall, P. Preservation and storage of biologic materials. U.S. Patent 3,677,024, July 18, 1972.
- Shoemaker, S., *Self-Knowledge and Self-Identity,* Ithaca: Cornell University Press, 1963.

- Sider, T., *Four Dimensionalism*, Oxford University Press, 2001.
- Spinoza, Baruch: "The Foundations of the Moral Life." *In Man and Spirit*, ed. S. Commins and R. N. Linscott, Random House, New York, 1947.
- Sternberg, H., P. E. Segall, H. Waitz, A. BenAbraham. Interventive gerontology cloning and cryonics: Relevance to life extension. Annual Spring Symposium on Biochemistry, George Washington University, 1988.
- Tandy, Charles (ed.) *The Philosophy of Robert Ettinger*, Ria University Press, 2002.
- Tandy, Charles (ed.) *Death and Anti-Death Volume 1,* Ria University Press, 2003.
- Tandy, Charles (ed.) *Death and Anti-Death Volume 2,* Ria University Press, 2004.
- Tipler, Frank J.*The Physics of Immortality,* Doubleday 1994.
- Vidigal, Diaulas, *Die, Freeze and be Back.* Minerva Press, London, 1997.
- "Vitalism." Def. 1a. Webster's. 2nd Ed. 1984
- Walford, R. L. *maximum Life Span.* Norton, New York, 1983.
- Watson, James D.: *Molecular Biology of the Gene.* W. A. Benjamin, New York, 1965.
- Yount, James. "An Advocation for Immortality," *Being Human: the techonological extensions of the body*, Apres-Coup, New York, 1999.
- Yount, James. "The State of Cryonics −2005," *The Prospect of Immortality*, Ria University Press, California.

INDEX

A

absolutism, 25, 147
absurdity, 30, 167, 178, 183, 355, 363, 365, 367, 368, 371, 373, 379, 380, 386, 387, 388, 389, 390, 391, 394, 395
aging, 17, 18, 25, 69, 121, 122, 125, 128, 130, 132, 143, 307, 308
Algerian Civil War, 26, 174, 189
ante-mortem person, 26, 205, 206
aphoristic politics, 26, 181
artificial intelligence, 20, 25, 28, 126, 137, 138, 432
assisted suicide, 20, 27, 227, 229, 230, 231, 232
assisted suicide law, 27
avoidable option, 25

B

barbarism, 23, 41, 43, 339
Baruchello, Giorgio, 7, 17, 23, 33, 45, 46, 416
Best, Benjamin P., 7, 17, 23, 53
bioartificial organs, 24
Bostrom, Nick, 13
brain, 23, 28, 55, 56, 58, 60, 61, 62, 63, 64, 65, 67, 68, 70, 71, 73, 74, 75, 77, 126, 138, 139, 217, 263, 296, 297, 298, 300, 306, 319, 359, 421, 422, 423, 425, 426, 427, 428, 429, 430, 431, 432, 434, 435, 437, 441, 442, 443
Buford, Thomas O., 7, 17, 24, 79, 90

C

caloric restriction, 25, 126, 128, 129, 130, 143
Camus, Albert, 3, 4, 5, 8, 11, 13, 17, 19, 21, 23, 26, 29, 30, 171, 172, 173, 174, 175, 176, 178, 179, 180, 181, 182, 183, 184, 185, 186, 187, 188, 189, 190, 191, 192, 193, 194, 195, 196, 197, 198, 333, 334, 336, 337, 338, 339, 340, 342, 343, 344, 345, 346, 347, 349, 350, 351, 352, 353, 355, 356, 357, 358, 359, 360, 361, 363, 364, 365, 367, 368, 371, 373, 376, 377, 378, 379, 380, 385, 386, 388, 389, 390, 391, 392, 393, 394, 398, 399
cause, 17, 28, 42, 55, 58, 60, 61, 86, 122, 125, 129, 131, 132, 158, 189, 239, 297, 310, 380
Chang, Alice, 72, 73, 91, 115
civilization, 23, 36, 40, 42, 43, 44, 45, 175, 185, 186, 193, 194, 274, 335, 336, 380, 396, 407
cognitive enhancement, 29, 306
Comfort, Alex, 198
computation, 28, 221, 236, 262, 293, 294, 298
consciousness, 26, 28, 40, 61, 63, 65, 79, 85, 157, 174, 223, 237, 262, 274, 298, 325, 327, 333, 363, 374, 375, 391, 419, 425
creativity, 19, 26, 174, 180, 181, 184, 192, 194, 218, 219, 220, 303, 395, 396
cryobiology, 23, 59
cryonic suspension, 31, 421, 423, 439, 441, 443
cryonics, 17, 20, 31, 53, 54, 55, 58, 61, 64, 65, 66, 67, 68, 69, 238,

309, 314, 315, 334, 335, 337, 340, 341, 347, 353, 355, 358, 359, 375, 381, 388, 396, 397, 412, 413, 414, 424, 437, 439
hormone replacement, 25, 127, 132
human rights, 20, 27

I

ice, avoidance of, 24
ideology, 30, 190, 195, 300, 389, 390, 395
immortality, 20, 28, 31, 50, 51, 149, 235, 238, 275, 289, 389, 390, 432
information theory, 31, 419, 425
intentional communities, 31, 409
interest, 20, 21, 23, 26, 33, 51, 92, 97, 122, 124, 147, 162, 194, 201, 204, 206, 209, 232, 260, 262, 264, 328, 402, 403, 413, 416, 421, 426

J

James, William, 7, 18, 19, 25, 147
justice, 26, 171, 173, 179, 187, 194, 398

K

kidney, rabbit, 24

L

law, 19, 23, 27, 37, 38, 39, 52, 84, 86, 87, 154, 161, 166, 173, 185, 188, 189, 225, 226, 229, 231, 232, 314, 350
LeBlanc, John Randolph, 8, 19, 26, 171
Lee, Jack, 8, 19, 26, 199, 416
living option, 25, 149, 151, 157, 165, 167
logical behaviorism, 28, 296

longevity, 18, 25, 125, 126, 139, 141, 142, 446
Lucas, J. R., 8, 19, 26, 211, 215, 221, 223, 224

M

machines, 27, 28, 138, 139, 214, 298, 424, 429, 431, 434
mathematics of immortality, 28
mathematics of resurrection, 28
meaning, 30, 109, 127, 181, 194, 204, 231, 260, 282, 297, 340, 353, 354, 365, 368, 373, 388, 390, 391, 397, 402
meaning of life, 30
memory, 31, 44, 75, 238, 262, 263, 293, 317, 368, 386, 419, 423, 424, 425, 427, 432, 441
mind, 13, 14, 15, 21, 23, 39, 52, 63, 64, 65, 84, 90, 148, 150, 156, 158, 160, 161, 163, 173, 176, 182, 185, 190, 191, 193, 202, 203, 212, 213, 214, 215, 216, 217, 218, 219, 221, 222, 294, 305, 316, 327, 352, 364, 383, 389, 393, 420, 422
minds, 27, 28, 36, 124, 148, 154, 155, 156, 160, 161, 169, 211, 213, 214, 215, 216, 219, 238, 296, 305, 316, 317, 318, 344, 363
Minelli, Ludwig A., 8, 20, 27, 225
modernism, 26, 177, 194, 384
modernity, 26, 175, 176, 178, 180, 182, 183, 193
molecular nanotechnology, 25, 137
momentous option, 25, 165
multiverse, 28, 235, 254, 255, 257, 262, 275, 388

N

Nagel, Thomas, 30, 223, 373, 374, 375, 377, 378
nanobots, 25, 138, 140